Biomimetic Adaptive Buildings

Biomimetic Adaptive Buildings

Guest Editors

Negin Imani
Brenda Vale
Derek Clements-Croome

MDPI

Basel • Beijing • Wuhan • Barcelona • Belgrade • Novi Sad • Cluj • Manchester

Guest Editors

Negin Imani
Centre for Sustainability
University of Otago
Dunedin
New Zealand

Brenda Vale
Wellington School of Architecture
Victoria University of Wellington
Wellington
New Zealand

Derek Clements-Croome
School of the Built Environment
University of Reading
Reading
UK

Editorial Office
MDPI AG
Grosspeteranlage 5
4052 Basel, Switzerland

This is a reprint of the Special Issue, published open access by the journal *Biomimetics* (ISSN 2313-7673), freely accessible at: https://www.mdpi.com/journal/biomimetics/special_issues/8V6VB4T647.

For citation purposes, cite each article independently as indicated on the article page online and as indicated below:

Lastname, A.A.; Lastname, B.B. Article Title. *Journal Name* **Year**, *Volume Number*, Page Range.

ISBN 978-3-7258-6906-0 (Hbk)
ISBN 978-3-7258-6907-7 (PDF)
https://doi.org/10.3390/books978-3-7258-6907-7

Cover image courtesy of Thomas Kinto

Contents

About the Editors

Negin Imani

Negin Imani is a Senior Research Fellow at Bodeker Scientific, New Zealand, and a Senior Lecturer in Built Environment at the Open Polytechnic of New Zealand. She leads research in biomimetic and bio-inspired architecture, focusing on energy-efficient and adaptive building design. Her work combines architecture, engineering, and biology, turning ideas from nature into innovative façades and sustainable building solutions. She has been leading New Zealand government-funded research on biomimetic buildings, supported by the Ministry of Business, Innovation and Employment. She has published books on nature-inspired approaches to building performance, including Biomimetic Buildings: Copying Nature for Energy Efficiency (2022) and Heating with Wolves, Cooling with Cacti: Thermo-bio-architectural Framework (ThBA) (2021). She has also taught at several universities and polytechnics in New Zealand and internationally. Before her academic career, she worked in architectural practice, gaining practical design experience that informs her research and teaching. Negin has served as a committee member of the National Association of Women in Construction (NAWIC), Wellington Branch, and on the Research Ethics Committee at the Open Polytechnic of New Zealand. She is the founder of Biomimicry New Zealand, a guest editor for *Biomimetics* (MDPI), and a frequent invited speaker on biomimetic design and sustainable buildings. Her work has been recognised through national awards, research grants, and coverage in international media.

Brenda Vale

Brenda Vale is a Professorial Research Fellow at the School of Architecture, Victoria University of Wellington, New Zealand. She is an architect and academic whose research interests include ecological footprinting, sustainable building design, zero energy housing design, and the history of prefabrication. Along with Robert Vale, she wrote one of the first books on sustainable design and architecture, "The Autonomous House" in 1975. She has been involved in the design of several award winning sustainable commercial buildings in the UK, including the UK's first autonomous house (1993) and the first zero-emissions settlement (1998), developed in partnership with Robert Vale. Their collaborative work received international recognition, including awards from the United Nations and the European Solar Energy Society. She also contributed, together with Robert Vale, to the development of the Australian government's National Australian Built Environment Rating System (NABERS). She has co-authored numerous influential publications with him including "Time to Eat the Dog? The real guide to sustainable living" which analysed the impact on the environment of a western lifestyle and everyday habits, "Living within a Fair Share Ecological Footprint" which has chapters written by many of their former postgraduate students and "Architecture on the Carpet" which explores the links between architecture and construction toys such as Meccano and Lego over the last hundred years. Recently, Brenda has co-authored many environmental books, including "The Environmental Impact of cities. Death by democracy and capitalism" (2023), "Heating with Wolves, Cooling with Cacti: Thermo-bio-architectural Framework (ThBA)" (2021), and "Everyday Lifestyles and Sustainability: The Environmental Impact Of Doing The Same Things Differently" (2018). Her latest book with Emilio Garcia explores the influence of wealth on the built environment.

Derek Clements-Croome

Derek Clements-Croome is an Emeritus Professor at the University of Reading, UK. His research interests include intelligent buildings; design of workplaces; logistic support analysis; individual environmental control; low-energy buildings; indoor air quality; alternative approaches to heating, ventilating, and air conditioning of buildings; and innovation sustainability in design, construction, and operation. He has founded and directed courses, including a BSc in building environmental engineering at Loughborough University in 1970 and an interdisciplinary Government-funded MSc in Intelligent Buildings at Reading University in 1996. He has also worked in architecture and building engineering at the University of Bath (1978-1988) with Sir Ted Happold. He founded the CIBSE Intelligent Buildings Group in 2006 and co-founded the CIBSE Natural Ventilation Group in 1992. He was Vice President for CIBSE in 2007–2009 and holds their bronze, silver, and gold medals. He edits the *INBI Journal*. His book Designing Buildings for People: Sustainable Liveable Architecture was published in 2020, and a third edition of Intelligent Buildings and Infrastructure in 2024. He contributed a chapter to the 2025 book Essential Lifestyle Medicine. He acts as a research project adviser to governments in New Zealand, Denmark, and Hungary, and BCO in the UK. He was awarded a team gold medal at the RHS Chelsea Flower Show 2022 for his work on flourishing biophilic classrooms. He is a member of the BSLM Advisory Committee.

Preface

This Reprint explores how principles drawn from nature can inform the development of buildings that actively respond to their environments. Focusing on adaptive façades, smart materials, and bio-based design strategies, it brings together research that demonstrates the potential of biomimicry to reduce energy use, improve comfort, and strengthen the relationship between buildings and their surrounding ecosystems.

Our motivation for compiling this work comes from the growing recognition that static construction methods are no longer adequate in the face of climate change and rising performance expectations. By examining solutions inspired by biological systems, we aim to highlight approaches that support more resilient and efficient buildings.

This volume is intended for researchers, practitioners, and students in architecture, engineering, and environmental design who are seeking practical and scientifically grounded pathways for applying biomimetic thinking in building practice. We hope it serves as a useful reference and a catalyst for further interdisciplinary work in this rapidly evolving field.

Negin Imani, Brenda Vale, and Derek Clements-Croome
Guest Editors

biomimetics

MDPI

Editorial

Editorial for Special Issue on Biomimetic Adaptive Buildings

Negin Imani [1,2,3,*], Brenda Vale [4] and Derek Clements-Croome [5]

1 Bodeker Scientific, Wellington 6012, New Zealand
2 Centre for Sustainability, University of Otago, Dunedin 9016, New Zealand
3 Open Polytechnic, Lower Hutt 5011, New Zealand
4 Wellington Faculty of Architecture and Design Innovation, Victoria University of Wellington, Wellington 6140, New Zealand; brenda.vale@vuw.ac.nz
5 School of the Built Environment, University of Reading, Reading RG6 6UR, UK; d.j.clements-croome@reading.ac.uk
* Correspondence: negin.imani@gmail.com

It seems that the future of building envelopes is moving towards adaptivity and self-regulation, reflecting the growing view that a vital strategy in addressing climate change is understanding buildings as living systems rather than static entities. This shift in perspective encourages the development of façades capable of dynamically responding to environmental conditions. Such systems can autonomously regulate temperature, daylight, and airflow, thereby reducing operational energy demand while enhancing indoor comfort.

Such a reduction is important, as buildings account for 30% of global energy use and 27% of emissions [1]. Biomimetic Adaptive Buildings (Bio-ABFs) can reduce building energy use and enhance thermal comfort [2–9] by 50% [10] and 37.1%, respectively [11]. Living building technologies rely on active or passive self-regulating mechanisms. Active systems utilise external components, such as sensors, processors, and actuators, to achieve adaptive behaviour. Recent innovations include self-adjusting shading systems, dynamic insulation, and responsive ventilation technologies that enable real-time, automated performance control. In contrast, passive systems rely on intrinsic material properties and structural characteristics that respond naturally to changing environmental conditions. As highlighted by the International Energy Agency's 2022 report [12], conventional materials with static behaviour fall short in response to changing climatic conditions.

As technology rapidly develops, contemporary buildings face increasingly complex environmental challenges. Architects and engineers are therefore called to rethink the relationship between the built environment and its surroundings. Conventional, static design approaches are often insufficient in effectively responding to fluctuating climatic conditions, resource limitations, and evolving occupant comfort and wellbeing standards. This Special Issue, "Biomimetic Adaptive Buildings," examines an emerging paradigm in which buildings emulate nature's capacity for adaptation, efficiency, and resilience. As most buildings last for decades, if not more, such an adaptation is crucial.

Biomimetic design has long inspired architecture, from the thermoregulatory efficiency of termite mounds to the structural ingenuity of plant stems. Recent advances in smart materials, sensor systems, and computational design now allow these biological principles to be implemented as functional architectural technologies. This Special Issue presents a range of contributions exploring how natural intelligence can inform the material, systemic, and ecological adaptability of built environments.

Our Special Issue brings together seven original research articles and one review paper that collectively explore the breadth of biomimetic adaptive building design. The contributions cover diverse aspects, including adaptive façades, bio-based materials, energy generation, environmental performance, and integrated design assessment frameworks.

A principal focus of this issue is adaptive façade systems, building envelopes capable of dynamically responding to environmental stimuli such as temperature, humidity, and solar radiation. Imani et al. [13], in "Biomimetic Shading Systems: Integrating Motorised and Moisture-Responsive Actuation for Adaptive Façades," present a two-stage investigation combining mechanical and passive adaptation. Nine motorised shading concepts inspired by plant stomata thermoregulation are developed, followed by a passive actuation prototype employing a polylactic acid (PLA) biocomposite and regenerated cellulose fibres. The findings indicate that such materials can autonomously alter geometry in response to humidity, offering a low-energy pathway toward climate-responsive façades.

In [14], "Biomimetic Adaptive Building Façade Modelling for Sustainable Urban Freshwater Ecosystems," Kahvecioğlu et al. [14] propose a sun-shading façade system inspired by lizard skin microstructures and leaf venation. The façade collects and channels rainwater to support urban freshwater ecosystems, extending its functionality beyond thermal regulation to an integrated ecological role. Similarly, Valinejadshoubi, Athienitis [15] present a dynamic photovoltaic façade combining shading control with energy generation, demonstrating significant improvements in thermal and visual comfort while reducing overall energy demand. The façade's responsiveness to changing solar conditions exemplifies how building envelopes can actively engage with their environment rather than remaining static barriers.

Sankaewthong et al. [16] investigate the Mimosa kinetic façade, inspired by the plant's adaptive movements, as a dynamic solution to improve indoor air quality and occupant comfort in densely populated buildings. Tests in Bangkok revealed that specific façade patterns excelled in terms of ventilation, while others optimised long-term comfort. Double-sided designs enhanced cross-ventilation and air purification, achieving airflow of up to 12 m/s compared to the 2.5 m/s seen with static façades. These findings highlight the potential of kinetic systems in transforming sustainable building practices; however, further research is needed to validate performance across diverse climates and real-world conditions.

AlAli et al. [17] explore nature-inspired design in laboratory buildings, emphasising biomimicry and biophilic principles to enhance sustainability and human well-being. They introduce the Nature-Inspired & Living Laboratory (NILL 1.0)™ Building Assessment Index, a framework for evaluating how buildings integrate sustainable practices, regenerative features, and occupant-focused design. By linking building systems to natural and human body analogies, the index guides the creation of laboratories that are resilient, environmentally responsible, and supportive of scientific innovation. Collectively, these studies illustrate how façades can evolve from static barriers into multifunctional, responsive, and ecologically engaged building skins.

The second thematic cluster bridges adaptive building concepts and sustainable, bio-based materials. Niculalde et al. [18] evaluate two Ecuadorian woods—Guayacan and oaks—as sustainable construction alternatives. Results show that Guayacan has lower mechanical strength and thermal conductivity, while oak properties vary by region; more specifically, Manabi oak exhibits slightly better thermal performance, while Loja oak shows superior elasticity. Multicriteria decision methods, such as the entropy-COPRAS approach, provide an objective framework for selecting timber based on varying mechanical and thermal properties, supporting informed building material choices.

El Haddaji et al. [19] explore Brewers' Spent Grains (BSGs) combined with cornstarch as a sustainable biocomposite, comparing the mixture with conventional hemp shives. While the BSG formulation exhibits higher thermal conductivity, it demonstrates superior hygrothermal performance, and its mechanical properties are comparable across all formulations. These findings indicate BSG waste as a promising construction material that supports circular economy principles; however, further research is needed to optimise formulations and assess durability.

Complementing the original research articles, this volume also includes a comprehensive review paper examining the role of biomimicry in architecture for improving energy efficiency and building performance. This review highlights how natural strategies, particularly those found in plants, can inspire adaptive and sustainable design solutions. It presents a synthesis of global case studies, demonstrating the versatility and effectiveness of biomimetic approaches across diverse climatic and architectural contexts [20].

Across these thematic strands, the contributions of this Special Issue collectively redefine building adaptation. From moisture-responsive shading modules and waste-derived composites to sensor-driven façades, adaptive design requires the integration of material behaviour, technological intelligence, and ecological connectivity. This collection aims to serve as a reference for future interdisciplinary research, fostering collaboration among architects, engineers, material scientists, and biologists. As biomimetic thinking continues to evolve, the ultimate objective is not merely to replicate natural forms but to embed processes of adaptation and regeneration into the built environment itself.

The guest editors extend their sincere gratitude to all contributing authors for their innovative research, to the reviewers for their thoughtful evaluations, and to the *Biomimetics* (MDPI) editorial team for their support and professionalism. Recognition is also given to the academic institutions, funding organisations and research centres advancing biomimetic design through experimentation, teaching, and cross-disciplinary collaboration.

Funding: This research received no external funding.

Conflicts of Interest: Author Negin Imani was employed by Bodeker Scientific, University of Otago and Open Polytechnic. The remaining authors declare that the research was conducted in the absence of any commercial or financial relationships that could be construed as a potential conflict of interest.

References

1. IEA. *Buildings*; International Energy Agency: Paris, France, 2023.
2. Francesco, S.; Lidia, B.; Gigliola, A. A critical review of biomimetic building envelopes: Towards a bio-adaptive model from nature to architecture. *Renew. Sustain. Energy Rev.* **2022**, *169*, 112850. [CrossRef]
3. Imani, N.; Vale, B. *Heating with Wolves, Cooling with Cacti: Thermo-Bio-Architectural Framework (ThBA)*; CRC Press: Milton, ON, Canada, 2021.
4. Attia, S.; Lioure, R.; Declaude, Q. Future trends and main concepts of adaptive facade systems. *Energy Sci. Eng.* **2020**, *8*, 3255–3272. [CrossRef]
5. Al-Masrani, S.M.; Al-Obaidi, K.M. Dynamic shading systems: A review of design parameters, platforms and evaluation strategies. *Autom. Constr.* **2019**, *102*, 195–216. [CrossRef]
6. Hosseini, S.M.; Mohammadi, M.; Guerra-Santin, O. Interactive kinetic façade: Improving visual comfort based on dynamic daylight and occupant's positions by 2D and 3D shape changes. *Build. Environ.* **2019**, *165*, 106396. [CrossRef]
7. Imani, N.; Vale, B. Biomimetic Design for Building Energy Efficiency 2021. *Biomimetics* **2022**, *7*, 106. [CrossRef] [PubMed]
8. Charpentier, L.; Cruz, E.; Nenov, T.; Guidoux, K. Pho'liage: Towards a Kinetic Biomimetic Thermoregulating Façade. In *Bionics and Sustainable Design*; Springer Nature: Singapore, 2022; pp. 367–401.
9. Imani, N.; Vale, B. *Biomimetic Buildings: Copying Nature for Energy Efficiency*; MDPI: Basel, Switzerland, 2022.
10. Webb, M. Biomimetic building facades demonstrate potential to reduce energy consumption for different building typologies in different climate zones. *Clean. Technol. Environ. Policy* **2022**, *24*, 493–518. [CrossRef] [PubMed]
11. Kuru, A.; Oldfield, P.; Bonser, S.; Fiorito, F. Performance prediction of biomimetic adaptive building skins: Integrating multifunctionality through a novel simulation framework. *Sol. Energy* **2021**, *224*, 253–270. [CrossRef]
12. Birol, F. *World Energy Outlook 2022*; International Energy Agency: Paris, France, 2022.
13. Imani, N.; Le Guen, M.-J.; Bedggood, N.; Betteridge, C.; Gauss, C.; Barbier, M. Biomimetic Shading Systems: Integrating Motorised and Moisture-Responsive Actuation for Adaptive Façades. *Biomimetics* **2025**, *10*, 711. [CrossRef]

14. Kahvecioğlu, B.; Avinç, G.M.; Selçuk, S.A. Biomimetic Adaptive Building Façade Modeling for Sustainable Urban Freshwater Ecosystems: Integration of Nature's Water-Harvesting Strategy into Sun-Breakers. *Biomimetics* **2024**, *9*, 569. [CrossRef]
15. Valinejadshoubi, M.; Athienitis, A.K.; Bagchi, A.; Abtahi, M. Integrated Dynamic Photovoltaic Facade for Enhanced Building Comfort and Energy Efficiency. *Biomimetics* **2024**, *9*, 463. [CrossRef]
16. Sankaewthong, S.; Miyata, K.; Horanont, T.; Xie, H.; Karnjana, J. Mimosa Kinetic Façade: Bio-Inspired Ventilation Leveraging the Mimosa Pudica Mechanism for Enhanced Indoor Air Quality. *Biomimetics* **2023**, *8*, 603. [CrossRef] [PubMed]
17. AlAli, M.; Beheiry, S.; Atabay, S. Strategies for the Design and Construction of Nature-Inspired & Living Laboratory (NILL 1.0) TM Buildings. *Biomimetics* **2024**, *9*, 441.
18. Nicolalde, J.F.; Martínez-Gómez, J.; Dávila, P.; Medrano-Barboza, J.; Molina-Osejos, J.V. Ecuadorian woods: Building material selection using an entropy-COPRAS comparative analysis based on the characterization of Ecuadorian Oak and Guayacan timber. *Biomimetics* **2024**, *9*, 443. [CrossRef] [PubMed]
19. El Haddaji, B.; Benzaama, M.-H.; Quiertant, M.; El Mendili, Y. Insight into the use of brewers' spent grain as a low-carbon aggregate in building materials. *Biomimetics* **2024**, *9*, 781. [CrossRef] [PubMed]
20. Bijari, M.; Aflaki, A.; Esfandiari, M. Plants Inspired Biomimetics Architecture in Modern Buildings: A Review of Form, Function and Energy. *Biomimetics* **2025**, *10*, 124. [CrossRef] [PubMed]

biomimetics

MDPI

Article

Biomimetic Shading Systems: Integrating Motorised and Moisture-Responsive Actuation for Adaptive Façades

Negin Imani [1,2,3,*], Marie-Joo Le Guen [4], Nathaniel Bedggood [2], Caelum Betteridge [4], Christian Gauss [5] and Maxime Barbier [4]

1 Bodeker Scientific, Wellington 6012, New Zealand
2 Centre for Sustainability, University of Otago, Dunedin 9016, New Zealand; ngbedggood@gmail.com
3 Open Polytechnic, Lower Hutt 5011, New Zealand
4 Scion Group, New Zealand Institute for Bioeconomy Science Ltd., Rotorua 3046, New Zealand; mariejoo.leguen@scionresearch.com (M.-J.L.G.); caelum.betteridge@scionresearch.com (C.B.); maximebbr@gmail.com (M.B.)
5 School of Engineering, University of Waikato, Hamilton 3240, New Zealand; christian.gauss@waikato.ac.nz
* Correspondence: negin@bodekerscientific.com or imane807@otago.ac.nz or negin.imani@openpolytechnic.ac.nz

Abstract

A biomimetic adaptive façade applies natural principles to building design using shading devices that dynamically respond to environmental changes, enhancing daylight, thermal comfort, and energy efficiency. While motorised systems offer precision through sensors and mechanical actuation, they consume energy and are complex. In contrast, passively actuated systems use smart materials that respond to environmental stimuli, offering simpler and more sustainable operation, but often lack responsiveness to dynamic conditions. This study explores a sequential approach by initially developing motorised shading concepts before transitioning to a passive actuation strategy. In the first phase, nine mechanically actuated shading device concepts were designed, inspired by the opening and closing behaviour of plant stomata, and evaluated on structural robustness, actuation efficiency, ease of installation, and visual integration. One concept was selected for further development. In the second phase, a biocomposite made of polylactic acid (PLA) and regenerated cellulose fibres was used for Fused Deposition Modelling (FDM) to fabricate 3D-printed modules with passive, moisture-responsive actuation. The modules underwent environmental testing, demonstrating repeatable shape changes in response to heat and moisture. Moisture application increased the range of motion, and heating led to flap closure as water evaporated. Reinforcement and layering strategies were also explored to optimise movement and minimise unwanted deformation, highlighting the material's potential for sustainable, responsive façade systems.

Keywords: biomimetic; adaptive building façade; façade technology; automated shading devices; 3D printing; biocomposite; energy efficiency; automation

1. Introduction

Building operation accounts for 30% of global energy consumption [1] and 27% of energy sector emissions [2]. As the impacts of climate change intensify, reducing energy consumption in buildings has become an urgent priority. Evidence suggests that adaptive façades can reduce energy consumption, improve thermal performance, and enhance occupant well-being.

Conventional materials are failing to respond to our changing climate, as evidenced by the International Energy Agency's 2022 report calling for higher-performance building façades. This while responsive facades outperform traditional solutions, achieving up to 65% higher efficiency [3].

Biomimetic Adaptive Building Façades (Bio-ABFs) are inspired by plants' adaptation to changing weather and can mitigate operational energy use while enhancing occupants' comfort [4–11]. The existing adaptive facades that operate at a real scale are motorised and use mechanical equipment for activation. The research shows that maintenance remains a critical challenge for such responsive facades.

The Institut du Monde Arabe (1981–1987) featured a complex system of 240 motor-controlled apertures, but due to high maintenance demands, it gradually failed, becoming non-functional within six years [12]. Similarly, the GSW Headquarters (1999) employs a double-skin façade with actuated blinds, which, while still operational after 25 years, requires more maintenance than conventional buildings [13]. Other examples, such as the Kiefer Technik Showroom (2007) and St. Ingbert Town Hall (2010) [14], also struggle with maintenance-intensive, bespoke systems. To ensure long-term viability, adaptive façades must be designed for durability, ease of maintenance, and standardisation.

While several studies have been conducted on kinetic façade systems (Table 1), there remains a lack of focus on the kinematic design of individual façade modules (shading systems) and how they can influence façade operation when integrated into the façade. A more holistic approach is required, one that considers the existing challenges associated with motorised systems. Although it is not an easy task to develop a systematic design framework for developing façade shading systems, it is possible to evaluate the kinematics of shading modules concerning the practical challenges they pose when integrated into the building envelope.

As identified by [5] when translating plant thermoregulation strategies into architectural equivalents, it is essential to consider the thermoregulation variables present in the heat transfer equation. A similar approach in architecture could yield comparable results. Interestingly, regardless of the specific thermoregulation solution used by plants, the main variables in thermoregulatory mechanisms and the heat transfer equation include either temperature gradient, solar heat coefficient (absorption or transmission), surface area, and heat transfer coefficient. These variables can be categorised under either spatial or morphological principles. This insight highlights the need to shift the biomimetic design approach from a focus on form to one centred on function. It is not the form of the plant organs or the specifics of their observed kinetics that directly impact heat transfer, but rather that these mechanisms ultimately depend on the fundamental principles and variables outlined above. That said, in the design of shading systems, the primary focus should not be on replicating the morphological forms or kinematic movements observed in biological analogues. Rather, emphasis should be placed on the underlying principles of these systems in thermal adaptation. The form and kinematics of the design of shading systems are significant only to the extent that they facilitate construction and operational efficiency, including the ease of fabrication, automation, maintenance, durability, and seamless integration within the building's broader management framework.

Existing Bio-ABFs rely on expensive, high-maintenance mechanical systems for adaptive façade geometries [8,10,15–19] and smart material responses attached to moving elements of the shading device [20–27]. 4D printing with biomaterials offers more cost-effective solutions, enabling the creation of dynamic, multifunctional structures [28,29].

Table 1. Synthesis of kinetic façade systems uses and findings in buildings drawn by past studies.

Building	Year	Prototype Scale	Fabrication Tool	Shape Change	Material of the Adaptive Component	Regulation	Design	Software Used	Authors
Buckling system	2022			Buckling, electricity	Shape Memory alloy.	Light	Bistable mechanism	Finite element analysis (FEA)	[30]
Microalgae bio-reactive facade	2022	Small scale	Lab tools	Algae growth, mechanical component	Algae	Light & heat	Control of bioreactor (microalgae medium flow)	-	[31]
Kinetic biomimetic Facade	2022	Small scale	Punch press industrial tools	Difference in the coefficient of expansion, temperature change of the material	Thermobimetal	Light & heat	Flowerlike triangular petals Open & close mechanism	Rhino, grasshopper. Finite element analysis (FEA)	[10]
La Seine Musicale	2017	Full scale-France		Pneumatic system mounted on a rail	Solar Panels	Energy	Honeycomb structure & Trussed steel framework		[32]
Ocean Pavilion	2012	Full scale	Labour work	Torsional buckling, mechanical actuator	Laminated sheet. glass fibre reinforced polymer (GFRP)	Light	Analog control of a mechanical System + buckling	-	[33]
Dancing Pavilion	2016	Full scale-Rio, Brazil	-	Swivelling metal panels, sensors	Metal disc controlled by a stepper Motor	-	-	Animation software	[32]
***Strelitzia reginae-* Flctofin**	2015	Single-component prototype	Labor fabrication	Punctually applied load, mechanical control	Glass fibre reinforced polymer (GFRP)	Heat & light	Rectangular frames	FAM, CAD	[34]
Hugroskin	2014	Full scale	7-axis industrial robot to CNC mill, vacuum moulding	Anisotropic structural and Hygroscopic characteristics of wood, ambient humidity	Elastically bent plywood. Wood veneers	Air flow	Triangular shape	Finite element analysis (FEA)	[35]
Al-Bahr Tower	2012	Full scale	-	Linear actuators, building management System	Stainless steel supporting frame, aluminium dynamic frame, polytetrafluo-roethylene (PTFE) coated Fibreglass panels	Light	Honeycomb-inspired structure for the curtain wall. Triangular unit for kinetic structure	CATIA Tool, Rhino Scripts integrated into BIM	[36]
MEDIA-ICT-Building	2011	Ful scale-Barcelona	Laser-cut machine	Pneumatic system, Lux meters trigger Of pneumatic shading or fog system	ETFE (ethylene Tetrafluoroethylene)	Light & temperature	Lattice structure	Finite element analysis (FEA)	[37]
Oval Cologne Offices	2010	Full Scale—Germany	-	Sensors	Glass	Heat & light	Rectangular frames	-	[38]
Ingbert Town Hall	2009	Full scale-Germany	-	Motorised pushing rod	Electropolished expanded stainless Steel sheet	Light	-	-	[14]
Milsertor centre	2008	Full-scale Austria		18 Electrical motor, rail and gliding rod, programmed System	Plexiglass panel	Light	Rectangular frame		[14]
Kiefer Technic Showroom	2007	Full-scale Germany	-	Electrical motor, rail and roller, a programmed System	Aluminium panel	Light	Rectangular frame	Programmable BUS/PLC system.	[39]

The development of 4D-printed building materials can enhance architectural design freedom and increase the environmental efficiency of adaptive façades [40]. The long-term benefits of these preprogrammed materials could include self-assembly and self-repair capabilities and addressing current technical challenges such as high maintenance and reliance on complex computer-controlled systems.

Early experiments in this field focused on harnessing the material properties of various metals to enable movement in response to temperature changes. Initially, research centred on shape memory alloys (SMA) [41]. However, due to their high cost and low mechanical strength, alternative materials had to be explored [42]. This led to increased interest in bimetallic systems over the past decade. These systems consist of rigidly connected metal sheets with differing thermal expansion coefficients [43]. When heated, one sheet expands more than the other, causing the structure to bend in one direction; as it cools, it returns to its original shape.

Sung [43] leveraged these properties to develop innovative systems for ventilation and shading. One example is the Bloom pavilion, composed of 414 stacked, hyperbolic paraboloid-shaped bimetallic panels. These panels automatically open at a specific temperature to release excess heat and close when cooled to retain warmth. Another system Sung [43] developed was the Pivot Shade System, which integrates bimetallic strips into a double-skinned façade to function as a shading device. Similar to the GSW Headquarters, these strips bend when heated, adjusting their angle to provide optimal shading for the building's interior. Collectively, these examples demonstrate the potential of bimetallic systems to regulate indoor temperature, shading, and ventilation while offering greater durability than traditional adaptive systems.

Inspired by flower petals Charpentier, Cruz [10] explored the use of bimetallic systems at a more advanced level. The rigid sections of the petals incorporated photovoltaic panels to generate passive energy while mitigating solar gain. Despite these advancements, the use of bimetallic panels was not considered in later development stages. Although bimetallic materials provide increased rigidity compared to other shape-changing materials, their high cost, limited availability, and lack of recyclability make them less viable for large-scale adoption [10]. These factors present a significant barrier to the widespread use of bimetallic systems in adaptive architecture. While they offer durability and robustness, their expensive and complex manufacturing process hinders their practicality.

Other similar projects have used layered copper and polyethene. However, due to cost considerations, these materials were replaced by cross-laminated timber veneer, taking advantage of the difference in expansion rates of timber running parallel and perpendicular to the grain. Coating was also used to modify the expansion speed [44].

The HygroSkin pavilion utilised the hygromorphic properties of timber. Developed by Achim Menges with the Institute for Computational Design at the University of Stuttgart, the pavilion responded to humidity changes, with timber absorbing moisture and expanding to actuate shutters constructed from veneer layers [35]. A different system developed by Mazzucchelli, Alston [45] integrated thin-film solar cells with a hygromorphic layer composed of two wooden slats from different wood types and trunk cuts. The deformation resulted from the differing expansion coefficients of the two wooden slats, causing them to expand and contract at different rates in response to temperature and humidity changes.

Achieving shape-changing capabilities is not restricted to architecture and has already been demonstrated in various other fields during the last decade. Table 2 compares examples of various shape morphing 4D printed composites, including PLA, chitosan-based polymers, hydrogels, soybean oil-based polymers, wood-fill filaments, and bacterial films, based on their stimuli responses, technical characteristics, fabrication methods, and applications. These materials respond to heat, pH, moisture, light, and humidity. Fabricated mainly through 3D printing, inkjet, or laser techniques, their transition times range from immediate to hours. Applications span biomedical fields (drug delivery, scaffolds, adaptive interfaces), engineering (solar tracking, filtration, acoustic wave control), and soft robotics, demonstrating their potential in self-morphing structures, responsive surfaces, and bio-compatible innovations.

Table 2. Synthesis of 3D printing materials/processes findings in passive actuation drawn by past studies.

Material	Stimuli	Technical Characteristics	Variable	Method of Fabrication	Transition Time	Application	Authors
PLA	Temperature (Hot water at 80 °C) Bending	Tg: 60~65 °C, Tm: 180 °C	Geometric shapes & patterns, Stored strain (by Print parameters; feed rate and stretching of filament, print temperature & direction)	Fused deposition moulding (FDM)printer	Not mentioned	Soft robotics, grippers & self-morphing structures	[46]
Chitosan hydrogel ink & Pretrained polystyrene (PS) film	Heat absorbed through light at 3.5 megawatts per mm Bending	Tg of shrink film is 105 °C	spatial pattern	Ink 3D printed on film then dried at 50 °C	30 to 60 s	Responsive truss structure	[42]
Hydrogel-based ink	Moisture and temperature Bending	Temperature above 32 °C, tensile strength: 642 KPa	Polymer chains in the ink formulations	3D printed as ink then UV polymerised	Immediate transition	Biomedical applications	[47]
PLA Filament	Temperature at 105 °C Bending	Elastic modulus: 50 MPa, Tg: 60~65 °C	Thickness Stored stress during printing, released upon heating.	FDM 3D printer.	18 s	Engineering applications such as filtering, localising and guiding acoustic waves, shape-changing solar concentrators	[48]
Soybean oil epoxidized acrylate	Temperature Fixed temporary shape at −18 °C, recovered original shape at human body temperature (37 °C)	Tg: 37 °C, UV curable	printer infill density, laser frequency and printing speed.	Custom-developed multilateral 3D printer system, laser printing technique	1 min	Biomedical scaffolds	[49]
Wood-fill fine filament (includes PLA, PHA & wood fibres)	Moisture	Young's modulus:—0° compressed composite: ~4000 MPa with ~0.15% swelling—90° compressed composite: ~3500 MPa with ~0.43% swelling—90° high-speed printing: ~900 MPa with ~0.5% swelling		Prusa I3 Rework 3D printer equipped with a 0.4 mm nozzle	1 day to 1 h	Passive devices for solar tracking systems, responsive building skins	[50]
***Bacillus subtilis* natto cell**	Moisture	50% swelling. Resists up to 2 min flush of water, 470° curvature		Inkjet 3D printer	A few seconds	Adaptive interface and wearable applications	[51]
Polyester wood filled filament and ABS	Moisture	Mutli phase motion based on direction of prints	Geometric shapes & patterns,	FDM	Min- to hours	Not specifically mentioned	[52]
Wood-fill (15, 25%) filament (PLA and arbocel)	Moisture	Degree and direction of bending depend on the PLA-to-wood-PLA layer ratio		FDM			[53]
Wood-fill (40%) PLA filament (Laywood)	Moisture and temperature	Cantilever actuator achieved a curvature of 120° at 120 min and progressed to 270° at 270 min under controlled humidity conditions	Porosity/infill, patterns, layers	FDM	Min-hours	Speculative: soft robotics, adaptive architecture, and responsive textiles,	[54]

While most research on adaptive building systems has focused on either passive or active actuation strategies, there remains a knowledge gap in understanding how these approaches can be effectively integrated. In this context, passive actuation refers specifically to the intrinsic shape-changing behaviour of the shading component itself—excluding supporting mechanisms such as hinges or linkages, which have already been widely studied. A sequential approach that combines passive and active strategies offers a promising pathway to overcome current limitations, enabling façade systems that are both energy-efficient and highly responsive. Advancing this sequential model is essential to improve the performance, reliability, and scalability of next-generation climate-adaptive building technologies.

While kinetic façade systems have been extensively explored, few studies have addressed the practical integration of motorised shading systems into building envelopes from a kinematic design perspective. Many instead focus on biomimetic movements or forms. As discussed earlier, this research takes a more structured approach, aiming to develop and evaluate formal and kinematic solutions that directly address the construction and operational challenges of kinetic adaptive façades. The research builds on existing work in modularity, customisation, and active automation within solar shading systems integrated into Bio-ABFs.

The proposed design components are inspired by stomatal regulation in plants, where opening and closure control water exchange and thermal balance, serving as a direct analogy for the adaptive shading response.

2. Materials and Methods

2.1. Design Development of Motorised Shading Systems with Mechanical Actuation (Using 3D Printing Prototyping)

The key criteria for the development process were chosen based on the core engineering principles of product development to ensure a practical outcome. Each judging criterion balances the principles of functionality (Actuation, Occlusion), longevity (Simplicity and Robustness, Repairability), practicality (Mounting) and also with the crucial element of form (Design Appeal) [55].

Our evaluation criteria are supported by prior studies: ease of maintenance and repairability [56,57], mounting feasibility [58], actuation method [57], practicality [58] occlusion ratio [59,60], design simplicity [7,60,61], durability and robustness [62] have all been identified as key considerations in façade and shading system design.

(1) Simplicity and Robustness

Minimise the number of joints and moving parts;
Avoid overly complex slats, plates, or geometries that could compromise reliability.

(2) Actuation Method

Evaluate the suitability of servos versus linear actuators;
Consider practical methods for mounting and integration

(3) Repairability

Explore design strategies that allow for easy maintenance and future repairs.

(4) Mounting Method

Assess the feasibility of mounting the device on buildings, both in retrofits and new constructions;
Minimise interference from the structural frame or mounting points.

(5) Occlusion Ratio

Aim for a dynamic range of 0% to 100% occlusion between fully open and closed states.

(6) Design Appeal

Ensure the design not only performs its function but also complements the architectural aesthetics;

Consider the visual impact of the device when repeated in a façade grid, will it appear cohesive or visually cluttered?

A range of concepts was developed, each demonstrating distinct mechanical behaviours, aesthetic qualities, and feasibility constraints. The following is a summarised overview of nine design concepts that were explored during the design process (Figure 1):

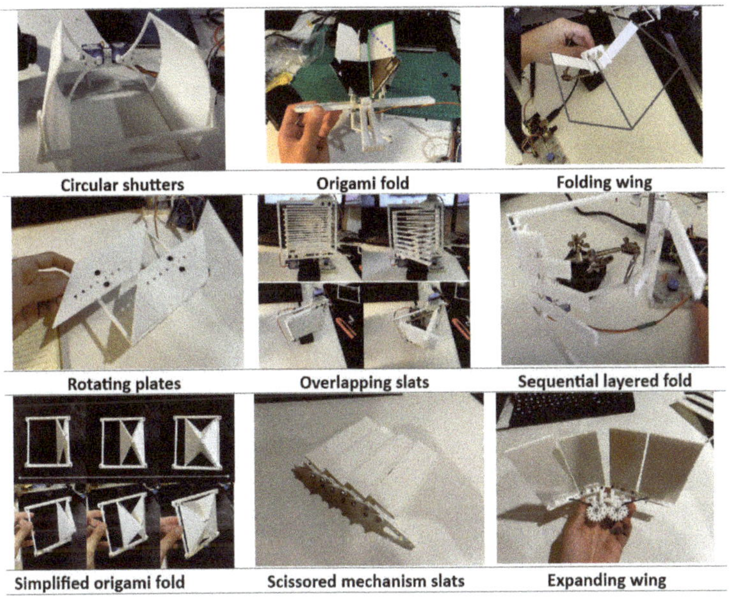

Figure 1. Design concepts for adaptive façade shading systems.

Circular Shutters: This concept consisted of two quarter-cylinder panels that rotated to form a semicircular closure. Each panel was rotated using two servos, although a single motor with interlinked gears could also have been used. A key limitation was the depth required within the frame to house the retracted shutters.

Origami Fold: Inspired by early origami folds, this design introduced complex motion through multiple hinges and moving parts. Two corners of the square panel travelled along a linear rail, while the other two remained free-floating. Actuation was considered via a linear actuator or a motor with linkages. A variation of the original origami design, this version modified the panel shapes to influence the system's overall form. In the closed position, the arrangement resembled a tulip, demonstrating how geometric changes could affect visual outcome.

Folding Wing: This concept used a compact linear actuation system to drive a larger folding movement. A central hinged section moved along a straight path, while the side panels were guided by rotating points. Although simple to actuate and mount, the design required further optimisation to reduce frame depth and improve occlusion performance.

Rotating Plates: This design used two flat shutters that rotated via dual servos. It investigated whether overlapping shutters that extended into adjacent modules could conceal the structural frame beneath, aiming to produce a more seamless appearance across the façade. This variation repositioned the rotating plates to alter their overlap behaviour. While lateral overlap with neighbouring modules was retained, vertical integration was reduced. Making the plates removable was also considered to facilitate maintenance and support a variety of material finishes.

Overlapping Slats: This design featured slats actuated by two servos, creating a phasing visual effect as they moved past one another. Although not fully opaque in the closed state, the design effectively blocked sunlight at common angles. As the slats transitioned between states, they formed a dynamic zig-zag texture across the surface.

Sequential Layered Fold: This concept was built upon the use of two servos to produce a cascading movement. The first slat was designed to move initially, then hook onto the next one, initiating a sequence that created a fanning effect. While actuation was simple, the slats needed to be robust and effective at varying angles, which introduced further design challenges.

Simplified Origami Fold: This design simplified the original origami-inspired concept by reducing the number of hinges. It retained the key folding motion while offering easier fabrication and a straightforward linear actuation path.

Scissored Mechanism Slats: his concept involved slats attached to a scissor mechanism. Although it occupied significant space when compressed, it allowed light penetration through controlled gaps. The design potentially shared actuation components with other linear rail systems and benefited from established mechanical principles.

Expanding Wing: This design employed gears and linkages to produce a single, smooth folding motion. The plates stretched and retracted to reveal or conceal the façade. While it was the most visually dynamic, it was also the most complex to implement as a full-scale, robust system.

Across these 9 concepts (Figure 1), most shared two primary actuation strategies: (1) rotational via dual motors/servos, or (2) linear via actuators on a central rail. These shared mechanisms suggest potential for standardised frames supporting multiple design types, though this could complicate frame design due to differing spatial and mechanical demands.

Based on six key criteria and following a comparative analysis of the nine design concepts (Results and Discussion section), three concepts were shortlisted, with the concept 'Rotating Plates' ultimately selected for further development (Figure 2). This design was subsequently used for 4D printing in the second phase of the research.

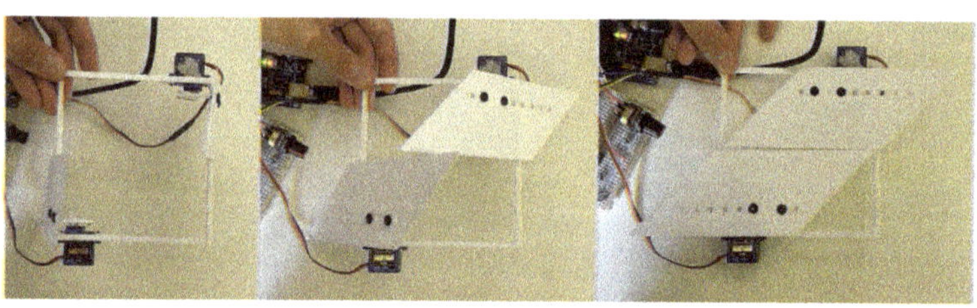

Figure 2. Rotating plates kinematic design concept.

The three shortlisted design concepts were modelled and rendered in Blender to visualise the scale of each module in the context of a human and a building (Figure 3). It was determined that each module would be sized at 800 mm by 800 mm, with advanced prototyping to be conducted at a 1:5 scale.

Figure 3. Design alternatives: Simplified Origami Fold, Overlapping Slats, and Rotating Plates from left to right.

To determine the appropriate size of the prototypes, two main aspects had to be considered: first, to build the prototypes to a scale that could realistically represent a full-size building and produce meaningful data; and second, to serve as an interactive display to help communicate the design vision. The 1:5 scale was a practical choice because it balanced the limits of the 9 g servos, which could move the 3D-printed parts without stalling, and allowed enough modules to be built and tested.

PLA was selected for the fabrication of motorized systems due to its excellent combination of printability, mechanical performance, and environmental safety. The eSUN Matte PLA+ used offered high rigidity, low shrinkage, minimal warping, and a lighter weight than standard PLA, reducing both material cost and overall mass while remaining durable and easy to process.

2.2. *Design Development of Non-Motorised Shading Systems with Passive Actuation Based on 4D Printing*

Building upon this design the parallel step was to investigate the addition of a passive actuation based on the materials composition. The actuated design concept comprises two plates mounted on a fixed frame (Figure 4 left). The opening mechanism of the flaps happens in response to variation in water content and temperature. The design of the shading device was simplified by dividing each flap into two triangular surfaces to enhance control over monitoring the shape-change mechanisms of individual flaps (Figure 4 right).

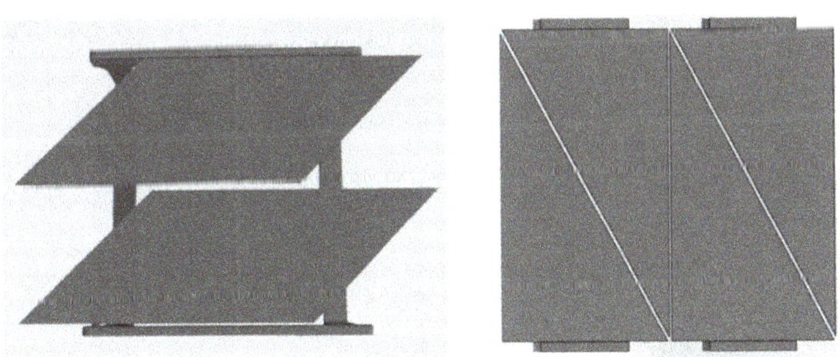

Figure 4. Initial design concept (**left**); modified design concept as assembly (**right**).

2.2.1. Compounding and 3D Printing

A composite material consisted of a PLA matrix (Ingeo 2003D, Natureworks, Blair, NE, USA) with 30 wt.% Lyocell fibres (FCP fibres with norminal length of 0.3 mm, supplied by Lenzing, Lenzing an der Ager, Austria) was used as an active/responsive layer in the bi-material structures. First, the polymer and fibres were melt compounded using a Process 16 Parallel Twin-Screw Extruder, (Thermo Fisher Scientific Inc., Waltham, MA,

USA), processed with a temperature profile of 165 °C, 180 °C, and the rest 185 °C at 70 rpm. The obtained material was pelletised and used for the production of 3D printing filaments similarly to what has been previously done [63]. The pelletised composite (<4 mm particles) was processed in a Filabot EX2 single screw extruder at 180 °C. The filament was air-cooled and spooled in a custom-made spooling device synchronised with the extrusion speed to obtain filaments of 1.75 ± 0.1 mm. For the passive layers, only PLA was used. The PLA filament was produced using a 3Devo filament maker and at a temperature ranging from 175–190 °C (Feed to die).

The two different types of material were thoroughly dried in a vacuum oven at 45 °C overnight (a period of at least 12 h) before being printed. The printing was performed using a Voron 350 FDM printer fitted with a 0.8 mm nozzle. The filament was printed at 220 °C onto a 60 °C bed. The layer height was set at 0.2 mm, and the part was printed at 20 mm/s.

Two types of design were used. A rectangular shape was adopted to explore three layouts of active-to-passive layers, while assemblies with a triangular shape were used to visualise opening and closing in a scaled-down assembly. In both rectangular and triangle designs, the first layer consisting of the composite active material was printed with 100% infill at raster angle of 90° and without perimeter. The second layer of PLA only was printed with a 50% infill and with a raster angle of 0°. A perimeter was printed for a stronger outline. Table 3 shows the slicer settings for the 3D-printed composites. The rigid support frame that holds the panels together within the assembly, was 3D printed using a commercially available filament (Guen, ColorFabb, Belfeld, The Netherlands).

Table 3. General slicer settings for the 3D printed composites both triangular and rectangular shapes.

Parameters	Slicer Settings
Line thickness [mm]	0.9
Layer height [mm]	0.2
Print speed [mm/s]	20
Nozzle temperature [°C]	220
Build plate temperature [°C]	60
Active density [%]	100
Active raster angle [°]	90
Passive density [%]	50
Passive raster angle [°]	0

The different layouts were used to assess their influence on the shape-change characteristics of the bi-material strips/triangles. A passive-to-active layer ratio of 1:1 is commonly used for hygromorphic structures, whereas, for PLA-lyocell composites, a lower ratio of passive to active layers has been shown to produce a more pronounced shape change or curvature [64]. In this research, the aim was to evaluate whether thicker samples (layout 2) would exhibit similar transformations to thinner samples (layout 1) by altering the passive-to-active layer ratio. Layout 3 was defined after initial trials with the other layouts revealed excessive curvature at the tips of the samples.

2.2.2. Layer Variation Based on Rectangular Shape

Rectangular shapes were 3D printed to quantitively evaluate the deformation and compare three different layouts (Figure 5). The rectangles were 80 × 10 mm in length and width (with thickness based on the layout), and five samples of each design were printed and tested at once.

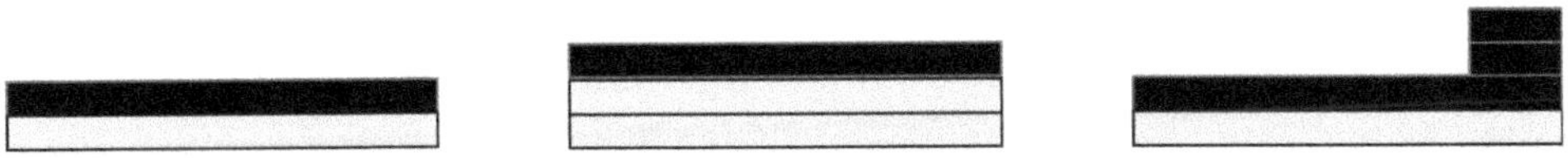

Figure 5. Schematic side view of the layout 1 (**left**), layout 2 (**middle**), layout 3 (**right**). Black represents the passive layer; grey represents the active composite layer.

a. Layout 1: One active layer and one passive layer (total thickness of 0.4 mm),
b. Layout 2: Two active layers and one passive layer (total thickness of 0.6 mm),
c. Layout 3: One active layer and one passive layer with 2 extra passive layers of 100% infill at one end that were 20 mm in length.

The environmental experiment on the samples was conducted at room temperature. These conditions were

a. Water spray from a spray bottle (until thoroughly wet) to activate the shape change transformation;
b. Monitoring/drying for 24 h at 23 °C and 50% RH;
c. Drying at 40 °C oven for 1 h.

The samples were allowed to move freely without any mechanical constraints. Photographs were taken at 0.1, 1, 2, 4, and 24 h to measure the height and radius of curvature. For each photograph, the sample was positioned on its side, and the tip height was measured using a ruler (Figure 6). The radius of curvature was determined by drawing base and height lines in ImageJ (Version 1.54p) 1.54 m and applying a macro based on the chord length formula [63].

$$K = 8d/(c^2 + 4d^2) \quad (1)$$

where K is curvature in mm^{-1}, d is the height in mm, and c is the chord length in mm.

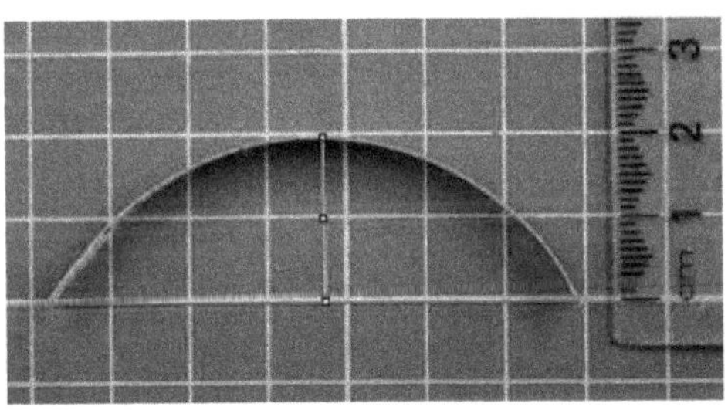

Figure 6. Base (chord length) and height lines drawn in ImageJ for radius calculation.

The chord method is a well-established approach to approximate curvature, particularly in contexts where arc measurements may be impractical or less precise [28,63].

2.2.3. Assemblies (Triangular Shapes)

The triangle assemblies were composed of four 3D printed intermeshing triangles. Two assembly layouts were tested based on Layout 1 and 3 of the rectangular shapes. The thicker section of Layout 3 was added to the narrowest part of the triangle (Figure 7). A tab was also added to the triangle shape to the slit onto the overall frame. This thicker section was made of extra layers of PLA. Glue was used to fix the triangles assemblies into the slits of the frame.

The triangular structure performances were tested under the following the same environmental conditions as the rectangular samples:

a. Water spray from a spray bottle (until thoroughly wet at room temperature) to activate the shape change transformation,
b. Monitoring/drying for 24 h at 23 °C and 50% RH,
c. Drying at 40 °C oven for 1 h

Photographs were taken throughout the process to monitor the shape variation.

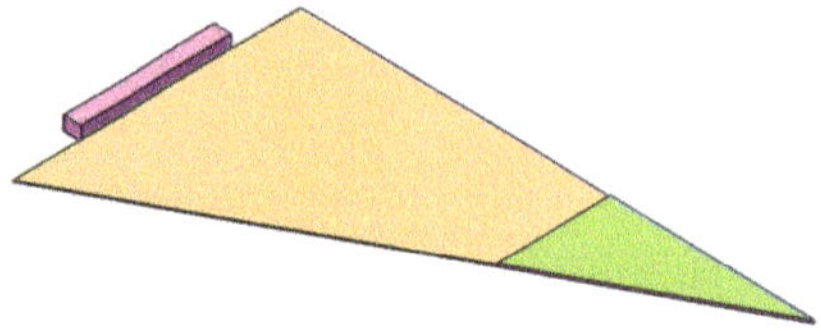

Figure 7. Triangular assembly 2 based on Layout 3. Further modified composite (orange) triangle including a tab for assembly (purple) and reinforced tip (green).

3. Results

3.1. Motorised Shading System with Mechanical Actuation (3D Printing)

A comparative analysis of the nine developed kinematic shading systems was conducted based on six key evaluation criteria: Simplicity and Robustness, Actuation Method, Repairability, Mounting Feasibility, Occlusion Ratio, and Architectural Integration (Table 4). From this analysis, three designs—Rotating Plates, Simplified Origami Fold, and Overlapping Slats—were shortlisted for further refinement, based on their balance of mechanical performance, scalability, and integration potential.

Table 4. Comparative evaluation of nine kinematic shading design concepts.

Design Concept	Simplicity & Robustness	Actuation Method	Repairability	Mounting Feasibility	Occlusion Ratio
Circular Shutters	Moderate—simple geometry but required depth for retraction	Servo or motor-geared	Moderate—moving parts enclosed in the frame	Limited—requires depth for installation on the facade	Moderate—partial occlusion
Origami Fold	Low—complex with many hinges	Linear actuator or motor-linkages	Low—intricate parts are harder to access	Complex—required rail system	High—near full closure
Folding Wing	High—simple folding with few parts	Linear actuator	High—accessible and modifiable	High—minimal structural impact	Moderate—needed tuning
Rotating Plates	High—few moving parts	Dual servos	High—removable shutters	High—minimal structural impact	High—full occlusion is achievable
Overlapping Slats	Moderate—phasing required precise alignment	Dual servos	Moderate—Posing complexity for maintenance	Moderate—required careful planning	Low—not fully opaque
Sequential Layered Fold	Moderate—cascading motion increased complexity	Dual servos	Low—difficult to access, maintain and repair	Moderate—required stable support	High—multi-stage occlusion
Simplified Origami Fold	High: reduced hinges, improved reliability	Linear actuator	Moderate—minimal complexity	Good—modular and scalable	Moderate—less occlusion than the original origami
Scissored Mechanism Slats	Moderate—known mechanism but spatially inefficient	Linear actuator	Moderate—layout, components and spatial arrangement lack efficiency and are hard to maintain and integrate.	Poor—large footprint even when compressed	Low—gaps remained in a closed state
Expanding Wing	Low—intricate linkages prone to failure	Gears and linkages	Low—difficult to service reliably	Low—challenging to integrate	Moderate—smoother transitions

Similar qualitative comparative analysis have been applied in the evaluation of kinetic facades and deployable structures [60].

(1) **Rotating Plates** performs strongly across all criteria with minimal compromises. Mechanically simple, visually clean, and architecturally feasible—making it a strong candidate for real-world application and scaling.
   - **Simplicity & Robustness**: Rated high—uses few moving parts, reducing potential failure points.
   - **Actuation**: Dual servos offer reliable, precise control with relatively simple implementation.
   - **Repairability**: Rated high—shutters are removable, allowing easy maintenance or replacement.
   - **Mounting Feasibility**: Rated high—design integrates easily with minimal impact on the building structure.
   - **Occlusion Ratio**: Rated high—capable of full closure, supporting effective shading.

(2) **Simplified Origami Fold** offers a good trade-off between mechanical novelty and practical deployment. It's visually interesting, functionally adaptable, and easier to manufacture than the original fold.
   - **Simplicity & Robustness**: Rated high—fewer hinges than the original origami design means greater reliability and easier fabrication.
   - **Actuation**: Linear actuator is straightforward and compatible with many control systems.
   - **Repairability**: Rated moderate—simple geometry aids maintenance, though less accessible than Rotating Plates.
   - **Mounting Feasibility**: Rated good—modular and scalable; fits well into façade grid systems.
   - **Occlusion Ratio**: Moderate—lower than the original origami fold, but still functional.

(3) **Overlapping Slats** Despite not achieving full occlusion, it offers dynamic surface texture and shading performance with simpler construction than more complex biomimetic or multi-layered options. Its aesthetic appeal in modular repetition adds architectural value.
   - **Simplicity & Robustness**: Moderate—requires precise alignment, but manageable.
   - **Actuation**: Uses dual servos—reliable and compact.
   - **Repairability**: Moderate—complexity in alignment may pose some maintenance challenges.
   - **Mounting Feasibility**: Moderate—integration into a grid is possible with thoughtful planning.
   - **Occlusion Ratio**: Low—does not achieve full opacity, but performs well for typical solar angles.

The three shortlisted concepts—Rotating Plates, Simplified Origami Fold, and Overlapping Slats—were prototyped at a 1:5 scale to assess their mechanical feasibility and design integration.

Although testing was conducted at reduced scale, consideration was given to the challenges of full-scale implementation to ensure the most appropriate concept was selected. The Simplified Origami Fold demonstrated potential for compact deployment and minimal spatial footprint. However, its fold-and-slide mechanism proved complex to fabricate, even at 1:5 scale. Issues with hinge alignment and actuator integration highlighted concerns regarding scalability and reliability. Despite these challenges, its flexibility and spatial efficiency remain promising for 1:1 deployment, provided further refinement is undertaken. In contrast, Rotating Plates and Overlapping Slats were mechanically more straightforward.

Their rotation-based mechanisms involved fewer components, simplifying fabrication and improving reliability. While these designs lacked the formal complexity and visual appeal of the Origami concept, their simplicity, robustness, and ease of maintenance position them as strong candidates for practical application in adaptive façade systems (Table 5).

Table 5. Key strengths and challenges of the three shortlisted shading concepts based on 1:5 scale prototyping.

Design	Strengths	Challenges
Simplified Origami Fold	Minimal spatial footprint	Complex hinging and alignment requirements Difficult to integrate actuators at prototype scale (1:5) The design at a scale of 1:1 could be compromised by the increased thickness of materials A more complex fold and slide mechanism
Overlapping Slats	Simple rotational mechanism Ease of fabrication Reliable and mechanically robust	Occupies more space when open, as both wings cover the length of the frame Gaps may be visible even if fully closed
Rotating Plates	Simple rotational mechanism Ease of fabrication Good structural stability Potential for creating a variety of patterns	No significant challenge was observed compared to the other two designs

Rotating plates were chosen because they offer a simple rotational mechanism that is easy to fabricate and provides good structural stability. Unlike overlapping slats, which can occupy more space when open and may show gaps when closed, rotating plates present no significant challenges and also allow for the creation of various patterns.

The Rotating Plates concept was selected for physical prototyping at a 1:5 scale to evaluate mechanical feasibility, modular integration, and readiness for upscaling. The prototype array was constructed using 2020 aluminium extrusion as the structural frame, with 3D-printed servo mounts and SG90 micro servos to actuate individual shutters.

Each module consisted of a pair of rotating plates mounted to a shared aluminium rail using a single bolt (Figure 8), allowing for rotational adjustment and ease of alignment. The design enabled the top shutter of a lower module and the bottom shutter of the upper module to share a mounting rail, reducing overall material use and visual obstruction. The extrusion framework also served as a routing channel for servo wiring, with grooves integrated into the mounts for cable concealment (Figure 9).

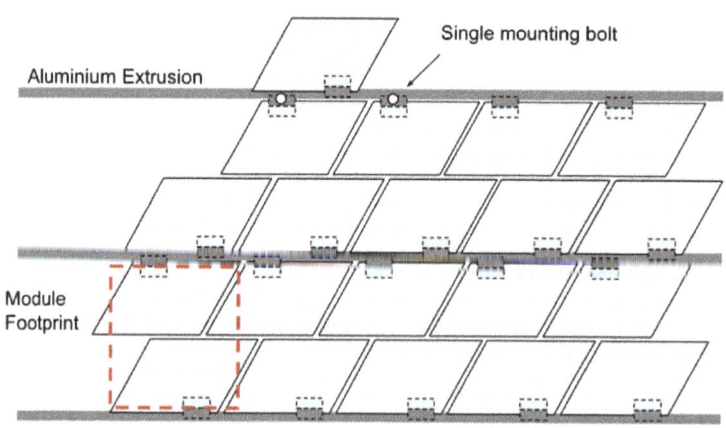

Figure 8. Design of the 1:5 assembly.

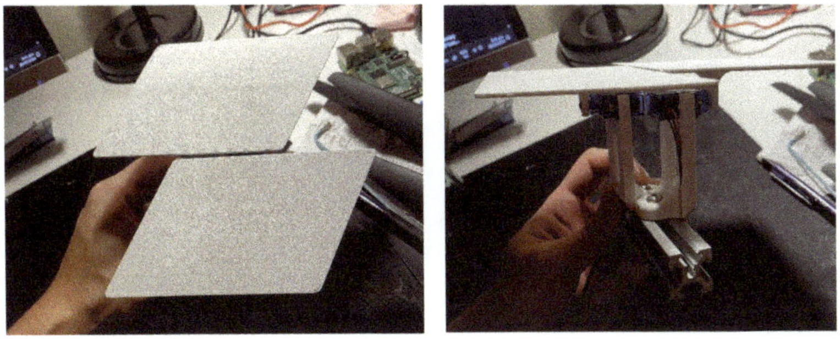

Figure 9. Top and bottom shutters from above view (**Left**) and side view attached to the frame (**Right**).

Sensor zones were mapped across the façade and included temperature and light sensors to simulate future feedback-based control but were not installed (Figure 10 Top). The servo mounts were designed to support lateral adjustment and to conceal servo wiring within integrated grooves, enhancing visual clarity and protecting cables during operation (Figure 10 Bottom). The SG90 servos proved sufficient for low-load actuation at this scale, but limitations were evident, including mechanical noise, restricted torque capacity, and a minimum step size of approximately one degree. These constraints could pose challenges for precision and durability in a full-scale deployment.

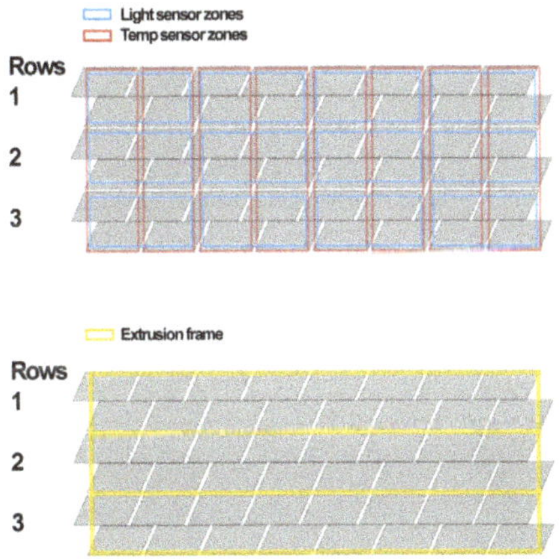

Figure 10. Design of sensor placement (**Top**) and support frame (**Bottom**).

The façade prototype was fabricated with the following dimensions: 675 cm (W) × 400 cm (D) × 250 cm (H), scaled down to 1:5. A complete module fitted a 160 mm square footprint with a 10mm gap around its perimeter. Additionally, the fixed distance between the horizontal frames required the shutter mounts to ensure sufficient clearance between actuated shutters to prevent interference. The complete 1:5 façade prototype (Figure 11) consisted of 27 servo-controlled modules. Each was pre-mounted to vertical aluminium rails before being installed on the frame. This modular approach allowed for efficient installation and wiring. The mounting strategy included a single bolt per shutter, which enabled small rotational adjustments during alignment. Each module occupied a footprint of 160 mm by 160 mm, with a 10 mm buffer zone to prevent interference between rotating parts. The system demonstrated strong repeatability and structural stability, validating the underlying mechanical concept.

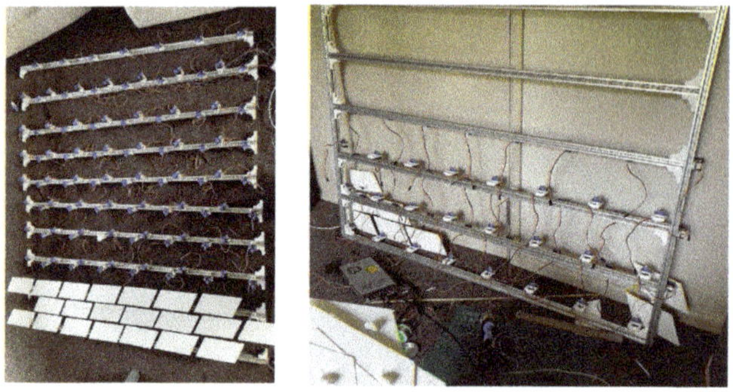

Figure 11. Scale prototype of the Rotating Plates concept being built, arrays (**left**) and arrays mounted onto a vertical frame (**right**).

At 1:5 scale, SG90 micro servos provide sufficient torque to rotate lightweight 3D-printed panels. However, in a full-scale (1:1) system, the mass, inertia, and wind load on each panel would increase significantly. The transition from model to full scale will require re-evaluation of actuation power, mechanical durability, and environmental resilience. The elegant simplicity of the rotating mechanism remains valid, but would be paired with industrial-grade components, stronger materials, and more sophisticated control systems.

3.2. Non-Motorised Shading Systems with Passive Actuation (4D Printing)

3.2.1. Rectangular Shape Characterisation

As reported by others, both the actuation amplitude and response rate are influenced by the geometry and material composition [28,63]. Layouts 1 and 3 showed similar height variation with asymmetrical curving; however, the asymmetrical reinforcement in Layout 3 prevented over-curling on one side of the sample (Figure 12). In contrast, Layout 2—with a double active layer—exhibited a reduction in movement despite the increased volume of actuation material. This behaviour can be attributed to an approximate 50% increase in overall thickness, which markedly enhances bending stiffness, given that flexural rigidity scales with the cube of the section thickness [65].

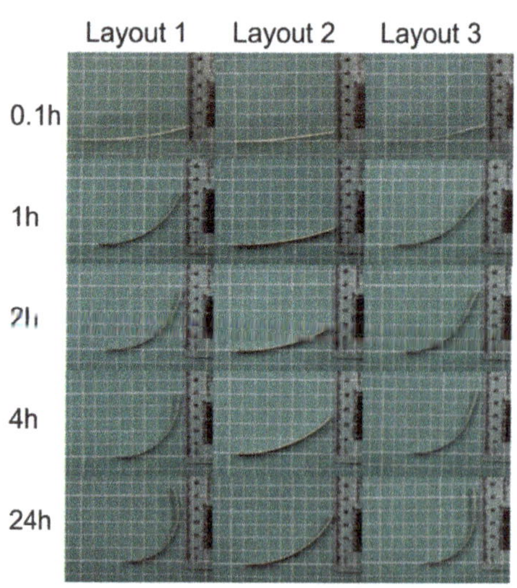

Figure 12. Photographs of the three rectangular layouts over time displaying deformation due to moisture.

Layout 1 with one passive layer and one active layer measured approximately 0.340 mm, while the layout 2 with two active layers and one passive layer measured around 0.570 mm. Layout 3 had the same thickness of layout 1 but with and extra 0.340 mm of passive material at the end of the samples.

When plotting the measurement variation as a function of time, Layout 2 (thickest samples) exhibited the slowest response and produced curvatures approximately 30% smaller than those observed for Layouts 1 and 3 (Figure 13). In all layouts, the fastest deformation occurs within the first hour and slowed down within 24 h. Layouts 1 and 3 showed no noticeable difference in height performance; however, reinforcing one end of the sample altered the curvature profile and resulted in an approximate 10% decrease in K. This prevented the "over-curling" effect observed in Layout 1.

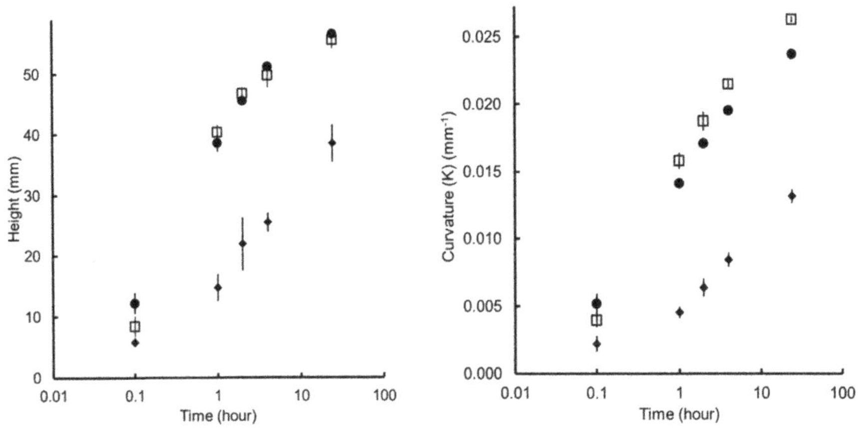

Figure 13. Variation in the height (left) and the curvature (right) as a function of time for Layout 1 (Square), Layout 2 (Black diamond) and Layout 3 (Black circle). The error bar represents the standard deviation for n = 5. The figure does include error bars; however, the errors are so small that they are contained within the size of the marker.

The plots indicate that all three layouts exhibit rapid actuation within the first hour of exposure. In both height and curvature measurements, a steep increase is observed between 0.1 and 1 h, highlighting that most of the shape change occurs within this early window. This suggests that the material–geometry system responds quickly to moisture ingress. Notably, the Layout 1 (circle) and Layout 3 (square) achieve over 80% of their final curvature and displacement before the 4-h mark, with only marginal changes, thereafter, indicating a fast approach to equilibrium. In contrast, the sample represented by Layout 2 (diamond) displays slower kinetics and lower overall actuation, with a more gradual increase in both height and curvature over time. This is expected as Layout 2 is thicker, and water diffusion is expected to be slower.

3.2.2. First Assembly

The passive testing of the triangle relied on fixing one end to a frame in cantilever mode and observing the curvature as moisture entered the composite, with photos taken at 1, 2, 4, and 24 h (Figure 14). The stiffness of the prints was reduced by the presence of random holes, which still provided partial shading but also decreased structural integrity. Similar to the rectangular-shaped composites, the triangular prints appeared to "over-curl" within 24 h, indicating that the movement was too pronounced. Some twisting was also observed, particularly toward the shorter leg of the triangle. This torsional behaviour is likely due to geometric and fabrication-related asymmetries. The triangular layout, with one side fixed and uneven leg lengths, creates an imbalance in stiffness that promotes twisting during activation. Additionally, the anisotropic properties inherent in the FDM printing process—especially differences in raster orientation between the triangle's edges—may add to uneven strain distribution. Residual stresses from printing, such as those caused by uneven cooling or layer deposition, may also relax over time and contribute to the observed out-of-plane twisting. While the primary goal is bending-driven actuation, this torsional response adds complexity to the motion and need to be accounted for in future design refinements.

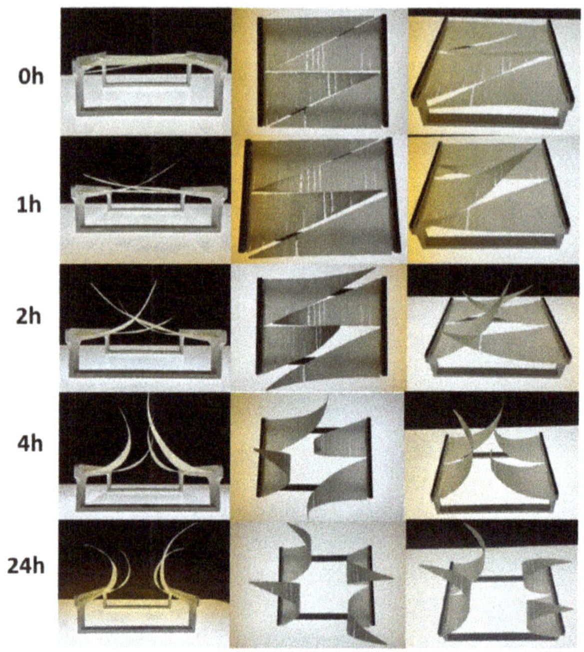

Figure 14. Evolution of the Assembly One over time under moisture stimulus.

3.2.3. Second Assembly

Four triangles were 3D printed using the same design with the exception of the tip reinforcement section that was increased in length from 25% of total length of the triangles (Figure 15). The slicer settings were the same as Table 3, with the addition of 2 passive layers with 100% infill at the tip. These composite triangles were then glued into the frame, sprayed with water and left for 24 h at ambient conditions. The reinforcement at the apex (pointed tip) locally increased stiffness and altered the overall deformation pattern. As expected, the reinforced tip resisted curling, leading to an obvious reduction in curvature compared to the unreinforced layout of the first assembly. This was consistent with a measured decrease in the curvature coefficient (K) by approximately 10% in the rectangular samples. Interestingly, while the overall bending was less pronounced, the sample exhibited more pronounced torsional behaviour. The deformation was accentuated compared to the first assembly toward the shorter edge of the triangle. This suggests that the interaction between the fixed base, the asymmetric geometry, and the local reinforcement creates an uneven stress distribution during actuation that was exacerbated by an increase in stiffness. As the reinforced tip constrains movement at one end, the free edges—especially the shorter leg—become the dominant paths for curvature and twist. The result is a less uniform, more complex motion that combines bending and asymmetric torsion, highlighting the sensitivity of the system to even minor stiffness modifications.

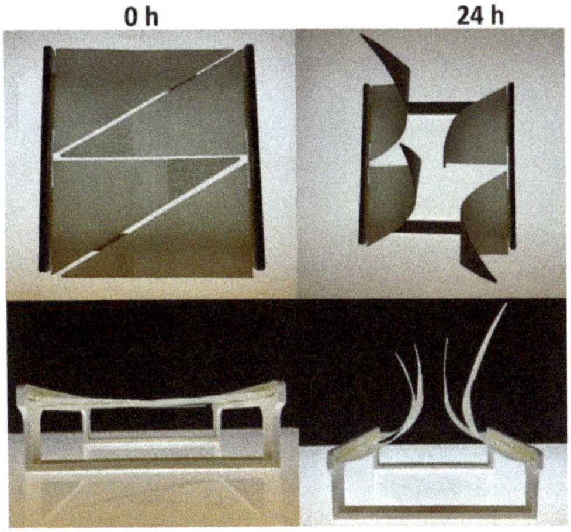

Figure 15. Second assembly of the structure in closed and opened positions.

4. Discussions and Conclusions

The comparative evaluation and 1:5 scale prototyping revealed the trade-offs between mechanical simplicity, fabrication ease, and architectural performance in kinetic façade systems. While visually expressive designs like the Simplified Origami Fold offered compactness and aesthetic potential, their complexity in actuation and fabrication raised concerns about scalability and long-term reliability. In contrast, the Rotating Plates and Overlapping Slats concepts, though less visually complex, demonstrated mechanical robustness and practical feasibility, attributes critical for deployment at architectural scale.

The selection of the Rotating Plates concept for physical prototyping underscored the importance of modularity, ease of integration, and structural clarity in kinetic shading systems. Although limitations in small-scale components such as servo torque and sensor implementation were identified, the overall mechanical concept proved sound.

This project also demonstrated in parallel the feasibility of using hygromorphic composite materials to create passive structures capable of dynamic shape change in response to environmental humidity. Specifically, bi-layered PLA/Lyocell composites, printed in combination with pure PLA, showed strong and consistent bending behaviour when exposed to moisture, confirming their suitability for adaptive applications.

Different materials have been used in hygromorphic systems. Wood-based approaches rely on differences in hygroexpansion between laminates with different grain orientations [66]. Although highly responsive, they face limitations in uniformity, manufacturability, and durability. Other studies use natural-fibre yarns or short fibres in a polymer matrix, opening up additional manufacturing routes, including additive manufacturing, which enables effective fibre alignment in predefined orientations.

Lyocell fibres are advantageous due to their uniform dimensions and predictable mechanical properties, enhancing processability and facilitating 3D printing of biocomposites. They also exhibit strong hygroscopic behaviour. In PLA composites, comparable moisture uptake yields ~3% hygroexpansion for lyocell versus ~0.7% for wood fibre [50,64]. Even relative to composites with 70 wt% continuous flax fibres in a PP matrix [67], which achieved a maximum hygroexpansion of ~3.4% in the transverse direction, the ~3% observed for lyocell is comparatively high. In addition, using short fibres rather than continuous fibres offers broader manufacturing options. Together, these factors suggest a higher potential for shape-change/curvature than systems based on other fibres.

The triangular actuator design successfully produced curvature and height changes over time, and partial design optimisation—such as tip reinforcement and multi-layer configurations—helped evaluate the influence of stiffness distribution and geometry. However, several opportunities for further refinement remain. Replacing the triangular shape with a rectangular profile may reduce unwanted torsion and improve packing density by allowing closer spacing and better shading coverage. Extending the composite length to create slight overlap across the frame could further increase functional coverage. Adjustments to the stiffened tip—either by increasing its area or removing active material entirely—could better control the onset and magnitude of curling, ensuring that the actuation reaches its target state at the desired time.

The PLA-based formulations were primarily used for research and prototyping purposes. It is important to mention the limitations of PLA due to its low UV resistance and biodegradability, which would restrict its use in real-life applications. However, its high moisture absorption is not considered a drawback in this context, since it is essential for the system to function, the composites must absorb water and expand for the shape transformation to occur. The main aim of this study was to demonstrate the concept and produce a prototype using the selected design. In practical applications, other stable and durable polymers could be used with lyocell fibres, such as polyketone (PK) or high-density polyethylene (HDPE), with the addition of UV stabilisers.

This study on hygromorphic composite materials has certain limitations that warrant acknowledgment. First, the cyclability of the developed materials was not examined in detail. While preliminary investigations were performed, this aspect was not pursued further, as it lies beyond the scope of the present work. Future studies should address long-term performance and repeatability, which are critical for practical applications.

Another important consideration relates to thickness effects. It is well established that increasing thickness raises bending stiffness, which in turn slows the hygromorphic response and water migration within the samples. Prior studies [64,66] have shown that both the passive-to-active layer ratio and the overall thickness of bilayer hygromorphic structures govern curvature. This behavior can be described using a modified Timoshenko equation, originally developed for bimetallic strips with differing thermal expansion coefficients and later adapted for differential hygroexpansion. Such relationships provide a means to tune both the magnitude and rate of transformation. Although highly relevant to the potential applications of the structures presented here, this line of inquiry is outside the scope of the current manuscript and represents a promising direction for future investigation.

In addition, the materials used in this study were selected primarily for proof-of-concept purposes. While the incorporation of hygromorphic materials demonstrates feasibility, the scalability of 3D printing with PLA/lyocell remains uncertain. Factors such as the manufacturing process and the overall thickness of the structures may influence performance, including the risk of delamination. These challenges underscore the need for further research into alternative material systems and process optimization before practical deployment can be realized.

The development of 3D-printed motorised shading systems highlighted several technical and practical challenges that will need to be addressed in future stages. These include improving sensor integration, refining control and communication systems, selecting appropriate materials and fabrication methods, and developing effective mounting strategies for both new and existing buildings. Considerations around scalability, system reliability, and ease of installation will be key to ensuring the viability of the design in real-world architectural applications.

Future work should focus on refining material design and actuation strategies to balance responsiveness, stability, and durability, while also standardising testing methods for consistent evaluation across studies. To enable real-world application in adaptive façades, challenges related to manufacturability, long-term performance, and modelling of moisture–structure interactions must be addressed through advanced optimisation and predictive tools.

Author Contributions: This paper represents the results of teamwork. Conceptualisation, N.I. and N.B.; Methodology, N.I., N.B., M.-J.L.G., C.G. and M.B.; Software, N.B. and M.-J.L.G.; Validation, N.I. and M.-J.L.G.; Investigation, N.I. and M.-J.L.G.; Resources, C.G. and M.B.; Writing—Original Draft Preparation, N.B. and C.B.; Writing—Review and Editing, N.I. and M.-J.L.G.; Visualisation, N.B. and C.B.; Supervision, N.I. and M.-J.L.G.; Funding Acquisition, N.I. and M.-J.L.G. All authors have read and agreed to the published version of the manuscript.

Funding: This project formed part of a larger research programme funded by the Ministry of Business, Innovation and Employment of New Zealand under the National Science Challenge—Science for Technological Innovation, led through the University of Otago (Grant E4203), the programme was titled 'Biomimetic Buildings for Climate Change Adaptation: Future Generation of Facades.' The work builds on earlier research within the Challenge, which focused on additive manufacturing and the development of 3D and 4D printed bio-composites led by Scion (Grant 2019-S5-CRS).

Institutional Review Board Statement: Not applicable.

Informed Consent Statement: Not applicable.

Data Availability Statement: The datasets presented in this article are not readily available because the data are part of an ongoing study. Requests to access the datasets should be directed to corresponding author.

Conflicts of Interest: Authors Marie-Joo Le Guen, Caelum Betteridge, and Maxime Barbier were employed by the company New Zealand Institute for Bioeconomy Science, and Christian Gauss was employed by the University of Waikato. The remaining authors declare that the research was conducted in the absence of any commercial or financial relationships that could be construed as a potential conflict of interest.

Abbreviations

The following abbreviations are used in this manuscript:

PLA	Polylactic acid
FDM	Fused Deposition Modelling
Bio-ABFs	Biomimetic Adaptive Building Façades
SMAs	Shape memory alloys

References

1. IEA. *Energy Efficiency Policy Toolkit 2025*; IEA: Paris, France, 2025; Licence: CC BY 4.0.; Available online: https://www.iea.org/reports/energy-efficiency-policy-toolkit-2025 (accessed on 28 September 2025).
2. Globalabc. IEA Tracking Report—Buildings. 2023. Available online: https://globalabc.org/index.php/resources/publications/iea-tracking-report-buildings (accessed on 30 September 2022).
3. Dewidar, K.; Mahmoud, A.H.; Magdy, N.; Ahmed, S. The role of intelligent façades in energy conservation. In Proceedings of the International Conference on Sustainability and the Future: Future Intermediate Sustainable Cities (FISC 2010), Elshourouq, Egypt, 23–25 November 2010.
4. Francesco, S.; Lidia, B.; Gigliola, A. A critical review of biomimetic building envelopes: Towards a bio-adaptive model from nature to architecture. *Renew. Sustain. Energy Rev.* **2022**, *169*, 112850. [CrossRef]
5. Imani, N.; Vale, B. *Heating with Wolves, Cooling with Cacti: Thermo-Bio-Architectural Framework (ThBA)*; CRC Press: Boca Raton, FL, USA, 2021.

6. Attia, S.; Lioure, R.; Declaude, Q. Future trends and main concepts of adaptive facade systems. *Energy Sci. Eng.* **2020**, *8*, 3255–3272. [CrossRef]
7. Al-Masrani, S.M.; Al-Obaidi, K.M. Dynamic shading systems: A review of design parameters, platforms and evaluation strategies. *Autom. Constr.* **2019**, *102*, 195–216. [CrossRef]
8. Hosseini, S.M.; Mohammadi, M.; Guerra-Santin, O. Interactive kinetic façade: Improving visual comfort based on dynamic daylight and occupant's positions by 2D and 3D shape changes. *Build. Environ.* **2019**, *165*, 106396. [CrossRef]
9. Imani, N.; Vale, B. Biomimetic Design for Building Energy Efficiency 2021. *Biomimetics* **2022**, *7*, 106. [CrossRef]
10. Charpentier, L.; Cruz, E.; Nenov, T.; Guidoux, K.; Ware, S. Pho'liage: Towards a kinetic biomimetic thermoregulating façade. In *Bionics and Sustainable Design*; Springer Nature: Singapore, 2022; pp. 367–401.
11. Imani, N.; Vale, B. *Biomimetic Buildings: Copying Nature for Energy Efficiency*; MDPI: Basel, Switzerland, 2022.
12. Meagher, M. Designing for change: The poetic potential of responsive architecture. *Front. Archit. Res.* **2015**, *4*, 159–165. [CrossRef]
13. Alemdağ, E.L.; Beyhan, F. A research on construction systems of double skin facades. *Gazi Univ. J. Sci.* **2017**, *30*, 17–30.
14. Schumacher, M.; Schaeffer, O.; Vogt, M.-M. *Move: Architecture in Motion-Dynamic Components and Elements*; Walter de Gruyter: Berlin, Germany, 2012.
15. Sheikh, W.T.; Asghar, Q. Adaptive biomimetic facades: Enhancing energy efficiency of highly glazed buildings. *Front. Archit. Res.* **2019**, *8*, 319–331. [CrossRef]
16. Hosseini, S.M.; Mohammadi, M.; Rosemann, A.; Schröder, T.; Lichtenberg, J. A morphological approach for kinetic façade design process to improve visual and thermal comfort. *Build. Environ.* **2019**, *153*, 186–204. [CrossRef]
17. Körner, A.; Born, L.; Mader, A.; Sachse, R.; Saffarian, S.; Westermeier, A.; Poppinga, S.; Bischoff, M.; Gresser, G.; Milwich, M. Flectofold—A biomimetic compliant shading device for complex free form facades. *Smart Mater. Struct.* **2017**, *27*, 017001. [CrossRef]
18. Badarnah, L.; Zolotovsky, K. Morphological differentiation for the environmental adaptation of biomimetic buildings: Skins, surfaces, and structures. In *Biomimicry for Materials, Design and Habitats*; Elsevier: Amsterdam, The Netherlands, 2022; pp. 439–466.
19. López, M.; Rubio, R.; Martín, S.; Croxford, B. How plants inspire façades. From plants to architecture: Biomimetic principles for the development of adaptive architectural envelopes. *Renew. Sustain. Energy Rev.* **2017**, *67*, 692–703. [CrossRef]
20. Mi-jin, K.; Baek-gyeom, K.; Je-sung, K.; Hwang, Y. Flexural biomimetic responsive building façade using a hybrid soft robot actuator and fabric membrane. *Autom. Constr.* **2023**, *145*, 104660. [CrossRef]
21. Saidam, M.W.; Al-Obaidi, K.M.; Hussein, H.; Ismail, M.A. The application of smart materials in building facades. *Ecol. Environ. Conserv.* **2017**, *23*, S8–S11.
22. Holstov, A.; Farmer, G.; Bridgens, B. Sustainable materialisation of responsive architecture. *Sustainability* **2017**, *9*, 435. [CrossRef]
23. Smith, S.I. Superporous intelligent hydrogels for environmentally adaptive building skins. *MRS Adv.* **2017**, *2*, 2481–2488. [CrossRef]
24. Villegas, J.E.; Gutierrez, J.C.R.; Colorado, H.A. Active materials for adaptive building envelopes: A review. *J. Mater. Environ. Sci.* **2020**, *11*, 988–1009.
25. Yi, H.; Kim, D.; Kim, Y.; Kim, D.; Koh, J.-S.; Kim, M.-J. 3D-printed attachable kinetic shading device with alternate actuation: Use of shape-memory alloy (SMA) for climate-adaptive responsive architecture. *Autom. Constr.* **2020**, *114*, 103151. [CrossRef]
26. Reichert, S.; Menges, A.; Correa, D. Meteorosensitive architecture: Biomimetic building skins based on materially embedded and hygroscopically enabled responsiveness. *Comput. Aided Des.* **2015**, *60*, 50–69. [CrossRef]
27. Imani, N.; Donn, M.; Balador, Z. Bio-inspired materials: Contribution of biology to energy efficiency of buildings. In *Handbook of Ecomaterials*; Springer: Berlin/Heidelberg, Germany, 2018; pp. 2213–2236.
28. de Kergariou, C.; Demoly, F.; Perriman, A.; Le Duigou, A.; Scarpa, F. The design of 4D-printed hygromorphs: State-of-the-art and future challenges. *Adv. Funct. Mater.* **2023**, *33*, 2210353. [CrossRef]
29. Barbier, M.; Le Guen, M.J.; McDonald-Wharry, J.; Bridson, J.H.; Pickering, K.L. Quantifying the shape memory performance of a three-dimensional-printed biobased polyester/cellulose composite material. *3D Print. Addit. Manuf.* **2021**, *8*, 193–200. [CrossRef]
30. Khezri, M.; Rasmussen, K. Functionalising buckling for structural morphing in kinetic façades: Concepts, strategies and applications. *Thin-Walled Struct.* **2022**, *180*, 109749. [CrossRef]
31. Talaei, M.; Mahdavinejad, M.; Azari, R.; Haghighi, H.M.; Atashdast, A. Thermal and energy performance of a user-responsive microalgae bioreactive façade for climate adaptability. *Sustain. Energy Technol. Assess.* **2022**, *52*, 101894.
32. Schumacher, M.; Vogt, M.-M.; Krumme, L.A.C. *New Move: Architecture in Motion-New Dynamic Components and Elements*; Birkhauser: Boston, MA, USA, 2019.
33. ArchDaily. One Ocean, Thematic Pavilion EXPO 2012/Soma. Available online: https://www.archdaily.com/236979/one-ocean-thematic-pavilion-expo-2012-soma (accessed on 13 July 2025).
34. Schleicher, S.; Lienhard, J.; Poppinga, S.; Speck, T.; Knippers, J. A methodology for transferring principles of plant movements to elastic systems in architecture. *Comput. Aided Des.* **2015**, *60*, 105–117. [CrossRef]
35. Krieg, O.D. HygroSkin—Meteorosensitive Pavilion. In *Advancing Wood Architecture*; Routledge: Oxfordshire, UK, 2016; pp. 125–138.

36. Attia, S. Evaluation of adaptive facades: The case study of Al Bahr Towers in the UAE. *QScience Connect.* **2017**, *2017*, 6.
37. Glynn, R.; Sheil, B. (Eds.) *Fabricate 2011: Making Digital Architecture*; UCL Press: London, UK, 2017; Volume 1.
38. Loonen, R.C.; Rico-Martinez, J.; Favoino, F.; Brzezicki, M.; Ménézo, C.; La Ferla, G.; Aelenei, L.L. Design for façade adaptability: Towards a unified and systematic characterization. In Proceedings of the 10th Conference on Advanced Building Skins, Bern, Switzerland, 3–4 November 2015; Economic Forum: Munich, Germany, 2015; pp. 1284–1294.
39. Wigginton, M. *Intelligent Skins. Butterworth-Heinemann Linacre House*; Jordan Hill: Oxford, UK, 2002.
40. Cheng, T.; Tahouni, Y.; Sahin, E.S.; Ulrich, K.; Lajewski, S.; Bonten, C.; Wood, D.; Rühe, J.; Speck, T.; Menges, A. Weather-responsive adaptive shading through biobased and bioinspired hygromorphic 4D-printing. *Nat. Commun.* **2024**, *15*, 10366. [CrossRef]
41. Koh, J.-S.; Kim, S.-R.; Cho, K.-J. Self-folding origami using torsion shape memory alloy wire actuators. In *International Design Engineering Technical Conferences and Computers and Information in Engineering Conference*; American Society of Mechanical Engineers: New York, NY, USA, 2014; Volume 46377, p. V05BT08A043.
42. Zolfagharian, A.; Kaynak, A.; Khoo, S.Y.; Kouzani, A. Pattern-driven 4D printing. *Sens. Actuators A Phys.* **2018**, *274*, 231–243. [CrossRef]
43. Sung, D. Smart geometries for smart materials: Taming thermobimetals to behave. *J. Archit. Educ.* **2016**, *70*, 96–106. [CrossRef]
44. Foged, P. Sense Envelope IV. 2016. Available online: https://pasoldfoged.com/sense-iv (accessed on 28 September 2025).
45. Mazzucchelli, E.S.; Alston, M.; Brzezicki, M.; Doniacovo, L. Study of a BIPV adaptive system: Combining timber and photovoltaic technologies. *J. Facade Des. Eng.* **2018**, *6*, 149–162.
46. Alshebly, Y.S.; Mustapha, K.B.; Zolfagharian, A.; Bodaghi, M.; Mohamed Ali, M.S.; Almurib, H.A.; Nafea, M. Bioinspired pattern-driven single-material 4D printing for self-morphing actuators. *Sustainability* **2022**, *14*, 10141. [CrossRef]
47. Naficy, S.; Gately, R.; Gorkin III, R.; Xin, H.; Spinks, G.M. 4D printing of reversible shape morphing hydrogel structures. *Macromol. Mater. Eng.* **2017**, *302*, 1600212. [CrossRef]
48. Zhang, Q.; Zhang, K.; Hu, G. Smart three-dimensional lightweight structure triggered from a thin composite sheet via 3D printing technique. *Sci. Rep.* **2016**, *6*, 22431. [CrossRef] [PubMed]
49. Miao, S.; Zhu, W.; Castro, N.J.; Nowicki, M.; Zhou, X.; Cui, H.; Fisher, J.P.; Zhang, L.G. 4D printing smart biomedical scaffolds with novel soybean oil epoxidized acrylate. *Sci. Rep.* **2016**, *6*, 27226. [CrossRef] [PubMed]
50. Le Duigou, A.; Castro, M.; Bevan, R.; Martin, N. 3D printing of wood fibre biocomposites: From mechanical to actuation functionality. *Mater. Des.* **2016**, *96*, 106–114. [CrossRef]
51. Yao, L.; Ou, J.; Cheng, C.-Y.; Steiner, H.; Wang, W.; Wang, G.; Ishii, H. BioLogic: Natto cells as nanoactuators for shape changing interfaces. In Proceedings of the 33rd Annual ACM Conference on Human Factors in Computing Systems, Seoul, Republic of Korea, 18–23 April 2015.
52. Correa, D.; Poppinga, S.; Mylo, M.D.; Westermeier, A.S.; Bruchmann, B.; Menges, A.; Speck, T. 4D pine scale: Biomimetic 4D printed autonomous scale and flap structures capable of multi-phase movement. *Philos. Trans. R. Soc. A* **2020**, *378*, 20190445.
53. Krapež Tomec, D.; Straže, A.; Haider, A.; Kariž, M. Hygromorphic Response Dynamics of 3D-Printed Wood-PLA Composite Bilayer Actuators. *Polymers* **2021**, *13*, 3209.
54. Tahouni, Y.; Krüger, F.; Poppinga, S.; Wood, D.; Pfaff, M.; Rühe, J.; Speck, T.; Menges, A. Programming sequential motion steps in 4D-printed hygromorphs by architected mesostructure and differential hygro-responsiveness. *Bioinspiration Biomim.* **2021**, *16*, 055002. [CrossRef]
55. Eppinger, S.D.; Ulrich, K. *Product Design and Development*; McGraw-Hill: New York, NY, USA, 1995.
56. Mangkuto, R.A.; Koerniawan, M.D.; Apriliyanthi, S.R.; Lubis, I.H.; Atthaillah; Hensen, J.L.; Paramita, B. Design optimisation of fixed and adaptive shading devices on four façade orientations of a high-rise office building in the tropics. *Buildings* **2021**, *12*, 25. [CrossRef]
57. Al Dakheel, J.; Tabet Aoul, K. Building Applications, opportunities and challenges of active shading systems: A state-of-the-art review. *Energies* **2017**, *10*, 1672.
58. Roshan Kharrat, R. Dynamic and Adaptive Photovoltaic Shading Systems: Design and Architectural Integration for Energy Production and Indoor Comfort. Ph.D. Thesis, Politecnico di Torino, Turin, Italy, 2024.
59. Tabadkani, A.; Haddadi, M.; Rizi, R.A.; Tabadkani, E. A hierarchical multi-purpose roller shade controller to enhance indoor comfort and energy efficiency. In *Building Simulation*; Springer: Berlin/Heidelberg, Germany, 2023.
60. Konstantoglou, M.; Tsangrassoulis, A. Dynamic operation of daylighting and shading systems: A literature review. *Renew. Sustain. Energy Rev.* **2016**, *60*, 268–283. [CrossRef]
61. Wu, H.; Zhang, T. Optimal design of complex dynamic shadings: Towards sustainable built environment. *Sustain. Cities Soc.* **2022**, *86*, 104109. [CrossRef]
62. Brzezicki, M. A Systematic Review of the Most Recent Concepts in Kinetic Shading Systems with a Focus on Biomimetics: A Motion/Deformation Analysis. *Sustainability* **2024**, *16*, 5697. [CrossRef]

63. Gauss, C.; Pickering, K.; Barbier, M.; Miller, T. Additive manufacturing of hygromorphic structures using regenerated cellulose/PLA biocomposites. *Mater. Today Proc.* **2023**. [CrossRef]
64. Bachmann, P.A.; Pickering, K.; Gauss, C. Combining Hydromorphic and Shape-Shifting Effects for Programmable Passive Actuation in 4D-Printed Biocomposites. *Compos. Part. C Open Access* **2025**, *18*, 100632. [CrossRef]
65. Timoshenko, S. Analysis of bi-metal thermostats. *J. Opt. Soc. Am.* **1925**, *11*, 233–255. [CrossRef]
66. Holstov, A.; Bridgens, B.; Farmer, G. Hygromorphic materials for sustainable responsive architecture. *Constr. Build. Mater.* **2015**, *98*, 570–582. [CrossRef]
67. Le Duigou, A.; Keryvin, V.; Beaugrand, J.; Pernes, M.; Scarpa, F.; Castro, M. Humidity responsive actuation of bioinspired hygromorph biocomposites (HBC) for adaptive structures. *Compos. Part. A Appl. Sci. Manuf.* **2019**, *116*, 36–45. [CrossRef]

biomimetics

MDPI

Article

Biomimetic Adaptive Building Façade Modeling for Sustainable Urban Freshwater Ecosystems: Integration of Nature's Water-Harvesting Strategy into Sun-Breakers

Berkan Kahvecioğlu [1,2,*], Güneş Mutlu Avinç [3] and Semra Arslan Selçuk [1]

1 Department of Architecture, Faculty of Architecture, Gazi University, 06560 Ankara, Turkey; semraselcuk@gazi.edu.tr
2 Department of Construction and Technical, Erciyes University, 38280 Kayseri, Turkey
3 Department of Architecture, Faculty of Engineering and Architecture, Mus Alparslan University, 49250 Mus, Turkey; g.avinc@alparslan.edu.tr
* Correspondence: berkankahvecioglu@gmail.com

Abstract: Urban freshwater ecosystems have many critical functions, such as providing water to all living things and supporting biodiversity. Factors such as water pollution, increased water consumption, habitat loss, climate change, and drought threaten the health of urban freshwater ecosystems. Looking for solutions to these challenges, this article aims to recycle water and return it to its life cycle using a climate-sensitive water collection strategy. The model focuses on the biomimetic method as a basic strategy. In this regard, the concept of water-harvesting has been examined in detail by conducting a deep literature review, including architecture and engineering disciplines. With all these data obtained, a synthesis/integration study was carried out by developing a model proposal based on adaptive building façade elements to solve the water problems experienced in cities. The model proposal, which is directly related to the titles of "Clean Water and Sanitation (SDG 6)" and "Sustainable Cities and Communities (SDG 11)", which are among the Sustainable Development Goals (SDGs), aims to provide different perspectives on the disciplines with its superficial and functional features. In this context, it is anticipated that the article will become an indispensable resource for other researchers working on the subject.

Keywords: adaptive building; biomimetic façade modeling; clean water and sanitation; sun-breaker; sustainable cities and communities; water-harvesting; urban freshwater ecosystem

1. Introduction

The world's future depends on the cooperation of scientists, researchers, policymakers, practitioners, and controllers from different disciplines and the development of more sustainable living standards and strategies. For example, with the Annual Environment Policy Review reports organized every year with the initiatives of the European Commission, the current environmental developments at the European Union and national levels are monitored, and future risks and potentials are discussed together. As stated in the report prepared by the European Commission in 2009, the increasing environmental pressures indicate the need for innovative alternatives to the ecosystem-destroying technologies of today's industrial age; a solution [1]. So much so, that it is stated that approximately 7 trillion US dollars are invested annually by both the public and private sectors against activities that damage the ecosystem globally [2]; this number is more critical when considering interrelated crises such as climate change, biodiversity loss, damage to aquatic ecosystems, and land degradation.

The combination of many problems, such as rapid urbanization on a worldwide scale, disasters, climate change, and resource consumption, poses severe challenges for

21st-century cities. Along with urban pressures, especially the rapidly changing demographic structure of megacities, increasing transportation problems, resource depletion, and various natural phenomena (flooding, overheating, etc.) occurring frequently, the urban freshwater ecosystem, along with biodiversity, is being damaged, and alternative solutions are being sought [3–9]. According to the United Nations, approximately half of the world's population lives in urban areas [10], and this proportion is projected to increase to approximately 70% by 2050 [11]. This also emphasizes the need for severe measures to be taken. Therefore, it is clearly understood that living standards in cities will be more complex, and all disciplines should create different solutions regarding the environment and protecting clean water ecosystems.

Arid and semi-arid belts, which arise due to the uncontrolled consumption of water in most of the world, irregular rainfall, and the loss of most of the valuable water as surface runoff in a short period, also bring serious risks globally [4,9,12,13]. Therefore, recent water scarcity, increased consumption, and a decrease in rain put nature and humanity at significant risk [4,6,8,9,12,14,15]. Against these water-related problems, this study considered the water-harvesting strategy developed by living things in nature as a critical resource, with the reference that science receives from nature. Collecting and re-evaluating the flow of water without causing erosion constitutes one of the most essential objectives of sustainability [16,17]. The concept of biomimicry is a nature-inspired approach. It is considered a response to the increasing demands to find alternatives to the ecologically destructive technologies, systems, and methods of the current industrial age, which are expected to solve today's unsustainable human–nature relationship [18]. This strategy and approach can enable ecosystems and built environments to adapt to climate change by mimicking nature through durable, sustainable, and adaptable natural regeneration [19]. In this respect, the adaptive reuse of buildings is essential for sustainability and the future [20].

There are many new and different approaches to addressing global environmental problems in various disciplines. For example, risks such as climate change and damage to the ozone layer have brought the concept of energy to the forefront, and engineering and architectural disciplines have also brought the subject under the spotlight [21–24]. Shading systems and sun-breakers have gained popularity, especially considering strategies related to façades, and they have been ensured to control the energy used in buildings [23,24]. From this point of view, this study offers an alternative perspective in addition to the energy-efficient façade designs of sun-breaker building elements from shading systems, with an approach that includes architectural and engineering disciplines in perspective. At the same time, the importance of the study is revealed by the fact that the field is given a new dimension apart from the focused perspective and that the water-harvesting strategy, considered and determined to be necessary, contributes to the freshwater ecosystem in urban areas.

Being inspired by nature is an essential subject within architecture, like other disciplines. Various biomimetic design strategies have been developed, and attempts have been made to ensure the integration of biomimetics with architecture [25–28]. However, these integration initiatives still face difficulties [29–31]. Future studies are expected to provide perspectives and contributions if these difficulties are overcome. From this point of view, this study contributes to the literature by revealing the development and deciphering potential of the existing system; the subject is considered and approached in the context of engineering, architecture, climate adaptability, and the sustainability of the connections between urban and ecological systems. This article presents a system initiative, with perspectives emphasizing the importance of an interdisciplinary approach to urban freshwater ecosystems and the achievements obtained from joint analyses.

2. Materials and Methods

This study, which deals with an integration initiative, aims to provide an architectural solution based on the main problems of global environmental issues such as drought, desertification, water scarcity, and water loss. The model's integration initiative and development process started with a deep literature review. After the data collection and analysis processes, thematic analysis was used to categorize and explore the development stages of the model. From this point of view, the concept of biomimicry was first considered, and the current studies in the literature were evaluated. After that, within the architecture discipline's framework, a literature review was conducted, primarily on water harvesting/acquisition; studies and approaches in the field were put forward. The literature about sun-breakers, which are taken as the main design element of this study and can be preferred as aesthetic building elements, is presented alongside their functional architectural value in buildings. The focus of both research topics on the literature on a global scale has provided critical perspectives and references in terms of opening horizons for the study. This study has formed an essential basis for the development of the model. The bio-sun-breaker model, designed as a result of the development stages, proposes a different integration initiative that can solve water loss problems with nature's strategic water-harvesting approach and contribute to urban freshwater ecosystems along with functional and aesthetic purposes in the design of buildings.

The literature review was considered to have two stages. As a strategy, the approaches to water-harvesting and sun-breaker building elements from the selected shading systems are discussed in detail. The points that can improve the model proposal are indicated more in the analysis and results. In this context, the research and studies conducted on sun-breaker structural elements in the literature have been examined in detail, and gaps in the field have been identified. After that, the structural component of the sun-breaker was re-examined through the concept of biomimicry, which is the primary approach of this study, and alternatives that can be solutions to problems were developed by presenting perspectives. With the biomimicry approach, sun-breakers have been integrated with the water-harvesting strategy, and a new model proposal has been introduced to the field (Figure 1).

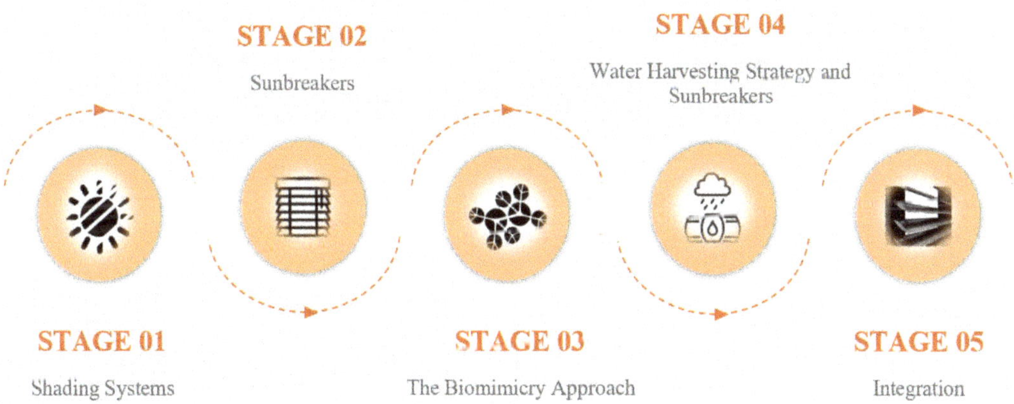

Figure 1. Model proposal development diagram.

3. Literature Review and Theoretical Background

The nature-inspired/learning/application approach, defined by concepts such as biomimicry, biomimetics, bio-knowledgeable, biomimesis, bio-imitation, and bio-inspired, constitutes the primary approach of this study because it offers different perspectives and prioritizes sustainability. Biomimicry (bio—life, and mimesis—to imitate) is a design approach inspired by the functional concepts of an organism or a natural ecosystem [32]. According to Benyus [33], biomimicry imitates the processes occurring in nature to create innovative solutions with sustainable design. Bio-imitation generally refers to using ecological criteria to evaluate sustainability by addressing it and to create models and designs inspired by nature within the framework of forms, processes, and ecosystems [33,34]. It has been observed in the literature review that the biomimicry approach is an interdisciplinary research field in which experts from different backgrounds, such as architecture, education, health, philosophy, computer science, biomedicine, physics, and chemistry, are involved in biology and engineering to create highly sustainable products [35–38].

The biomimicry approach offers inspiration to disciplines in three ways: organism (the imitation of nature), behavior (the imitation of natural processes), and ecosystem levels (the imitation of ecosystem functioning principles) [39]. At the organism level, design and architecture come to the foreground, and inspiration is taken from the shape of a building, the form of the design, or the structure of the model. At the behavioral level, the unity and interaction between the ecosystem and its environment inspire its design. At the ecosystem level, the main focus is on the overall holistic relationships with other parts of the organism, and how they interact with each other is of interest. Each stage constitutes an essential approach in terms of its characteristics, the fact that it can be interpreted in many disciplines, and the fact that it provides broad perspectives.

Biomimicry against rapidly occurring climate changes and environmental degradation is critical today. The biomimicry approach, which has recently increased in popularity as a concept, has also been adopted in recent centuries and is inspired by nature. For example, in 1482, Leonardo da Vinci invented a flying machine inspired by the flight mechanisms of birds and bats, and this study is considered the first example of biomimicry [40,41]. Therefore, the need to study and imitate nature to find practical solutions to human needs is not a new strategy, but a process that continues and develops from the past to the present.

The approaches used for biomimicry in different disciplines are broad and involve various solutions. Thus, using nature-inspired strategies provides excellent benefits for each area. For example, although large in volume, a bionic car prototype inspired by the box fish (Ostracon Meleagris) has a different vehicle design, with a small wheelbase [42]. The bionic car consumes less fuel because the volume is more aerodynamic as a result of the imitation of a box fish. The tree development model is also supported, using the minimum necessary materials in the car's structure. Therefore, a new perspective has been created in the considered vehicle design, which is more sustainable and includes comfort in a developmental sense. Other examples in different disciplines include hydrophobic and friction-reducing materials in textile production [43,44], velcro inspired by cocklebur seed shells [45], sea current turbines inspired by dryobalanops aromatica seeds [46], self-cleaning paints [47,48], and super-hard ceramics imitating mother-of-pearl pearls [49]. Similarly, there are studies produced using a biomimetic approach.

3.1. Water-Harvesting Strategy in Nature

The water-harvesting approach, one of nature's many strategies, offers one of the solutions to the most critical water problem today. Animals and plants are the choices for water-harvesting-based solutions in the literature because water shortages are becoming increasingly severe globally, as shown in Tables 1 and 2. It is observed that these studied species exhibit similar approaches in terms of their perspectives and working principles. The approach strategies of the animal and plant species included in the tables have enabled the development of the model proposed in this study and provided a meaningful perspective.

Table 1. Studies and animals inspired by water-harvesting strategy.

Latin Name	Image	Type/Location	Studies	Bio Inspiration	Latin Name	Image	Type/Location	Studies	Bio Inspiration
Crotalus cerastes laterorepens		Snake/Yuma Desert	[50,51]	Rainwater harvesting strategy	*Psammobates tentorius trimeni*		Turtle/SouthAfrica	[52]	Rainwater harvesting strategy
Physosterna Cribripes/stenocara gracilipes		Insect/Namib Desert	[51,53,54]	The strategy of collecting water droplets	*Onymacris unguicularis*	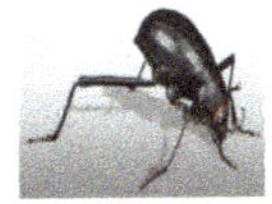	Insect/Namib Desert	[54–57]	Upside-down water collection strategy
Litoria caerulea		Frog/Australia	[58]	The skin cavity strategy	*Moloch horridus*		Lizard/Australia	[51,59,60]	Water collection strategy
Phrynosoma cornutum		Lizard/North America	[59]	Water collection strategy	*Uloborus walckenaerius*		Spider	[51,61]	Spider web strategy

Table 2. Studies and plants inspired by water-harvesting strategy.

Latin Name	Image	Type/Location	Studies	Bio Inspiration	Latin Name	Image	Type/Location	Studies	Bio Inspiration
Sekoya sempervirens		Tree/California	[62]	Fog collection strategy	*Pseudotsuga menziesii*		Tree	[62,63]	Fog collection strategy

Table 2. *Cont.*

Latin Name	Image	Type/ Location	Studies	Bio Inspiration	Latin Name	Image	Type/ Location	Studies	Bio Inspiration
Welwitschia mirabilis		Plant/ Namib Desert	[55]		*Trianthema 8hereroensis (Aizooceae)*		Plant/ Namib Desert	[55,57]	
Stipagrostis sabulicola		Plant/ Namib Desert	[55,57,64,65]	Water collection strategy	*Opuntia mikrodazisi*		Cactus	[66]	
Copiapoa haseltoniana		Cactus	[67]	Fog condensation strategy on thorn	*Discocactus horstii*		Cactus	[68]	Fog condensation strategy on thorn

When we look at the plants and animals that include the water-harvesting strategy in the tables, it is seen that their capacity to adapt to nature is quite strong. In addition, the fact that similar plant and animal species are often found in studies that address water-harvesting strategies indicates that species in nature have identical strategies. Tables 1 and 2, which summarize the sensitive cycle of nature in this area of water acquisition, are of critical value in structural and operational planning and are expected to reference other disciplines to be studied in the future.

When the animal and plant species in the tables are examined, it is seen that they have different structures and characteristics, but they apply similar strategies for the same purpose. When going into detail, it has been determined that the superficial features of plants and animals have an identical form and shape and consist of structures that direct water. Curved and bumpy structures are composed of the complex and soft shells of animals, and paths consist of needles and capillary structures in plants and direct water or moisture in the air by holding them. These morphological and behavioral characteristics have been referenced many times in past studies and have paved the way to the model proposal in this article.

3.2. Biomimicry Approach and Studies in Architecture

People resort to innovative methods by taking inspiration from nature to create different solutions to various challenges and maintain their lives. Architecture is also among the disciplines that fortify themselves from nature and often take reference from nature. For this purpose, the discipline of architecture is based on the passive, efficient, and cyclical design principles found in nature by learning from nature to achieve sustainable and conscious architectural design and minimize environmental impact [69]. Biology and architecture are also prerequisites for each other [70].

When we look at bio-inspired projects directly or indirectly related to architecture, systems that take different importance or provide advantages to climatic conditions have been integrated. For this reason, biomimicry has significant potential and benefits in architecture and design. For example, many strategies and problem-solving approaches, models, and projects have been developed, such as the shape and behavior of certain animals that adapt to the effects of the sun and wind; examples of nature approaches in construction materials; and different approach solutions related to heating, cooling, and lighting on building envelopes and façades, inspired by plants due to extreme changes in climatic conditions [37,71–74].

In architecture, the concept of inspiration from nature is generally seen to be carried out in research and development studies on topics such as light structures, water-harvesting, and thermoregulation. This study, which addresses the water problem, one of the world's leading climate crises, also analyzed various studies by assessing the current perspectives from the perspective of the architecture discipline. The studies developed as a result of the research and proposed as a model are given in Table 3. When Table 3 is examined, the links between the studies conducted and the proposed studies referencing nature are relatively strong. Models designed with similar and different approaches taken from nature are critical as a reference for subsequent studies. The increase in the quantity and quality of this approach and model studies leads to an unavoidable situation where they have positive effects both in disciplinary and global terms.

The projects in Table 3 show that the building shells or envelopes are inspired by nature and that the water-saving models have progressed to the application stage. When viewed as a whole, although the objectives and working principles of the projects seem to contain similar strategies, it is seen that the approach angles have a horizon-opening effect for future studies. Therefore, developing models with different perspectives regarding approach strategies is critical to the field.

Table 3. Examples of biomimicry projects and approach strategies.

<table>
<tr><td>Project name: Biomimicry Museum [32]
Bio-info: Cactus and camel
Purpose: canopy design with a bio-inspired strategy as a solution to the water needs of plants.
Working Principle: the upper cover of the floor has a large area and is designed inspired by the spongy bone structure of camels in their noses and the spines of cacti that harvest water from the air to reduce evaporation.</td><td>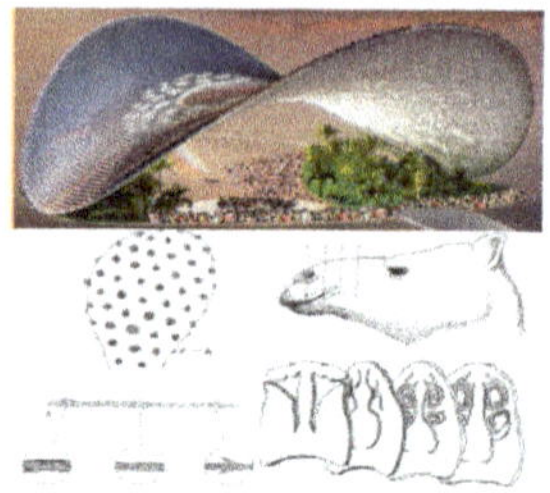</td></tr>
<tr><td>Project name: RainBellows [75]
Bio-Inspiration: Ice Plant (mesembryanthemum crystallinum) Purpose: to store rainwater by developing a façade system.
Working Principle: it is inspired by the ice flower plant, which can store water. There are façade elements that can increase the storage capacity thanks to the mechanisms that come out of the façade during the rainwater intake of the façade. After the necessary water intake, it is filtered and transmitted to where it is needed in the building.</td><td></td></tr>
<tr><td>Project name: AquaWeb [76]
Bio-Inspiration: honeycombs, spider webs, ice plant, and the mycelium plant.
Purpose: a system that can be adapted to the façades of buildings, which harvests and stores atmospheric water with a flexible and modular design and transmits it if necessary.</td><td>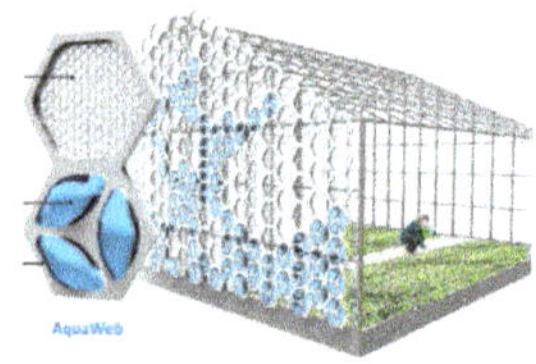</td></tr>
<tr><td colspan="2">Working Principle: Inspired by honeycombs, there is a web system that can harvest atmospheric water such as rain, fog, and humidity, and imitate the fibers of a spider web. The water that accumulates in the compartments that imitate the sacs of ice plants is stored and transmitted by pipes with the transport feature of mycelium plants.</td></tr>
</table>

3.3. Sun-Breakers

The etymology of sun-breakers is derived from the French language. The word 'brise' is expressed as 'breeze, break and breaking' in association with the word brise-soleil in the eighth edition of the Dictionary of the French Academy [77], while the word 'soleil' corresponds to the word 'sun'. In English, and within the scope of the literature, it is expressed with the words "sun breaker", "solar breaker", or "sunshade". The architectural element, whose primary function is to break sunlight and control the light coming into the interior, is essential in buildings' façade style and language.

Shading systems such as sun-breakers, blinds, and curtains (Figure 2) are often used in façades or landscape areas. In addition to its aesthetic value, it has an essential place in architecture and engineering in terms of containing functional importance regarding heat, light, sun, and privacy (the feature of interrupting outdoor and indoor visual communication). According to the literature, the elements found under the title of sun shading and shading systems are expressed with different words in their vocabulary, and other word origins are used together in combinations of concepts [78–82]. It has been found that terms such as sun breakers, sunscreens, sunshades, and blinds include the same, different, or similar systems in shading systems.

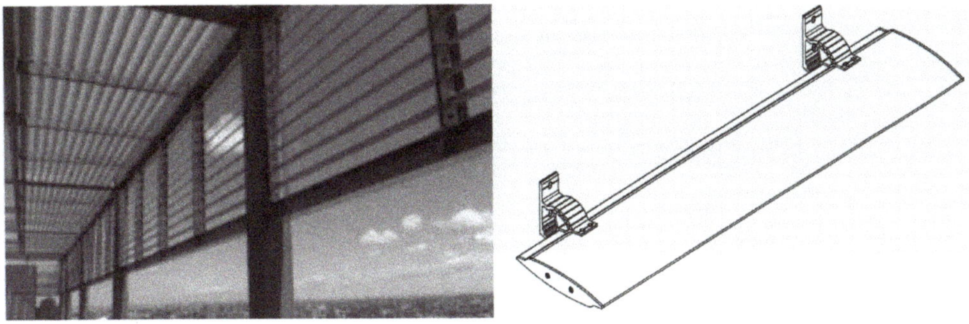

Figure 2. Classic details of the façade sun-breaker.

Sun-breakers have been designed in different ways depending on the style of positioning and the material type of the façade. They have added meaning to building façades because of the aesthetic value of the past and present shading systems. They are used in a linear (horizontal/vertical) or grid way with different materials (concrete, cement, wood, metal, aluminum, sheet metal, etc.), preferably in fixed or moving (manual/mechanic) forms. With the development of technology, different forms, purposes, and especially sun-breakers, which are constantly being developed, appear as an indispensable design element for structures today. For example, the sun control system and logic of the sun-breakers in Brazil and the Philippines, which were designed in 1948 (Nova Cintra Building) (Figure 3a) and 1961 (The Philamlife Building) (Figure 3b), as well as on the tower façades built in Abu Dhabi in 2012 (The Al-Bahar Towers 2024) (Figure 3c) as a result of the innovations and opportunities brought by technology with materials, forms, and usage purposes, clearly express the development over the years and changes in perspective approaches.

(a) (b) (c)

Figure 3. (**a**) Nova Cintra Building, 1948, Brazil [83]; (**b**) The Philamlife Building, 1961, Philippines [84]; (**c**) Al Bahar Towers, 2012, Abu Dhabi [85].

Sun control is of critical importance to minimize the adverse effects of the sun in the discipline of architecture. When we look at the sun- and wind-adapted façade systems, it is seen that the elements that serve as sun-breakers have different behaviors and characteristics. The CJ R&D Center (Yazdani Studio, Gyeonggi-do, South Korea, 2016), which is based on sun control, is one of these examples. A different strategy was developed according to the hole ratios on the panels to adjust the natural light levels of the sun-breaker strips on the façade of the building to the interior (Figure 4a). Another approach, the wind-adapted façade system, was designed to construct the San Francisco Public Utilities Commission Headquarters (SFPUC, KMD Architects and Stevens Architects, San Francisco, ABD, 2012). According to the study results, approximately 7% of the building's energy needs are met by renewable photovoltaic and wind sources integrated into the building (Figure 4b).

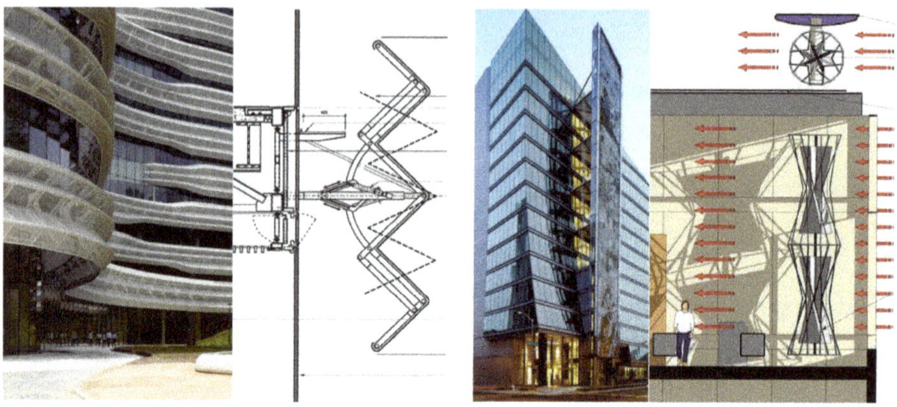

(**a**) (**b**)

Figure 4. (**a**) CJ R&D Center [86]; (**b**) SFPUC Building [87].

Some studies have investigated shading systems with a rainwater-harvesting strategy. The examples designed with the advantages of rainwater-harvesting and sun-breakers stand out as individual shading systems that generally work independently of the structure. For example, in Figure 5a, a rectangular tower for the water tank and a collection surface were created that also served as a shading tool for heating/cooling the water tank and then created particular conditions for the adjacent area [88]. Another example (Figure 5b) is a sun-breaker system in the form of a concave umbrella made by designer Fabrice Bardon, which can also be used as a fog and rainwater collector. The model is supported by a concave umbrella shape that collects morning mist and rainwater to meet garden watering needs [89].

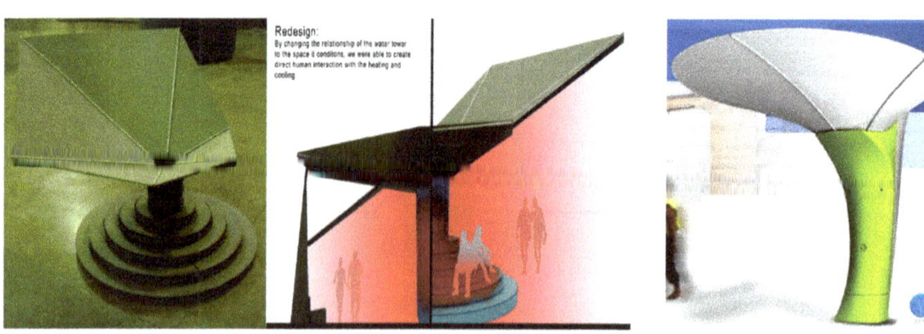

(**a**) (**b**)

Figure 5. (**a**) Shading systems; (**b**) rainwater-harvesting.

By including the different word groups identified in the literature review in the study, the latest advancements in global research and the perspectives considered, as well as the sun-breaker elements that are the subject of this research, have been chosen to form the foundation of the review. This literature review provides essential data for future studies to reveal the advantages and disadvantages of sun-breakers and shading systems. Therefore, while shading systems are found in many studies in the literature, the system has been developed using different analysis and research techniques. These studies show that sun-breakers and shading systems approached from various perspectives offer opportunities for other projects and research. The studies that deal with different perspectives are listed as follows:

- Light transmittance and illumination levels [79,90],
- Heat conduction and propagation [79,91],
- Acoustics characteristics [92,93],
- Movement characteristics [80,91,94,95],
- Relative humidity levels [94,95],
- Materials characteristics [79,80,90,91,94,95].

The developments in shading systems show that progress has been made in design and functional terms along with technology, and various practical studies have been carried out. It is understood that this situation will be used more in structures, especially as a result of the sensitivity of the systems to energy efficiency, and it will be in a position open to development. In this study, a new function was integrated into sun-breaker systems, and a model was proposed to develop the system from a different perspective.

4. Integration of Water-Harvesting Strategy to Sun-Breakers and Model Proposal

Water acquisition in architectural building shells involves four basic strategies: water collection, water transmission, water transportation, and storage [25]. According to the literature and case study analysis, the backbone of the proposed model and water-harvesting strategy were considered in three steps. For each step, different acquisition strategies in plant and animal species were considered, inspired by nature, and the stages were recorded with developable perspectives. These three steps are shown in Table 4, including the stage, the type of biological organism, and the strategies for handling it.

The strategies selected as biological organisms, and which are a reference for each stage, constitute the critical steps in the development of the model (Table 4). Similar methods for different animal and plant species observed in the literature were influential in developing the stages. So much so, that the first stage, the water collection stage, was based on the spiny structure of the Moloch Horridus animal species, which lives in the desert, has adapted to arid regions, and was integrated into the model. The natural strategies of the lizard species at the point of water acquisition also contributed to other phases of the proposed model (water transmission and water absorption). The natural capillary pathways and asymmetric structures formed by Moloch Horridus's spiny structure aim to ensure natural circulation at the model's water flow/transmission point. In the same way, the leaf layout in the water transmission phase constitutes another strategy that is referenced from another aspect of nature. In the leaf layout, like the lizard species, the natural transmission of water is ensured by the different lengths and thicknesses of the asymmetric capillary pathways. It constitutes an important step for storing water with minimal loss. During the water absorption phase, the waters collected on the surface were inspired by stoma, which plant species use effectively for water and air diffusion, and the model was developed together with all other strategies.

Table 4. Strategy approach in the model proposal.

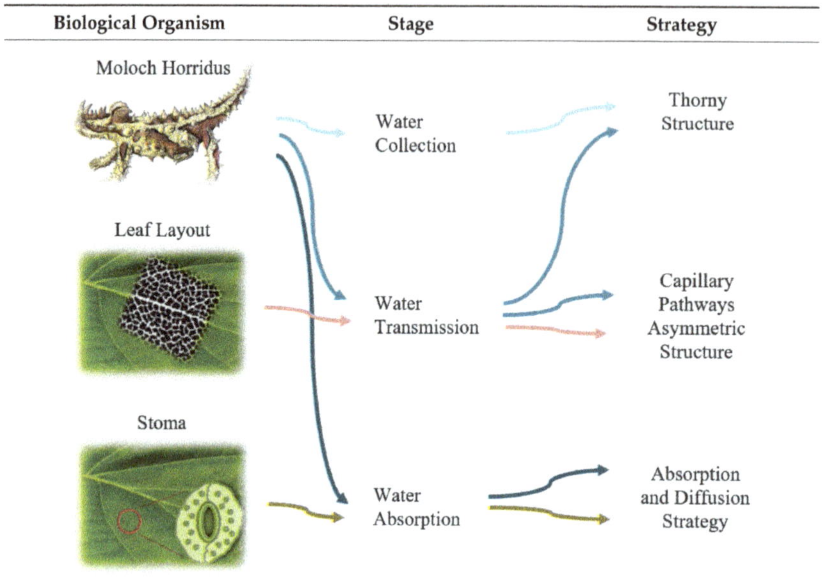

4.1. Water Collection Stage

Water is usually a limiting factor for nature and requires all living things to have special skills to survive in all conditions. For example, it has been found that some frog species have similar passive water collection and water transport capabilities using capillary skins [96]. Again, for lizards that collect moisture, fog, and rainwater, Withers evaluated five possible sources of water intake and considered them in the form of rain, puddles, water droplets caused by dew or fog condensation, moist soil, and thermally facilitated condensation [97]. Within the scope of this study, inspiration was taken from the lizard species "Moloch Horridus," which is one of the most remarkable animal species in water-harvesting and is known as the "Thorny Devil" (Figure 6).

Figure 6. Moloch Horridus.

The two large horned scales on the thorny devil's head resemble the figure of a dragon or devil due to the "false head". This illusion has, therefore, also been influential in its naming. Moloch was used in ancient times for a god often depicted as an ugly monster [98]. The skin of the thorny devil has protruding tips, and each item of water that touches these protrusions helps to collect it in the skin. Water collected through capillary channels between spiny structures is transported to the mouth (Figure 7). Lizard species also collect moisture in deserts by turning the dew condensation at and after extremely low temperatures to an advantage [99].

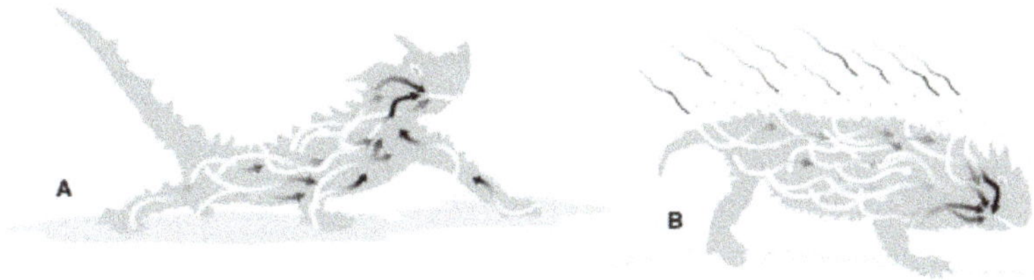

Figure 7. (**A**) Water directed to the mouth in motion; (**B**) water directed to the mouth in a stationary state [100].

Comanns et al. [101] evaluated the water acquisition strategy and skin capillary volumes, detailing he effectiveness of different water sources in filling capillary channels to provide drinking water for the Moloch Horridus (Figure 8). They found that an integrated water collection amount as low as 0.2% of the body weight is sufficient to prewet the skin surface and make the skin super hydrophilic. This situation shows that this ease and strategy for determining surface wettability is expected to be inspired by the skins of the spiny devil species, which can be used to provide a reference for other disciplines to be considered and to encourage the ability to collect water from various sources.

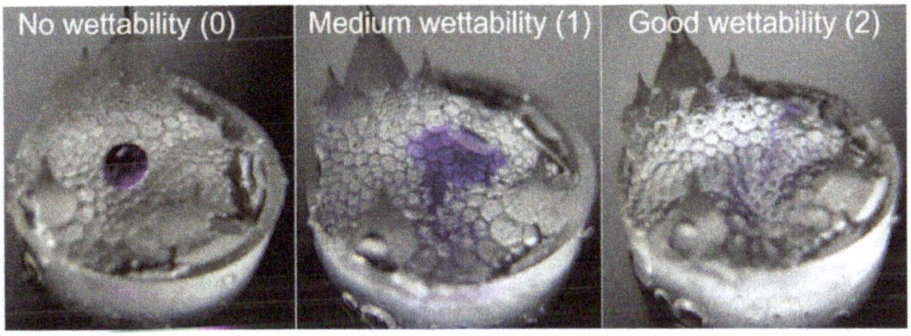

Figure 8. Moloch Horridus skin absorption distribution [101].

4.2. Water Transmission Stage

The water transmission phase of the model proposal was inspired by the thorny devil lizard of Australia, which is adapted to drought [102]. Water flow is provided to the hexagonal microstructures on the skin surfaces by the hydrophilicity of the skin surface upon contact with trace amounts of water with the skin (pre-wetting). The captured water elements are passively transported by capillary channels in an asymmetric and interconnected system that surrounds the entire body surface of the lizard (Figure 9). When the blisters and channels on the skin surface are directed toward the oral tract, they decrease in size and width, facilitating water flow and accelerating natural flow [60,99,103].

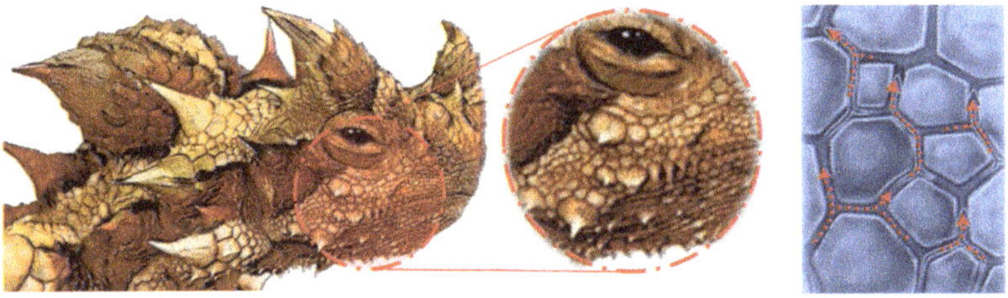

Figure 9. Skin layer (spines and capillary pathways, size change towards the mouth).

Leaf vein layout, another study strategy for water transmission, was used as another reference for this study. As a transport system, water is distributed to different aspects of the plant through channels in the leaf veins. Generally, the transport system is divided into three degrees in terms of the thickness and length of the vein [104] (Figure 10). Vein number 1 is the main vessel and constitutes the largest channel size. Vein 2 is thinner than the central vein and connects the other two channels. Vein number 3 is the thinnest capillary and constitutes the most extended vascular channel in the carrier system. Changes in vein thickness are the most crucial strategy for allowing water to flow passively. For this reason, many disciplines are bio-inspired by the leaf vein arrangement system. Considering the different perspectives, the leaf vein layout, transport system, and relationships with each other can improve the discipline from a biomimetic point of view [105–109].

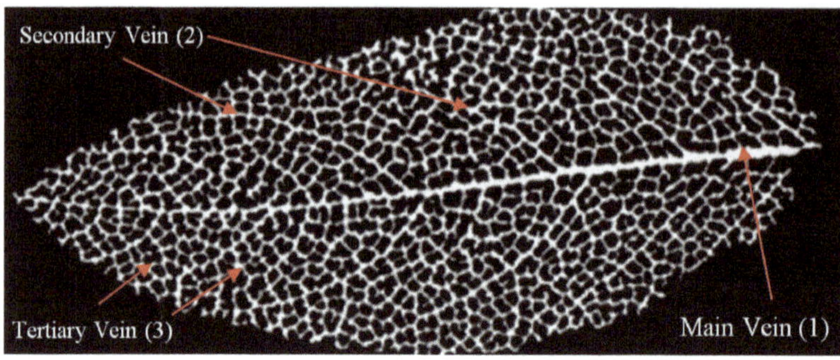

Figure 10. Leaf vein layout [110].

4.3. Water Absorption Stage

The absorption stage of water constitutes an important step, as does the collection and transportation of water. In nature, there are different strategies for water absorption on the surface. At this point, it is necessary to take the water from the surface before the water from the existing surface is transferred to the stored area. Within the scope of this study, the effective use of stoma by plant species for water and air diffusion was focused on, and inspiration was taken regarding water absorption and air permeability as a strategy.

Stomata control the amount of CO_2 absorbed and the changes associated with water. While many variables affect these processes, stoma features contain significant structural variations depending on the ecological balance or the natural environment [111–114]. The aperture variability and spatial size of the stoma pores resemble those of two beans, kidneys, or standing coffee beans side by side (Figure 11). The internal pressure of the cell (turgor) determines the shape changes. With increased turgor pressure, the cells bend outwards by stretching in the abdominal membrane, which hardens with increasing pressure. With increasing internal pressure, that is, turgor, the stoma becomes curved, and the stoma opens, and when the turgor decreases, it takes on a closed shape. However, research and discussions are still being conducted on the working mechanics of stomas, and their working principles continue to be analyzed in different ways [114–116]. Biomimetic studies have focused on stomata's transpiration and vapor permeability via this natural strategy [117–121]. Engineering, health, mechanics, and material sciences work on different principles by imitating the system with many other approaches. Model suggestions are seen in many disciplines, and inspiration from nature continues to be used.

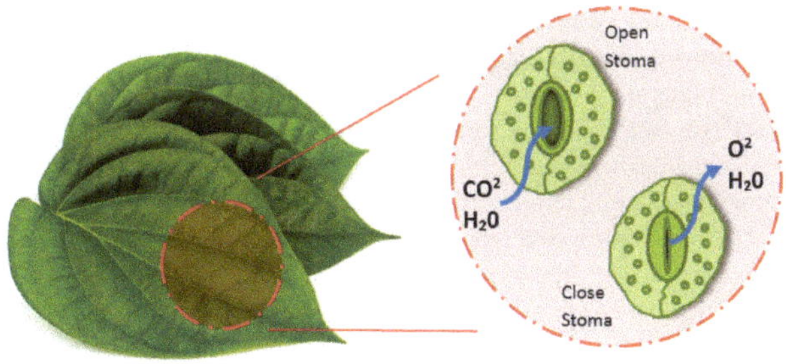

Figure 11. Stoma opening/closure conditions (by the author).

5. Model Proposal and Working Strategy

A standard model type is often used for the main proposed sun-breaker, and the appropriate production cost has been outlined (Figure 12). Therefore, the proposed stages have a significant advantage in adaptability and applicability to reality. The realization of the design with a foot-on-the-ground approach to the study has added value to the study. For this reason, the proposed sun-breaker model and form characteristics, like those of standard-type sun-breakers, undergo minor changes depending on the water-harvesting strategy. The form changes made, on the other hand, were designed in such a way as to provide an advantage in terms of water collection and accumulation, and the final form was given (Figure 13).

Figure 12. Selected standard sun-breaker.

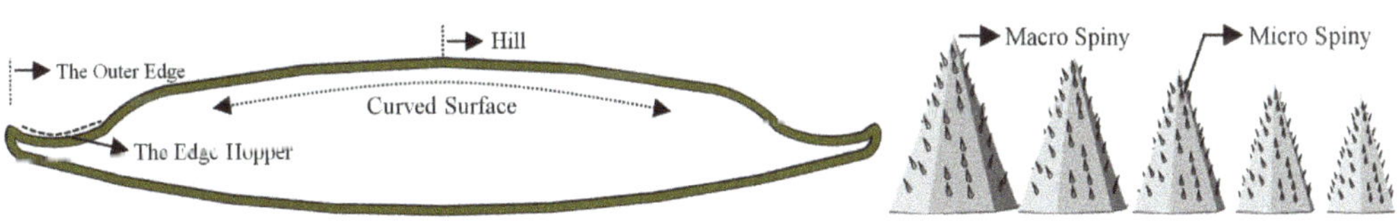

Figure 13. Recommended sun-breaker form and spiny protrusion forms.

The integration of the strategies used in the animal and plant species studied and inspired by the model has been realized in stages, and the design has been created. During the application of each stage to the model, the proposed layers and strategies were clearly expressed in step-by-step steps. In this context, a form design with areas depending on water collection and water accumulation strategies has been created to determine the sun-breaker form. The water-harvesting strategy, which is the first stage in acquiring water-harvesting for the created form, was inspired by the spiny skin layer of the Moloch Horridus lizard species, and work on the model began. This stage formed the starting point of the surface shape of the sun-breaker model. It aimed to facilitate moisture and fog collection by creating spiny protrusions on the spiny structures (Figures 6 and 9), observed on the skin layer of the thorny devil species, and are the most dominant feature of the water collection strategy (Figure 13). Spiny protrusions were systematically placed on the sun-breaker model in the following stages.

Different strategies were activated during the placement of spiny structures on the sun-breaker model, and systematic formation was achieved. In this direction, the form of the sun-breaker element was first considered during the formation of the sun-breaker surface. As shown in Figures 13 and 14, the design of the form in a curvilinear structure makes an essential contribution to the flow and collection of water. The sun-breaker element hill and the outer edge regions were formed for this design. Water is collected from the air, moisture, and fog, thanks to the spines on the back of the form; the natural flow of water towards the edge cavity is ensured by the shape of the sun-breaker. Moreover, considering the energy-efficient and water-harvesting gain designs of sun-breakers, the proposal of a mobile system has also been included in the model stage. In this context, the aim is to construct a sun-breaker model based on a moving system. Increasing the aesthetic value of the façade also reveals a significant meaning and advantage in sustainable achievements.

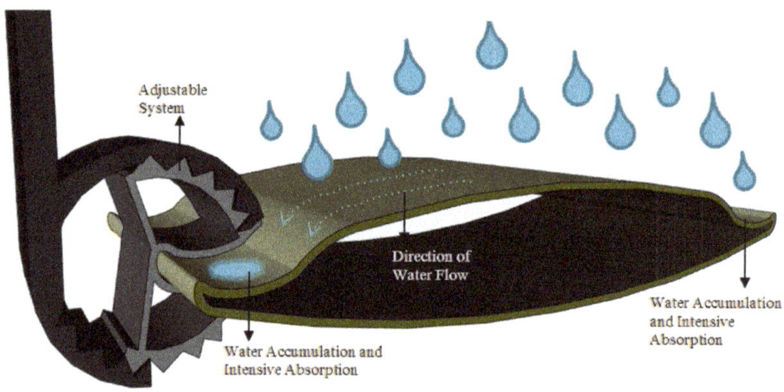

Figure 14. Recommended sun-breaker model and water/moisture/fog intake working principle.

As a result of the conditions imposed by the sun-breaker form, the moment the water from the air falls on the ridge surface or condenses, the flow of water towards the edge is expected to be fast, and the absorption is expected to be low. For this reason, different integrated strategies involving absorption centers, stoma, and channel (groove) formations have been developed. To direct the flow of incoming water grains due to the ridge and curvilinear surfaces brought by the formation of channels, curvilinear forms have been created that slow the water flow rate and facilitate absorption, rather than the perpendicular stripes of channels. Stomata were placed at the intersection points to absorb the water directed with the channels. Due to the low absorption in the back and the high water flow, the number of stomata remained low in the back regions and increased towards the edge cavity regions (Figure 15). Therefore, the possibility of directing the water flow and absorption and controlling the speed has been provided. Studies were completed by placing spiny structures inspired by the lizard species between the created channels and integrating them into the sun-breaker model (Figure 15).

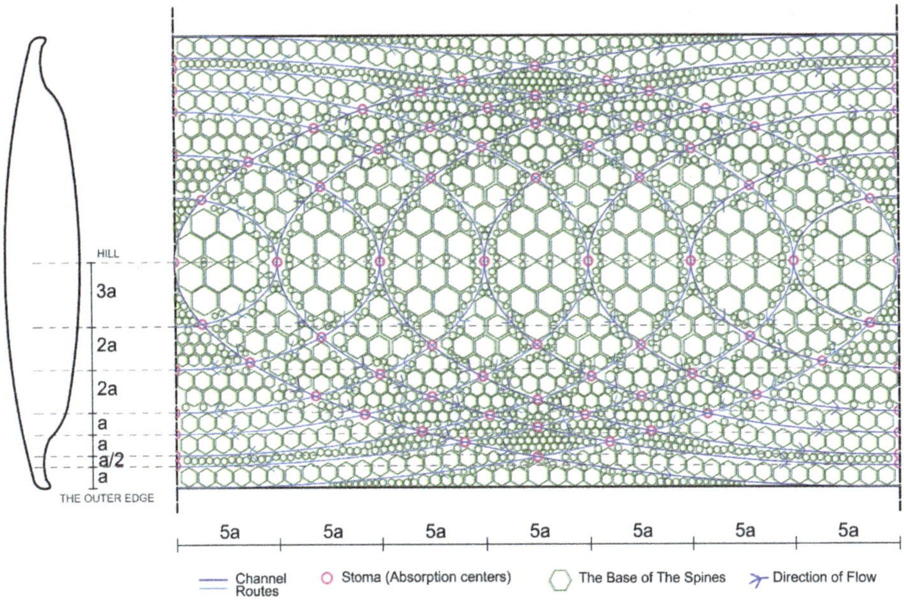

Figure 15. Recommended sun-breaker model: groove, protrusion, and stoma map.

Similar strategic approaches in the channels and stomata determined the placement of the spines by taking inspiration from the dimensional changes of the lizard species in the spiny structure moving toward the mouth (Figure 9). In this respect, support is provided for the passive water flow on the model. For this purpose, the spiny structure has been considered in three dimensions: the macro-, meso-, and micro-scales (Figures 16 and 17). Macro-spines constitute the largest spiny structure in the model, as the region in the back area will have the most contact with the air. Meso-spines include spines of different sizes that support macro-spines in a way that describes the route formed by the main channels. Conversely, micro-spines are located on macro- and meso-spines to facilitate water collection and condensation. The dimensional approaches considered were sized and placed depending on the position of the sun-breaker, the desired flow of water, the channel, and the stoma strategies created (Figure 18).

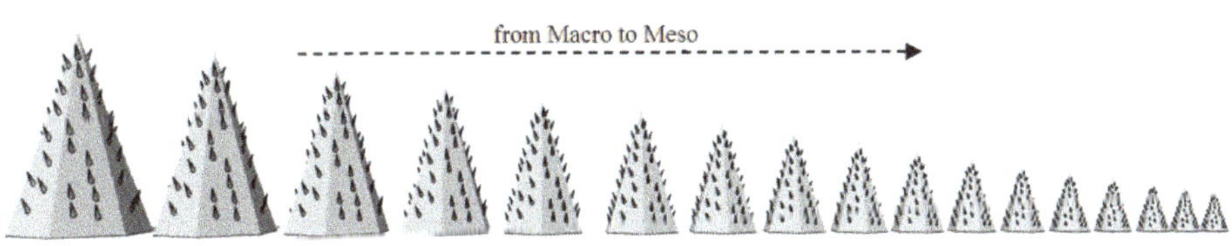

Figure 16. Recommended spiny sizes.

Within the scope of the proposed model, an additional strategy has been proposed that will facilitate the passive flow of water within the channels created on the sun-breaker and help with the water. Considering the thorny devil lizard species (Figure 9) and leaf-vein patterns (Figure 10), this strategy, which will be developed in channel structures, has been confirmed by examples from nature that it will also make an essential contribution to the collection, transmission, and absorption of water. In this context, access to natural pressure and speed will be provided with channel structures, lengths, widths, the attraction of water within itself, and thinning, shrinking, and elongating dimensions. From this point of view, the channels are sized in three ways: macro-, meso-, and micro-channels. Macro-channels are the first and largest channels and cover the channels that directly access the stomata. Meso-channels constitute the channels that are the means for storing water from the stomata. Conversely, micro-channels have been proposed as intermediate channels that provide access to the meso-channel for storing water from capillaries and stomas in the space between spines of different sizes (Figures 17 and 18).

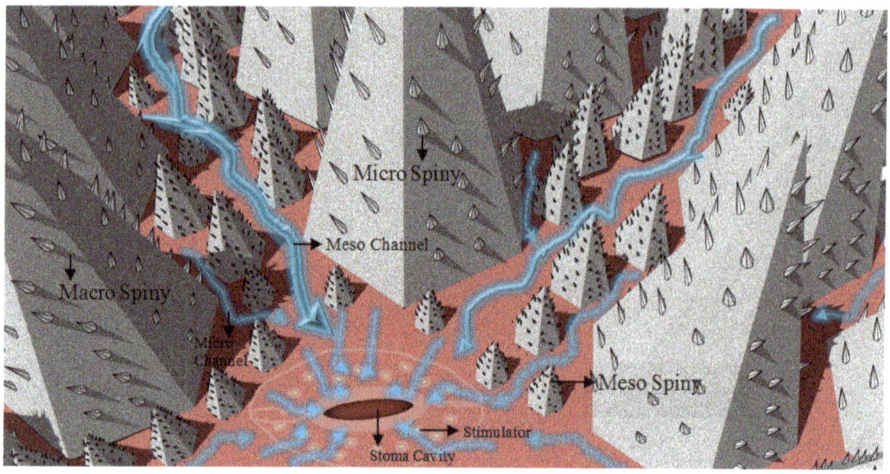

Figure 17. Passive water flows to the stomata.

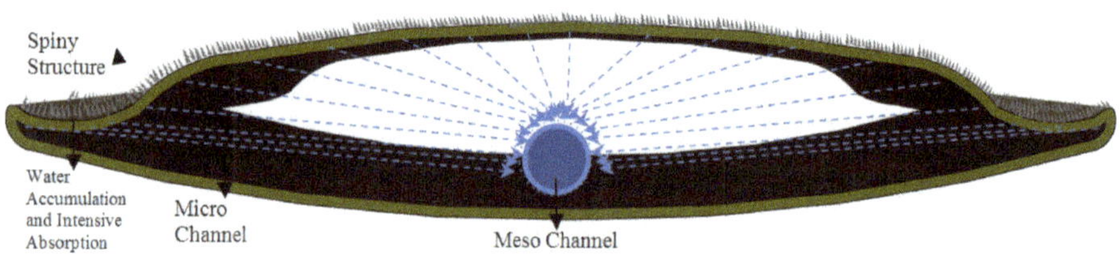

Figure 18. The proposed model perspective section.

The strategies recorded in stages and the references taken from nature are integrated into the model. Their relationship to the façade is shown in Figure 19. The diagram clearly shows the working principle and provides water transfer to the greywater storage and distribution center, with horizontal and vertical macro-channels offered as suggestions within the carrier system. At this point, offering alternative water use in buildings with water recycling provides an essential advantage for urban freshwater ecosystems.

The attempt of the model to exhibit a flexible and open-ended approach in terms of design and working principles, along with similar alternatives, can be introduced. This developmental feature also paves the way for the model to be used as an aesthetic building element on the façades of buildings with different architectural designs. In this aspect, the study contributes to the purpose of the proposed model integration and adds critical value. Therefore, this model integration proposal can be repeated and tested together with experimental work processes to ensure the applicability of the design and project initiative and to develop by evolving into different disciplines and dimensions.

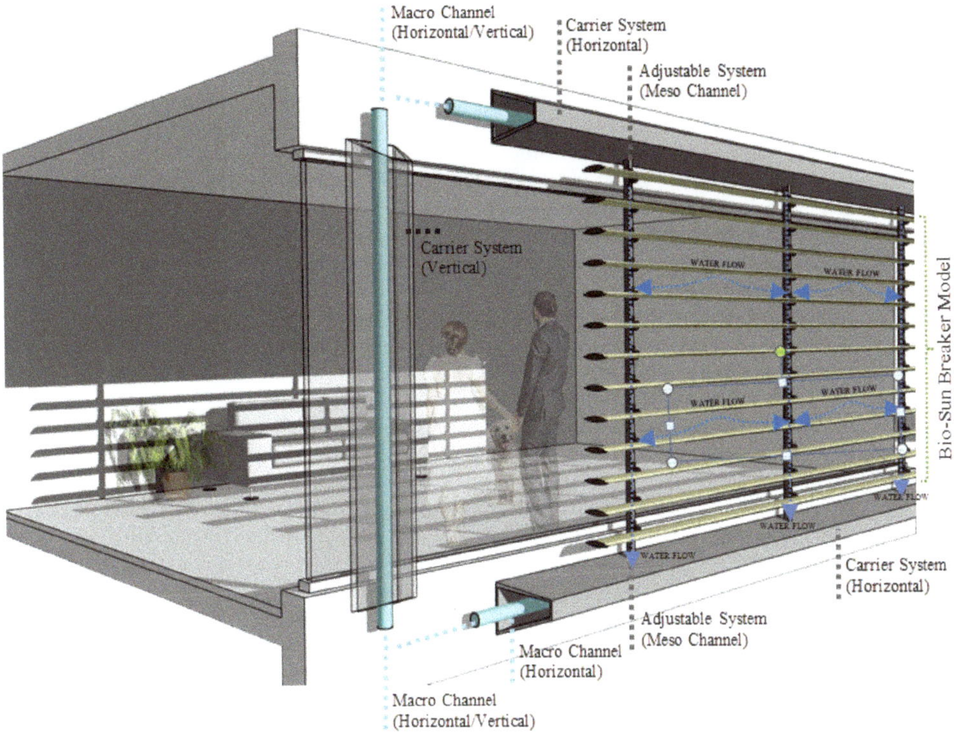

Figure 19. Working principle of the proposed bio-sun-breaker model with a façade.

6. Conclusions

Researchers, designers, policymakers, and investors worldwide are taking essential roles in improving and making sustainable living standards continuous, and different disciplines are trying to develop this issue with other strategies. Therefore, throughout history, designs inspired by nature have taken place in many fields, such as industrial design, health, medicine, materials science, nanotechnology, robotics, and engineering. The discipline of architecture also aims to reach literature and solutions with different perspectives on this subject. Similarly, the results of this study suggest that the discipline of architecture should adopt a method inspired by nature, the biomimicry method, which prevents environmental imbalance, not leaving building designs out of the ecosystem and making them a part of the ecosystem. Understanding and developing this situation, analyzing nature well, identifying its forms, systems, materials, functions, processes, aesthetics, and ecosystems, and discovering relevant solutions and new possibilities for disciplines, are becoming necessary in architecture.

Architectural façade elements play a complementary and additional role in managing the physical environmental conditions of buildings. When an element can control other physical environmental conditions besides the primary construction function, it has an essential advantage in contributing to the sustainability of the building and the environment. For example, daylight in architecture is critical in building design. Volumes with sufficient natural light in the relationship between people and space directly affect the mental and physical health of the user. In addition, considering the impact of buildings on energy consumption, the results also support the sustainability of buildings. However, in addition to these advantages, in this study, in addition to the tasks and characteristics of existing sun-breakers, nature's water acquisition strategy has been integrated on top of its current potential with a biomimetic approach.

It is envisaged that the inclusion of the proposed model in the project only during the preliminary design of the building to achieve the correct and targeted gains will provide a significant gain. Because of the purpose of use of the building, the shape, the climatic conditions in which the building is located, the size, color, and textural properties of the proposed model are related to criteria such as surface materials and section thickness; it is considered necessary to determine its separation and capacity for each building. The model proposal, which is directly related to the titles of "Clean Water and Sanitation (SDG 6)" and "Sustainable Cities and Communities (SDG 11)", which are among the Sustainable Development Goals (SDGs), aims to provide different perspectives on the disciplines with its superficial and functional features. In this context, it is anticipated that the article will become an indispensable resource for other researchers working on the subject.

The strategies integrated into the proposed model design constitute the most important privilege of this study to strengthen the urban freshwater ecosystem by providing a different and horizon-broadening perspective. In addition, model integration, which involves the use of a biomimetic approach, is expected to contribute to other biomimetic studies to be conducted in the future with integration methods to structure and space designs that are more sustainable, focused on urban problems in terms of the way they are handled and the model approach. It is necessary to make the subject developable as an essential reference for future research, for studies with other perspective approaches to be physically handled in research and development laboratories, and to complete the study by applying it; performance analyses should be performed. Again, it is necessary to select the building materials for the model's design, determine the sensitivity conditions at the junction points, test different scales and dimensions, choose the most efficient form and system, and reveal the negative and positive aspects for practical applications. In the same way, it is necessary to integrate the model and determine all the effects related to all economic, commercial, and climatic situations in the application stages to be included in the future-oriented strategies of the model. By conducting experimental studies, breakthroughs can be developed for the future, including other disciplines except for the discipline of architecture, as well as the development of this study and the continuation of its results. Therefore, this article is expected to pave the way for future work by drawing attention to other studies, such as prototype production, simulation tests, and material analysis, with this model initiative.

Author Contributions: Conceptualization, methodology, formal analysis, investigation, resources, B.K., G.M.A. and S.A.S.; writing—original draft preparation, visualization, B.K.; writing—review and editing, G.M.A. and S.A.S. All authors have read and agreed to the published version of the manuscript.

Funding: This research received no external funding.

Institutional Review Board Statement: Not applicable.

Data Availability Statement: Data shall be available from the corresponding author upon reasonable request.

Conflicts of Interest: The authors declare no conflicts of interest.

References

1. European Commission 2010. *2009 Environment Policy Review*; Office for Official Publications of the European Communities: Luxembourg, 2009; Available online: https://wayback.archive-it.org/12090/20221009012538/https://ec.europa.eu/environment/archives/pdf/policy/EPR_2009.pdf (accessed on 13 May 2024).
2. United Nations Environment Programme 2023. State of Finance for Nature: The Big Nature Turnaround Repurposing $7 Trillion to Combat Nature Loss—Technical Annex. Available online: https://wedocs.unep.org/20.500.11822/44706 (accessed on 13 May 2024).
3. Alcock, P.O.; Verster, E. Investigation into unconventional sources of water for a period of time in an urban/rural district of KwaZulu. *S. Afr. J. Sci.* **1987**, *83*, 348.
4. Falkenmark, M.; Lundquist, J.; Widstrand, C. Macro-scale water scarcity requires micro-scale approaches: Aspects of vulnerability in semi-arid development. *Nat. Resour. Forum* **1989**, *13*, 258–267. [CrossRef] [PubMed]
5. Perrings, C.; Folke, C.; Mäler, K.G. The ecology and economics of biodiversity loss: The research agenda. *Ambio* **1992**, *21*, 201–211.

6. Cosgrove, W.J.; Loucks, D.P. Water management: Current and future challenges and research directions. *Water Resour. Res.* **2015**, *51*, 4823–4839. [CrossRef]
7. Hoornweg, D.; Hosseini, M.; Kennedy, C.; Behdadi, A. An urban approach to planetary boundaries. *Ambio* **2016**, *45*, 567–580. [CrossRef]
8. Li, W.; Hai, X.; Han, L.; Mao, J.; Tian, M. Does urbanization intensify regional water scarcity? Evidence and implications from a megaregion of China. *J. Clean. Prod.* **2020**, *244*, 118592. [CrossRef]
9. Sun, S.; Zheng, X.; Liu, X.; Wang, Z.; Liang, L. Global pattern and drivers of water scarcity research: A combined bibliometric and geographic detector study. *Environ. Monit. Assess.* **2022**, *194*, 523. [CrossRef] [PubMed]
10. United Nations 2007. World Urbanization Prospects: The 2007 Revision Highlights, New York: Department of Economic and Social Affairs, United Nations. Available online: https://lternet.edu/wp-content/uploads/2016/07/2007WUP_Highlights_web.pdf (accessed on 17 May 2024).
11. United Nations 2011. World Urbanization Prospects: The 2011 Revision Highlights, New York: Department of Economic and Social Affairs, United Nations. Available online: https://www.un.org/en/development/desa/population/publications/pdf/urbanization/WUP2011_Report.pdf (accessed on 18 May 2024).
12. Mekonnen, M.M.; Hoekstra, A.Y. Four billion people facing severe water scarcity. *Sci. Adv.* **2016**, *2*, e1500323. [CrossRef]
13. Kreyling, J.; Arfin Khan, M.A.; Sultana, F.; Babel, W.; Beierkuhnlein, C.; Foken, T.; Jentsch, A. Drought effects in climate change manipulation experiments: Quantifying the influence of ambient weather conditions and rain-out shelter artifacts. *Ecosystems* **2017**, *20*, 301–315. [CrossRef]
14. Kummu, M.; Ward, P.J.; De Moel, H.; Varis, O. Is physical water scarcity a new phenomenon? Global assessment of water shortage over the last two millennia. *Environ. Res. Lett.* **2010**, *5*, 034006. [CrossRef]
15. Zeng, Z.; Liu, J.; Savenije, H.H. A simple approach to assess water scarcity integrating water quantity and quality. *Ecol. Indic.* **2013**, *34*, 441–449. [CrossRef]
16. Boers, T.M.; Ben-Asher, J. A review of rainwater harvesting. *Agric. Water Manag.* **1982**, *5*, 145–158. [CrossRef]
17. Li, F.; Cook, S.; Geballe, G.T.; Burch Jr, W.R. Rainwater harvesting agriculture: An integrated system for water management on rainfed land in China's semiarid areas. *AMBIO A J. Hum. Environ.* **2000**, *29*, 477–483. [CrossRef]
18. *ISO 18458:2015-05*; Biomimetics—Terminology, Concepts and Methodology. International Organization for Standardization (ISO) Copyright Office: Geneva, Switzerland, 2015. Available online: https://cdn.standards.iteh.ai/samples/62500/c1f61c3fb5024e5e91a8e4e1e2ac4fa0/ISO-18458-2015.pdf (accessed on 20 May 2024).
19. Pedersen Zari, M. Can biomimicry be a useful tool for design for climate change adaptation and mitigation? In *Biotechnologies and Biomimetics for Civil Engineering*; Pacheco Torgal, F., Labrincha, J., Diamanti, M., Yu, C.P., Lee, H., Eds.; Springer International Publishing: Cham, Switzerland, 2014; pp. 81–113. [CrossRef]
20. Kahvecioğlu, B.; Arslan Selçuk, S. Adaptive Reuse in the Realm of Architecture: Global Research Trends and Gaps for the Future Studies. *Sustainability* **2023**, *15*, 9971. [CrossRef]
21. Lindblad, P.; Lindberg, P.; Oliveira, P.; Stensjö, K.; Heidorn, T. Design, engineering, and construction of photosynthetic microbial cell factories for renewable solar fuel production. *Ambio* **2012**, *41*, 163–168. [CrossRef]
22. Creutzig, F.; Goldschmidt, J.C.; Lehmann, P.; Schmid, E.; von Blücher, F.; Breyer, C.; Fernandez, B.; Jakob, M.; Knopf, B.; Lohrey, S. Catching two European birds with one renewable stone: Mitigating climate change and Eurozone crisis by an energy transition. *Renew. Sustain. Energy Rev.* **2014**, *38*, 1015–1028. [CrossRef]
23. Loonen, R.C.; Rico-Martinez, J.M.; Favoino, F.; Brzezicki, M., Ménézo, C.; La Ferla, G.; Aelenei, L.L. Design for façade adaptability: Towards a unified and systematic characterization. In Proceedings of the 10th Conference on Advanced Building Skins, Bern, Switzerland, 3–4 November 2015; pp. 1284–1294.
24. Shahin, H.S.M. Adaptive building envelopes of multistory buildings as an example of high performance building skins. *Alex. Eng. J.* **2019**, *58*, 345–352. [CrossRef]
25. Badarnah, L. Water management lessons from nature for applications to buildings. *Procedia Eng.* **2016**, *145*, 1432–1439. [CrossRef]
26. Aslan, D.; Selçuk, S.A.; Avinç, G.M. A Biomimetic Approach to Water Harvesting Strategies: An Architectural Point of View. *Int. J. Built Environ. Sustain.* **2022**, *9*, 47–60. [CrossRef]
27. Arslan Selçuk, S.; Gülle, N.B.; Mutlu Avinç, G. Tree-Like Structures in Architecture: Revisiting Frei Otto's Branching Columns Through Parametric Tools. *Sage Open* **2022**, *12*, 21582440221119479. [CrossRef]
28. Di Bitonto, M.G.; Kutlu, A.; Zanelli, A. Fog Water Harvesting Through Smart Façade for a Climate Resilient Built Environment. In *International Conference on Technological Imagination in the Green and Digital Transition*; Springer International Publishing: Cham, Switzerland, 2023; pp. 725–734. [CrossRef]
29. Badarnah, L.; Kadri, U. A methodology for the generation of biomimetic design concepts. *Archit. Sci. Rev.* **2015**, *58*, 120–133. [CrossRef]
30. Garcia-Holguera, M.; Clark, O.G.; Sprecher, A.; Gaskin, S. Ecosystem biomimetics for resource use optimization in buildings. *Build. Res. Inf.* **2016**, *44*, 263–278. [CrossRef]
31. Badarnah, L. Form follows environment: Biomimetic approaches to building envelope design for environmental adaptation. *Buildings* **2017**, *7*, 40. [CrossRef]
32. Pawlyn, M. *Biomimicry in Architecture*, 2nd ed.; Riba Publishing: Newcastle, NSW, Australian, 2016; pp. 8–18,88–95.
33. Benyus, J.M. *Biomimicry: Innovation Inspired by Nature*, 1st ed.; Harper Collins Publishers: New York, NY, USA, 1997; pp. 1–21.

34. Ilieva, L.; Ursano, I.; Traista, L.; Hoffmann, B.; Dahy, H. Biomimicry as a sustainable design methodology—Introducing the 'Biomimicry for Sustainability framework. *Biomimetics* **2022**, *7*, 37. [CrossRef] [PubMed]
35. Feinberg, A.W.; Alford, P.W.; Jin, H.; Ripplinger, C.M.; Werdich, A.A.; Sheehy, S.P.; Grosberg, A.; Parker, K.K. Controlling the contractile strength of engineered cardiac muscle by hierarchal tissue architecture. *Biomaterials* **2012**, *33*, 5732–5741. [CrossRef]
36. Coban, M.; Coştu, B. Integration of biomimicry into science education: Biomimicry teaching approach. *J. Biol. Educ.* **2023**, *57*, 145–169. [CrossRef]
37. López, M.; Rubio, R.; Martín, S.; Croxford, B. How plants inspire façades. From plants to architecture: Biomimetic principles for the development of adaptive architectural envelopes. *Renew. Sustain. Energy Rev.* **2017**, *67*, 692–703. [CrossRef]
38. Mutlu Avinç, G.; Arslan Selçuk, S. Bio-design as a Basis for the Creation of New Architectural Materials: Experience of the Faculty of Architecture of Gazi University. *Civ. Eng. Archit.* **2024**, *12*, 643–667. [CrossRef]
39. Zari, M.P. Biomimetic approaches to architectural design for increased sustainability. In Proceedings of the SB07 NZ Sustainable Building Conference, Auckland, New Zealand, 14–16 November 2007; p. 033.
40. Taylor, R. Reflecting the impossible. *Nature* **2009**, *460*, 462. [CrossRef]
41. Gruber, P. *Biomimetics in Architecture: Architecture of Life and Buildings*, 1st ed.; Springer: Wien, Austria, 2011; pp. 16–18. [CrossRef]
42. Elsamadisy, R.; Sarhan, A.E.; Farghaly, Y.; Mamdouh, A. Biomimicry as a design approach for adaptation. *J. Al-Azhar Univ. Eng. Sect.* **2019**, *14*, 1516–1533. [CrossRef]
43. Smith, J. It's only natural. *Ecologist* **2007**, *37*, 52–55.
44. Guo, Z.; Liu, W.; Su, B.L. Superhydrophobic surfaces: From natural to biomimetic to functional. *J. Colloid Interface Sci.* **2011**, *353*, 335–355. [CrossRef] [PubMed]
45. Mueller, T. Design by nature. *Natl. Geogr.* **2008**, *213*, 68–91.
46. Chu, Y.J. A new biomimicry marine current turbine: Study of hydrodynamic performance and wake using software OpenFOAM. *J. Hydrodyn.* **2016**, *28*, 125–141. [CrossRef]
47. Vartan, S. The ultimate flattery. *E Environ. Mag.* **2006**, *17*, 44. [CrossRef]
48. Wanieck, K.; Fayemi, P.E.; Maranzana, N.; Zollfrank, C.; Jacobs, S. Biomimetics and its tools. *Bioinspired Biomim. Nanobiomaterials* **2017**, *6*, 53–66. [CrossRef]
49. Heintz, M. Biomimicry to the rescue. *Mater. Today* **2009**, *12*, 6. [CrossRef]
50. Rorabaugh, J. Apparent rain harvesting by a Colorado Desert sidewinder (Crotalus cerastes laterorepens). *Son. Herpetol.* **2007**, *20*, 128–129.
51. Bhushan, B. Design of water harvesting towers and projections for water collection from fog and condensation. *Philos. Trans. R. Soc. A* **2020**, *378*, 20190440. [CrossRef]
52. Auffenberg, W. A note on the drinking habits of some land tortoises. *Anim. Behav.* **1963**, *11*, 72–73. [CrossRef]
53. Parker, A.R.; Lawrence, C.R. Water capture by a desert beetle. *Nature* **2001**, *414*, 33–34. [CrossRef] [PubMed]
54. Nørgaard, T.; Dacke, M. Fog-basking behaviour and water collection efficiency in Namib Desert Darkling beetles. *Front. Zool.* **2010**, *7*, 1–8. [CrossRef]
55. Van Damme, P. Plant ecology of the Namib desert. *Afr. Focus* **1991**, *7*, 355–400. [CrossRef]
56. HAMILTON III, W.J.; Seely, M.K. Fog basking by the Namib Desert beetle, Onymacris unguicularis. *Nature* **1976**, *262*, 284–285. [CrossRef]
57. Seely, M.K. Irregular fog as a water source for desert dune beetles. *Oecologia* **1979**, *42*, 213–227. [CrossRef] [PubMed]
58. Tracy, C.R.; Laurence, N.; Christian, K.A. Condensation onto the skin as a means for water gain by tree frogs in tropical Australia. *Am. Nat.* **2011**, *178*, 553–558. [CrossRef]
59. Comanns, P.; Effertz, C.; Hischen, F.; Staudt, K.; Böhme, W.; Baumgartner, W. Moisture harvesting and water transport through specialized micro-structures on the integument of lizards. *Beilstein J. Nanotechnol.* **2011**, *2*, 204–214. [CrossRef] [PubMed]
60. Gans, C.; Merlin, R.; Blumer, W.F.C. The water-collecting mechanism of Moloch horridus re-examined. *Amphib. Reptil.* **1982**, *3*, 57–64. [CrossRef]
61. Zheng, Y.; Bai, H.; Huang, Z.; Tian, X.; Nie, F.Q.; Zhao, Y.; Zhai, J.; Jiang, L. Directional water collection on wetted spider silk. *Nature* **2010**, *463*, 640–643. [CrossRef]
62. Dawson, T.E. Fog in the California redwood forest: Ecosystem inputs and use by plants. *Oecologia* **1998**, *117*, 476–485. [CrossRef]
63. Ingwersen, J.B. Fog drip, water yield, and timber harvesting in the Bull Run municipal watershed, Oregon 1. *J. Am. Water Resour. Assoc.* **1985**, *21*, 469–473. [CrossRef]
64. Ebner, M.; Miranda, T.; Roth-Nebelsick, A. Efficient fog harvesting by Stipagrostis sabulicola (Namib dune bushman grass). *J. Arid. Environ.* **2011**, *75*, 524–531. [CrossRef]
65. Roth-Nebelsick, A.; Ebner, M.; Miranda, T.; Gottschalk, V.; Voigt, D.; Gorb, S.; Stegmaier, T.; Sarsour, J.; Linke, M.; Konrad, W. Leaf surface structures enable the endemic Namib desert grass Stipagrostis sabulicola to irrigate itself with fog water. *J. R. Soc. Interface* **2012**, *9*, 1965–1974. [CrossRef] [PubMed]
66. Ju, J.; Bai, H.; Zheng, Y.; Zhao, T.; Fang, R.; Jiang, L. A multi-structural and multi-functional integrated fog collection system in cactus. *Nat. Commun.* **2012**, *3*, 1247. [CrossRef]
67. Nobel, P.S. *Environmental Biology of Agaves and Cacti*, 1st ed.; Cambridge University Press: New York, NY, USA, 2003; pp. 1–50.
68. Schill, R.; Barthlott, W. Kakteendornen als wasserabsorbierende Organe. *Naturwissenschaften* **1973**, *60*, 202–203. [CrossRef]

69. Faragalla, A.M.; Asadi, S. Biomimetic design for adaptive building façades: A paradigm shift towards environmentally conscious architecture. *Energies* **2022**, *15*, 5390. [CrossRef]
70. Otto, F. *Occupying and Connecting*, 2nd ed.; Edition Axel Menges: Mitcham, UK, 2009; pp. 1–50.
71. Park, J.J.; Dave, B. Bio-inspired parametric design for adaptive stadium façades. *Australas. J. Constr. Econ. Build. Conf. Ser.* **2014**, *2*, 27–35. [CrossRef]
72. Varshabi, N.; Arslan Selçuk, S.; Mutlu Avinç, G. Biomimicry for Energy-Efficient Building Design: A Bibliometric Analysis. *Biomimetics* **2022**, *7*, 21. [CrossRef]
73. Reichert, S.; Menges, A.; Correa, D. Meteorosensitive architecture: Biomimetic building skins based on materially embedded and hygroscopically enabled responsiveness. *Comput. Aided Des.* **2015**, *60*, 50–69. [CrossRef]
74. Al-Masrani, S.M.; Al-Obaidi, K.M.; Zalin, N.A.; Isma, M.A. Design optimisation of solar shading systems for tropical office buildings: Challenges and future trends. *Solar Energy* **2018**, *170*, 849–872. [CrossRef]
75. Aasarchitecture. Available online: https://aasarchitecture.com/2017/05/winners-innovation-inspired-nature-elevens-international-biomimicry-competition/ (accessed on 14 March 2024).
76. Ajayi, T.O. Biomimicry: The nexus for achieving sustainability in the people-process-planet relationship. *Heliyon* **2023**, *9*, e16180. [CrossRef] [PubMed]
77. Dictionary of the French Academy. Available online: https://academie.atilf.fr/8/ (accessed on 17 March 2024).
78. Zhang, T.; Liu, J.P.; Zhang, Q.W. A New Navigate Pressure Deflector Louver Sun-Breaker Composed with Multi-Functions of Rainproof and Sun-Shading. *Adv. Mater. Res.* **2012**, *368*, 3737–3740. [CrossRef]
79. Raheem, A.A.; Issa, R.R.; Olbina, S. Solar transmittance analysis of different types of sunshades in the Florida climate. *Build. Simul.* **2014**, *7*, 3–11. [CrossRef]
80. Carl, T.; Schein, M.; Stepper, F. Sun Shades—About Designing Adaptable Solar Facades. In Proceedings of the 35th eCAADe Conference—Volume 2, Sapienza University of Rome, Rome, Italy, 20–22 September 2017; pp. 165–174. [CrossRef]
81. Yu, W.; Jiang, C. Study on intelligent adjustment blinds and building energy saving. In Proceedings of the IOP Conference Series: Earth and Environmental Science, Tokyo, Japan, 9–11 November 2018. [CrossRef]
82. Alkhater, M.; Alsukkar, M.; Eltaweel, A.; Su, Y. Parametric control of multiple blinds to enhance daylighting performance in the dome building: Case study of a mosque building in Saudi Arabia. *Solar Energy* **2023**, *266*, 112180. [CrossRef]
83. Mateus, D.; Henriques, G.C. Energy-Based Design: Improving Modern Brazilian Buildings Performance through Their Shading Systems, the Nova Cintra Case Study. *Buildings* **2023**, *13*, 2543. [CrossRef]
84. The Urban Roamer. Available online: https://www.theurbanroamer.com/ermita-icon-the-philamlife-building/ (accessed on 17 March 2024).
85. Livinspaces. Available online: https://livinspaces.net/ls-tv/discussing-the-design-an-indepth-look-at-the-design-of-the-al-bahar-towers-in-abu-dhabi-by-aedas-architects/ (accessed on 17 March 2024).
86. Yazdani Studio. Available online: https://yazdanistudio.com/portfolio/cj-corporation-cj-blossom-park/ (accessed on 17 March 2024).
87. Kmdarchitects. Available online: https://www.kmdarchitects.com/sfpuc (accessed on 13 March 2024).
88. Htdesignresearch. Available online: https://htdesignresearch.wordpress.com/2010/03/02/rainwater-harvestingsolar-shadingairplane-construction/ (accessed on 13 March 2024).
89. Designbuzz. Available online: https://designbuzz.com/solys-sun-shading-system-collects-rainwater-for-your-garden/ (accessed on 13 March 2024).
90. Zani, A.; Speroni, A.; Mainini, A.G.; Zinzi, M.; Caldas, L.; Poli, T. Customized shading solutions for complex building façades: The potential of an innovative cement textile composite material through a performance-based generative design. *Constr. Innov.* **2023**, *24*, 256–279. [CrossRef]
91. Baldinelli, G. Double skin façades for warm climate regions: Analysis of a solution with an integrated movable shading system. *Build. Environ.* **2009**, *44*, 1107–1118. [CrossRef]
92. Sakamoto, S.; Aoki, A. Numerical and experimental study on noise shielding effect of eaves/louvers attached on building façade. *Build. Environ.* **2015**, *94*, 773–784. [CrossRef]
93. Zuccherini Martello, N.; Aletta, F.; Fausti, P.; Kang, J.; Secchi, S. A psychoacoustic investigation on the effect of external shading devices on building façades. *Appl. Sci.* **2016**, *6*, 429. [CrossRef]
94. Nan, M.; Chen, Z.; Liu, L.; Baharlou, E. Hygrosensitive Kinetic Façade—A full-scale meteorosensitive shading system based on wood's self-actuated hygroscopic behavior, Anthropologic: Architecture and Fabrication in the cognitive age. In Proceedings of the 38th eCAADe Conference, TU Berlin, Berlin, Germany, 16–18 September 2020; pp. 133–142.
95. Vailati, C.; Bachtiar, E.; Hass, P.; Burgert, I.; Rüggeberg, M. An autonomous shading system based on coupled wood bilayer elements. *Energy Build.* **2018**, *158*, 1013–1022. [CrossRef]
96. Lillywhite, H.B.; Licht, P. Movement of water over toad skin: Functional role of epidermal sculpturing. *Copeia* **1974**, *1974*, 165–171. [CrossRef]
97. Withers, P.C.; Bradshaw, S.D. Water and energy balance of the thorny devil Moloch horridus: Is the devil a sloth? *Amphib. Reptil.* **1995**, *16*, 47–54. [CrossRef]
98. Pianka, E.R.; Pianka, H.D. The ecology of Moloch horridus (Lacertilia: Agamidae) in western Australia. *Copeia* **1970**, *1970*, 90–103. [CrossRef]

99. Comanns, P.; Esser, F.J.; Kappel, P.H.; Baumgartner, W.; Shaw, J.; Withers, P.C. Adsorption and movement of water by skin of the Australian thorny devil (Agamidae: *Moloch horridus*). *R. Soc. Open Sci.* **2017**, *4*, 170591. [CrossRef] [PubMed]
100. Sherbrooke, W.C.; Scardino, A.J.; de Nys, R.; Schwarzkopf, L. Functional morphology of scale hinges used to transport water: Convergent drinking adaptations in desert lizards (*Moloch horridus* and *Phrynosoma cornutum*). *Zoomorphology* **2007**, *126*, 89–102. [CrossRef]
101. Comanns, P.; Withers, P.C.; Esser, F.J.; Baumgartner, W. Cutaneous water collection by a moisture-harvesting lizard, the thorny devil (*Moloch horridus*). *J. Exp. Biol.* **2016**, *219*, 3473–3479. [CrossRef]
102. Witten, G. "Agamidae" (On-line). Fauna of Australia Series, 1993. Available online: https://www.dcceew.gov.au/science-research/abrs/publications/fauna-of-australia/fauna-2a (accessed on 17 March 2024).
103. Knight, K. How thorny devils tap damp sand to slake thirst. *J. Exp. Biol.* **2016**, *219*, 3366–3375. [CrossRef]
104. Roth-Nebelsick, A.; Uhl, D.; Mosbrugger, V.; Kerp, H. Evolution and function of leaf venation architecture: A review. *Ann. Bot.* **2001**, *87*, 553–566. [CrossRef]
105. He, J.; Mao, M.; Li, D.; Liu, Y.; Jin, Z. Characterization of leaf-inspired microfluidic chips for pumpless fluid transport. *J. Bionic Eng.* **2014**, *11*, 109–114. [CrossRef]
106. Alston, M.E.; Barber, R. Leaf venation, as a resistor, to optimize a switchable IR absorber. *Sci. Rep.* **2016**, *6*, 31611. [CrossRef]
107. Wang, W.; Wang, X.; He, X.; Wang, M.; Shu, H.; Xue, K. Comparisons of bioinspired piezoelectric wind energy harvesters with different layout of stiffeners based on leaf venation prototypes. *Sens. Actuators A Phys.* **2019**, *298*, 111570. [CrossRef]
108. Chen, G.; Chukwunenye, N.; Jones, G.F.; Li, C.H. Biomimetic structures by leaf vein growth mechanism for pool boiling heat transfer enhancements. *Int. J. Heat Mass Transf.* **2020**, *155*, 119699. [CrossRef]
109. Patino-Ramirez, F.; Arson, C. Transportation networks inspired by leaf venation algorithms. *Bioinspiration Biomim.* **2020**, *15*, 036012. [CrossRef] [PubMed]
110. Zhang, W.; Hansen, M.F.; Smith, M.; Smith, L.; Grieve, B. Photometric stereo for three-dimensional leaf venation extraction. *Comput. Ind.* **2018**, *98*, 56–67. [CrossRef]
111. Schönherr, J.; Bukovac, M.J. Penetration of stomata by liquids: Dependence on surface tension, wettability, and stomatal morphology. *Plant Physiol.* **1972**, *49*, 813–819. [CrossRef]
112. Spence, R.D.; Wu, H.; Sharpe, P.J.H.; Clark, K.G. Water stress effects on guard cell anatomy and the mechanical advantage of the epidermal cells. *Plant Cell Environ.* **1986**, *9*, 197–202. [CrossRef]
113. Willmer, C.; Fricker, M.; Willmer, C.; Fricker, M. The distribution of stomata. *Stomata* **1996**, *2*, 12–35. [CrossRef]
114. Lim, C.W.; Baek, W.; Jung, J.; Kim, J.H.; Lee, S.C. Function of ABA in stomatal defense against biotic and drought stresses. *Int. J. Mol. Sci.* **2015**, *16*, 15251–15270. [CrossRef]
115. Carter, R.; Woolfenden, H.; Baillie, A.; Amsbury, S.; Carroll, S.; Healicon, E.; Sovatzoglou, S.; Braybrook, S.; Gray, J.E.; Hobbs, J.; et al. Stomatal opening involves polar, not radial, stiffening of guard cells. *Curr. Biol.* **2017**, *27*, 2974–2983. [CrossRef]
116. Bidhendi, A.J.; Geitmann, A. Finite element modeling of shape changes in plant cells. *Plant Physiol.* **2018**, *176*, 41–56. [CrossRef]
117. Li, J.; Liu, C.; Xu, Z.; Zhanb, K.; Ke, X.; Wang, L. A microfluidic pump/valve inspired by xylem embolism and transpiration in plants. *PLoS ONE* **2012**, *7*, e50320. [CrossRef]
118. Kim, H.; Lee, S.J. Fabrication of triple-parted stomata-inspired membrane with stimulus-responsive functions. *Sci. Rep.* **2016**, *6*, 21258. [CrossRef] [PubMed]
119. Pan, D.; Cai, Z.; Ji, S.; Fan, S.; Wang, P.; Lao, Z.; Yang, L.; Ni, J.; Wang, C.; Li, J.; et al. Microtubes with complex cross section fabricated by C-shaped bessel laser beam for mimicking stomata that opens and closes rapidly. *ACS Appl. Mater. Interfaces* **2018**, *10*, 36369–36376. [CrossRef] [PubMed]
120. Durai Prabhakaran, R.T.; Spear, M.J.; Curling, S.; Wootton-Beard, P.; Jones, P.; Donnison, I.; Ormondroyd, G.A. Plants and architecture: The role of biology and biomimetics in materials development for buildings. *Intell. Build. Int.* **2019**, *11*, 178–211. [CrossRef]
121. Hosseini, S.M.; Mohammadi, M.; Schröder, T.; Guerra-Santin, O. Bio-inspired interactive kinetic façade: Using dynamic transitory-sensitive area to improve multiple occupants' visual comfort. *Front. Archit. Res.* **2021**, *10*, 821–837. [CrossRef]

biomimetics

MDPI

Article

Integrated Dynamic Photovoltaic Facade for Enhanced Building Comfort and Energy Efficiency

Masoud Valinejadshoubi *, Andreas K. Athienitis, Ashutosh Bagchi and Matin Abtahi

Centre for Zero Energy Building Studies, Department of Building, Civil, and Environmental Engineering, Concordia University, Montreal, QC H3G 1M8, Canada; andreask.athienitis@concordia.ca (A.K.A.); ashutosh.bagchi@concordia.ca (A.B.); matinabtahi@gmail.com (M.A.)
* Correspondence: valinejad.masood@gmail.com

Abstract: This simulation study explores the potential of a novel façade design with integrated control system comprising a dynamic photovoltaic (PV) facade integrated with dimming lighting control to enhance the work environment in office buildings and achieve energy-efficient solutions. Parametric modeling using the Grasshopper plug-in for Rhino software 7, coupled with energy simulation through the Honeybee environmental plug-in for the EnergyPlus program, are used in the methodology. The integrated control strategy was simulated to study in a single office space, utilizing the Daysim engine to assess indoor daylight quality and focusing on Daylight Factor (DF) and Daylight Glare Probability (DGP). Additionally, two artificial lighting control systems were examined for potential integration with the dynamic PV facade to minimize lighting load. The study employs the Galapagos evolutionary solver function embedded within Grasshopper to identify optimum solutions. The dynamic PV façade achieves substantial reductions in overall energy consumption, cutting it by 73% in June, 54% in July, 54.5% in August, and 52.55% in September. The results demonstrate substantial reductions in total energy consumption, with notable savings in heating and cooling due to the dynamic facade's ability to balance and control solar radiation during working hours. Moreover, the dynamic PV facade contributes to electricity generation, demonstrating its potential to improve visual comfort, decrease energy consumption, and generate electric energy through rotational adjustments and varying transparency levels.

Keywords: dynamic facades; BIPV/T; control systems; daylight factor; energy

1. Introduction

A substantial portion of global energy consumption and greenhouse gas emissions arises from buildings [1], surpassing even the energy usage in industry and transportation, constituting nearly one-third of the world's energy demand [2]. Among the various components of a building, electrical lighting, heating, and cooling contribute most to overall energy consumption. According to Waide and Tanishima [3], as reported by the International Energy Agency (IEA), grid-based electric lighting globally consumes 19% of total electricity production, with lighting representing an average of 34% of electricity consumption in the tertiary sector [4,5].

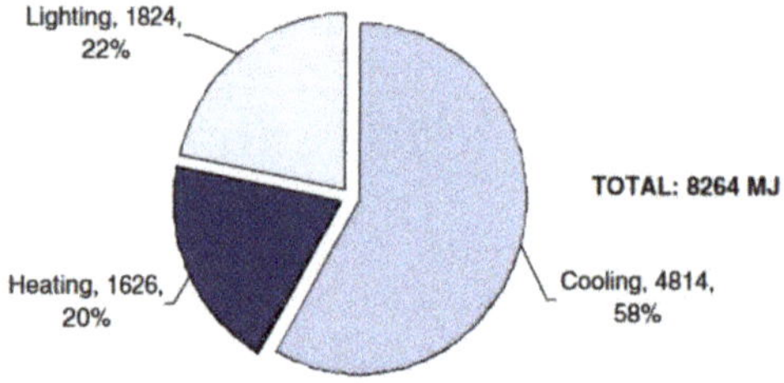

Figure 1. Energy consumption (MJ) for lighting, heating and cooling of a highrise office building in Montreal [6].

Figure 1 shows that lighting accounts for 22% of energy consumption of a typical highrise office building in Montreal, while space heating and cooling contribute 20% and 58%, respectively. By effectively utilizing and controlling daylight in office buildings, there is a significant potential for reducing energy consumption related to lighting, cooling in the summer, and heating in the winter. Additionally, this approach can enhance productivity among office workers [7]. However, buildings with extensive fenestration may experience fluctuating heating and cooling loads due to excessive solar gains, leading to glare issues, particularly on south-facing facades of office buildings. As a response, researchers and manufacturers are intensifying efforts toward designing energy-efficient buildings, incorporating renewable resources and passive design techniques to enhance energy efficiency and reduce operational energy demand [8–11]. Building Automation Systems (BASs) and Building Energy Management Systems (BEMSs) with enhanced environmental information offer exciting prospects to reduce building energy consumption and enhance comfort spaces. Given the significant influence of lighting, heating, and cooling on building energy consumption, particularly in office buildings, a holistic approach is necessary. This involves reducing energy demand, addressing lighting and thermal comfort, and deploying local efficient energy systems to meet additional demands [12]. The selection of energy and shading systems and control strategies, as highlighted by various sources [5,7], plays a crucial role in optimizing lighting and thermal performance in perimeter spaces.

In this context, dynamic building envelope technologies have garnered considerable attention due to their potential to enhance facade design, environmental responsiveness, and overall building energy performance, aiming to decrease energy demands and create comfortable office spaces [5,13,14]. Dynamic building envelopes must fulfill multiple requirements, such as light amount, uniformity, glare reduction, thermal comfort, and energy conservation. The concept introduces possibilities for creating a superior work environment in office buildings with energy-efficient solutions. Beyond controlling sunlight for heating, cooling, and glare, innovative facade systems may increase daylight levels and enhance uniformity within office spaces [15]. Dynamic facade concepts emerge as an effective strategy for managing outdoor and indoor interactions, with Building Integrated Photovoltaic (BIPV) systems and Semi-Transparent Photovoltaic (STPV) serving as integral components to bridge the gap between multifunctional dynamic facades and overall building energy balance. While BIPV and STPV are increasingly adopted as efficient solutions by designers and engineers [16,17], there remains a gap in research on optimizing such systems, considering factors such as Daylight Factor (DF), Daylight Glare Probability (DGP), artificial lighting control systems, heating, and cooling.

In this paper, DF and DGP are considered as the main indices to assess the visual comfort in a south-faced office unit, which is explained in the following:

i. Daylight Factor (DF): The Daylight Factor is an important index for evaluating the adequacy of natural light in an interior space, providing a measure of the natural light available compared to the light available outdoors under overcast sky conditions. According to CIBSE [18], Daylight Factor is defined as the ratio of indoor illuminance at a specific point to the simultaneous outdoor illuminance, expressed as a percentage.

This metric offers a simple yet powerful tool to gauge the natural lighting levels within a building.

Yose Rizala [19] stated that the Daylight Factor can be calculated using the following formula:

$$DF = \left(\frac{E_I}{E_O}\right) \times 100\% \tag{1}$$

where E_I is daylight illumination at a point on the workplane in the room, and E_O is simultaneous outdoor illuminance on the horizontal.

The Daylight Glare Index (DGI) is a principal metric for assessing discomfort glare from daylight, particularly when dealing with sources that have non-uniform luminance levels. Despite its importance, it does have some limitations and can be challenging to apply accurately. As defined by Hafiz [20], glare is the difficulty in seeing in the presence of bright light due to a direct or reflected light source within the visual field. This is typically quantified by comparing the size, location, and luminance of glare sources to the average luminance in the field of view, excluding the glare source itself.

Kurnia et al. [21] discussed that many studies use the concept of a glare index to measure indoor glare. A glare index is a quantitative measure indicating the level of perceived glare within a field of view. The general concept of the glare index can be expressed through a specific equation [21]:

$$\text{Glare} = \sum_{i=1}^{n} \frac{L_{s,i}^{\mathcal{F}}\, \omega_{s,i}}{L_b^{g}\, P_i^{h}} \tag{2}$$

While this equation provides a foundational concept, further research is necessary to refine it based on empirical data to determine an accurate glare index formula. Various experiments have led to the development of several glare indices, particularly for light sources such as daylight. In 1972, Hopkinson introduced the Daylight Glare Index (DGI), which accounts for the potential for significant glare sources, especially from the diffuse sky visible through a window.

The DGI remains a crucial index for evaluating daylight glare, especially from non-uniform luminance sources. However, its calculation is complex due to the need to determine geometric parameters, such as the solid angle and position index, and the luminance values perceived by the observer [22]. According to Hopkinson [23], the DGI formula is based on subjective perceptions. His research indicates that a DGI greater than 31 signifies intolerable glare, while a DGI less than 18 suggests that glare is barely perceptible. The DGI equation is as follows [23]:

$$DGI = 10 \times log_{10} 0.48 \sum_{i=1}^{n} \frac{L_{s,i}^{1.6}\, \omega_{s,i}^{0.8}}{L_b + \left(0.07\omega_{s,i}^{0.5}\, L_{s,i}\right)} \tag{3}$$

1.1. Adaptive Facade Systems

The widespread adoption of glazed exterior facades in contemporary office buildings has revolutionized workplace design, offering a plethora of physical and psychological health benefits to occupants. However, the deployment of these transparent structures has brought about challenges related to thermal resistance and visual discomfort [16]. In response to these challenges, there has been a concerted effort in recent years to develop energy-conscious buildings, with a particular focus on advanced adaptive facade systems [24]. Just as natural organisms exhibit adaptive behavior to survive and thrive in their habitats, adaptive facade systems can dynamically adjust their properties based on external factors such as sunlight intensity, temperature, and wind. By incorporating sensors and actuators, these systems can mimic the responsiveness of natural organisms, optimizing energy performance and occupant comfort. Among these innovations, dynamic PV facades have emerged as a promising solution capable of dynamically responding to climatic variations, thereby enhancing both visual comfort and energy performance [25]. In Canada, where lighting, heating, and cooling constitute the highest portions of electrical energy consumption in office buildings, ample opportunities exist for research [26].

Studies by Lee and Selkowitz [26] on adoptable shading devices and auto-dimming lighting systems revealed significant reductions in cooling and lighting energy consumption, showcasing the potential of dynamic shading systems. Athienitis and Tzempelikos [15] demonstrated 75% energy savings on overcast days and 90% on clear days through an advanced window system with controlled shading and light dimming. Additionally, research by Zhou et al. [27] emphasized the energy-saving benefits of auto-dimming lighting, reporting a 57% reduction in lighting energy consumption compared to non-dimming lighting [28].

1.2. Optimizing Visual Comfort and Energy Efficiency

In office environments, achieving optimal visual comfort is essential for productivity and well-being. DF is a metric used to quantify the amount of natural light available in a room relative to the illumination on a horizontal plane under an unobstructed sky hemisphere [29,30]. It provides insight into the distribution of daylight within a space, helping designers ensure adequate lighting levels without causing glare or discomfort. A high DF indicates a well-lit room with ample natural light, reducing the need for artificial lighting and consequently lowering energy consumption.

On the other hand, DGP assesses the likelihood of glare occurring in space due to excessive brightness contrasts between light and dark areas [29,30]. Glare can cause discomfort, eye strain, and reduced visual performance, negatively impacting occupant satisfaction and productivity. By analyzing DGP, designers can identify potential glare sources and implement mitigation strategies to enhance visual comfort.

In the context of office units, where sunlight exposure is more intense, balancing DF and DGP becomes paramount. Biomimetic-inspired STPV windows offer a unique opportunity to address this challenge by mimicking nature's adaptive strategies. Exploring STPV with varying transparency levels, Miyazaki et al. [31] found that optimum Semi-Transparent Photovoltaic (STPV) windows reduced total building electricity consumption by 55%, enhancing visual comfort and energy efficiency [32]. Drawing inspiration from natural systems, such as the self-regulating mechanisms found in plant leaves or animal eyes, designers can develop STPV windows and facades that dynamically adjust transparency levels based on incoming sunlight intensity. This biomimetic approach allows for optimized daylighting while minimizing glare, creating a harmonious and comfortable indoor environment reminiscent of natural ecosystems.

Despite the growing popularity of glazed exterior facades in modern office buildings due to their numerous benefits, including physical and psychological well-being, these structures present significant challenges. Issues related to thermal resistance and visual discomfort persist, prompting the need for innovative solutions. While extensive research has focused on developing energy-conscious buildings, particularly through adaptive facade systems, the integration of dynamic PV facades remains relatively unexplored. Furthermore, in regions like Canada, where lighting, heating, and cooling contribute significantly to electrical energy consumption in office buildings, a pressing need exists for sustainable and efficient building design solutions. Current studies highlight the potential of dynamic PV shading devices and auto-dimming lighting systems in reducing energy consumption, emphasizing the importance of implementing such technologies in building design. However, gaps remain in the literature regarding the integration of these systems and their impact on visual comfort and energy efficiency. This study aims to optimize visual comfort, reduce glare, and minimize energy consumption in glazed office buildings. By exploring the interplay between dynamic PV facades and lighting control systems, the research seeks to enhance the overall working environment and productivity while contributing to more efficient and sustainable building operations. Ultimately, this study will advance sustainable building design practices by providing insights and practical applications of adaptive building control systems, promoting the development of innovative and environmentally friendly building technologies. The usefulness of dynamic PV facades in this paper, inspired by biomimetics, lies in their ability to enhance sustainability and efficiency in building operations. By mimicking natural systems that adapt to environmental changes, these facades optimize visual comfort, reduce glare, and minimize energy consumption in glazed office buildings. They generate electricity, reduce the need for artificial lighting, and improve occupant comfort, thereby promoting innovative and sustainable building design practices rooted in the principles of biomimetics.

2. Research Methodology

The methodology section outlines the research approach, detailing the development of the dynamic PV facade system concept and its integration with artificial lighting control systems. The methodology encompasses several key components, starting with the conceptualization and parametric modeling of the dynamic PV facade system using Rhino/Grasshopper software 7. Rhino/Grasshopper, a graphical algorithm editor, enables rapid generation of parametric forms, bridging the gap between design concepts and energy performance considerations [33]. This computational tool facilitates the incorporation of biomimetic principles and adaptive behavior into the design process, allowing for the exploration of various facade configurations and lighting control strategies. The Rhino parametric software 7 integrates these design elements as multi-variable inputs in an optimization process aimed at achieving optimal outcomes, including high-quality daylight, minimal glare, and efficient electric energy consumption and generation [34]. Environmental simulations, encompassing visual comfort, electric lighting, heating, cooling, and internal energy analysis, are conducted using the Honeybee plug-in [35]. Integrated with Radiance and Daysim, Honeybee enables advanced grid-based daylight simulation and calculation of illuminance [9] and DGP through image-based simulation to assess the performance of the dynamic PV facade system.

2.1. Overall Concept and Design

Biomimetics, or biomimicry, is an interdisciplinary approach that draws inspiration from nature to solve human challenges and innovate in various fields, including architecture and technology. By studying biological systems, organisms, and ecosystems, biomimetics seeks to emulate their efficient structures, behaviors, and processes to design more sustainable, functional, and adaptive solutions. In the context of our research on dynamic PV facade systems, biomimetics informs the circular shape of the STPV and PV layers, mirroring natural forms observed in organisms. Additionally, biomimetic principles guide the system's adaptive function, reflected in the independent movement of PV layers to optimize solar energy capture and regulate daylight levels, akin to adaptive behaviors found in nature. The primary objective of this study is to create a dynamic facade solution capable of regulating daylight, reducing glare, and optimizing solar energy capture in office environments. By integrating biomimetic principles and advanced control systems, our dynamic PV facade model aims to balance natural lighting, energy consumption, and occupant comfort harmoniously.

The dynamic system comprises two Semi-Transparent Photovoltaic (STPV) layers, thin-film amorphous silicon glass with varying levels of transparency (Figure 2, No. 1 and 2), and one Polycrystalline PV layer (Figure 2, No. 3). Each PV layer possesses an independent structure (No. 5) connected to the central structural part (No. 4), allowing individual movement through dedicated centric servomotors (No. 6). A servomotor, serving as a rotary actuator, facilitates precise control over angular position, velocity, and acceleration [36]. The three PV layers, along with their structural and electrical components, collectively form a modular device. The central structural part is designed to connect to either the primary or secondary building beam, enabling the transfer of the system's load (No. 4). Figure 2 illustrates the simulated dimensions of one module of dynamic PV device in this research, and Figure 3 illustrates conceptual rendering of the dynamic PV facade on an office building. The rotations of the PV layers were specified in six steps (0°, 60°, 120°, 180°, 240°, 300°), as illustrated in Figure 2. In addition to the material properties of the PV layers within the dynamic system (Table 1), the materials used for the interior surfaces of the building (Tables 1 and 2) are paramount for controlling daylight and glare, thereby influencing visual comfort or discomfort.

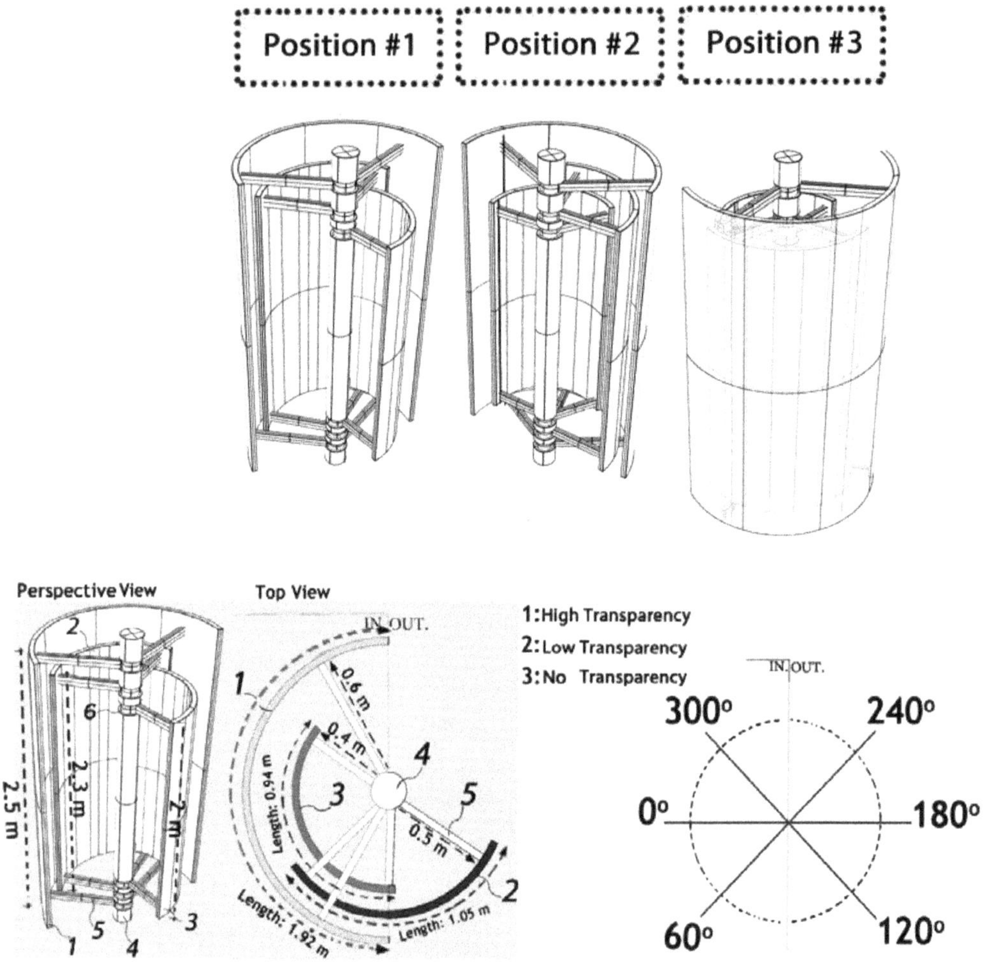

Figure 2. **Top**: The example of three positions of the dynamic PV system; **Bottom Right**: the diagram depicts the rotation degrees of the dynamic PV system consisting of layers 1, 2, 3, demonstrating the range of motion. **Left**: a detailed perspective and top view illustrate the components and structure of a module within the dynamic PV system.

Figure 3. Conceptual rendering of the dynamic PV facade on an office building.

The main design considerations for the dynamic PV facade system include:

- Biomimetic Design: Drawing natural inspiration to optimize system operation, structures, and behaviors.
- Modular Structure: Each PV layer has an independent structure, allowing for individual movement.
- Precise Control: Utilizing servomotors for accurate adjustment of PV layer positions.
- Material Selection: Choosing appropriate materials to balance energy generation and visual comfort.
- Rotational Adjustments: Specifying rotations to optimize solar energy capture throughout the day.

2.2. Lighting Control System Parameters

The dynamic PV facade functions as an automated system designed to regulate daylight within the necessary interior illuminance range (500 Lux) during office hours from 8 A.M. to 4 P.M. However, the DF provides an average lux value over broad surfaces, such as an entire room, which may not ensure uniform lighting throughout the office space [37]. Hence, artificial lighting becomes essential to achieve an appropriate lighting level in specific areas of the room. The artificial lighting control system is integrated with the dynamic PV facade to adopt the most effective hourly control strategy for reducing the lighting load.

Two artificial lighting control systems simulated in this research include (1) an Automated Switch-Off Occupancy Sensor and (2) an Always-On system during active occupancy working hours with auto-dimming. Following a lighting calculation, in simulation, six light sensors were placed in the further areas from the facade, receiving less daylight than the illuminance threshold (refer to Figure 4). These lighting sensors were simulated on the working table at a 1.40 m distance from each other, at a distance of 0.8 m from the floor, and 2.40 m from the facade. The lighting source chosen for this study was a light-emitting diode (LED) screw-base lamp with a power rating of 250 watts.

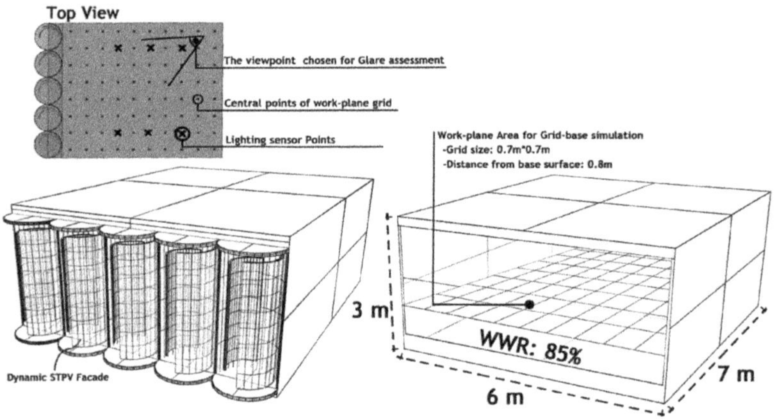

Figure 4. Right: The baseline office room model; **Left**: The office room model with dynamic PV facade.

2.3. Energy Analysis

In this research study, a simulation-based approach was employed to analyze the energy performance of a dynamic PV facade system implemented within an office space situated in Montreal, ASHRAE climate zone 6, characterized as Cold–Humid. The simulation focused on three key components: building material, operational schedule, and HVAC system. Table 1 details the properties of the PV layers, while Table 2 outlines the recommended wall, roof, and floor materials by ASHRAE for the specified climate zone, along with Table 3 interior surface characteristics relevant for lighting analysis. This simulation-based methodology comprehensively evaluated the dynamic PV facade system's energy performance under various environmental conditions and operational settings.

Table 1. Characteristics of PV layers in each analyzed module.

Element	Type	Max. Power (Wp/sqm) [38]	Visible Light Transmittance (%) [39]	SHGC (%) [39]	UV Transmittance (%) [39]	Exterior Reflection (%) [39]	Surface Area (Square Meter)
PV Layer 1	High Transparency	28	28	40	4	8.2	20
PV Layer 2	Low Transparency	40	10	29	0.1	8.3	10
PV Layer 3	Polycrystalline	200	0	0.5	0	0.7	8

Table 2. Characteristics of construction materials used in simulation.

Construction	Material (From Out to In)	U-Value	R-Value
Wall [7]	2.54 cm Stucco; 20.32 Concrete; Insulation; 1.27 cm Gypsum	0.425	2.349
Floor [7]	1.27 cm Gypsum; AtticFloor NonRes Insulation; 1.27 cm Gypsum	0.157	6.334
Roof [7]	Roof membrane; Roof insulation; Metal decking	0.18	5.36

Table 3. Characteristics of interior surfaces in simulation.

Interior Surfaces	Color	Roughness [40]	Secularity [40]
Interior Walls		0.03	0
Interior Ceiling		0.03	0
Interior Floor		0.01	0.05

To estimate electricity generation, we first evaluated the hourly average global solar radiation incident on the PV layers (Q) in kWh/m² using the Honeybee plug-in. The transmittance of glass (ŢG), as defined in Table 1, played a critical role in this calculation. The efficiency or yield of the PV panels (η) in this study varied depending on the PV type, also detailed in Table 1.

The performance ratio (PR), which is used to assess the system's overall performance, was established at 0.75, as referenced in previous studies [41,42]. With the radiation data in hand, we calculated the energy generation of the PV layers using the following equation [43]:

$$Electricity\ Generation = PV_{\eta_e}\,\tau_G\,A_{PV}\,Q \tag{4}$$

This approach allowed us to precisely estimate the energy output of each PV layer by taking into account the specific characteristics of the glass transmittance, PV panel efficiency, and the established performance ratio.

Figure 5 provides a comprehensive schedule for the office, encompassing occupants, lighting, equipment, and cooling and heating systems. Schedule input values are presented in fractional units within the range of 0 and 1, except for lighting, which is specified in lux. The regular workweek in Montreal spans 40 h, resulting in the office room being occupied from 8 A.M. to 4 P.M., with a density of 0.14 people/m² (six people for 42 m^2). The standard lighting level for office work is set at 500 lux, representing the sensor setpoint for lights during working hours [29]. Equipment usage was assumed to be 8 W/m^2, with some equipment operating continuously during operational hours.

Time	Mon-Fri Occupants (Fraction)	Sat, Sun Occupants (Fraction)	Mon-Fri Lighting (Lux)	Sat, Sun Lighting (Lux)	Mon-Fri Equipment (Fraction)	Sat, Sun Equipment (Fraction)	Mon-Fri Cooling (Fraction)	Sat, Sun Cooling Fraction) (	Mon-Fri Heating (Fraction)	Sat, Sun Heating (Fraction)
1a-6a	0	0	0	0	0.05	0.05	0	0	0.05	0.05
7a	0.2	0	0	0	0.05	0.05	0.2	0	0.2	0.05
8a-4p	0.9	0	500	0	0.9	0.05	0.5–0.9	0	0.8–0.9	0.05
5p	0.3	0	5	0	0.05	0.05	0	0	0.9	0.05
6p	0	0	6	0	0.05	0.05	0	0	0.05	0.05
7p	0	0	7	0	0.05	0.05	0	0	0.05	0.05
8p	0	0	8	0	0.05	0.05	0	0	0.05	0.05
9p	0	0	9	0	0.05	0.05	0	0	0.05	0.05
10p	0	0	10	0	0.05	0.05	0	0	0.05	0.05
11p	0	0	11	0	0.05	0.05	0	0	0.05	0.05
12a	0	0	12	0	0.05	0.05	0	0	0.05	0.05

Figure 5. The schedule of the office room investigated in this research.

Cooling and heating are assumed to be provided by Packaged Terminal Heat Pumps (PTHPs) with a coefficient of performance (COP) of 3.50. Night-time building setbacks were implemented (26 °C for cooling and 18 °C for heating) to reduce the temperature difference between ambient and indoor spaces, aiming to save energy. The cooling and heating systems were initiated at 5 AM, using 20% of the system capacity, to gradually balance the indoor air temperature within the setpoint range (24 °C for cooling and 20 °C for heating). Additionally, 5% of the heating system capacity was utilized to maintain an acceptable temperature during unoccupied periods, preventing equipment freezing in the office room in winter or overheating in summer.

2.4. Simulation Setup

The methodology is demonstrated by considering a south-facing single office space located in Montreal, Canada. Solar radiation, a non-controllable input, plays a significant role in the evaluation process. Figure 6 illustrates the annual variation in global solar radiation in Montreal, with peak values occurring in June, July, August, and September. Figure 7 illustrates the model parameters and performance criteria utilized in the evaluation process. The primary variable parameters included PV layer rotation, two lighting control systems, and four analysis periods during the year integrated into the Honeybee/Grasshopper algorithmic process.

Figure 8 provides a visual representation of the constant feedback loop during the exploration of control system scenarios, incorporating thresholds for illuminance, glare index, and lighting load. This exploration utilized a problem-solving exhaustive method through Galapagos, an evolutionary solver in Grasshopper known as a Genetic Algorithm (GA) [39]. Galapagos, functioning as a genetic multi-objective algorithm solver, operates based on a typical input–output relationship.

The optimization process involved defining three main objectives: achieving the required daylight illuminance threshold of 500 lux, reducing glaring risks in a specific occupant position, and minimizing the lighting load in an office room in Montreal, Canada. Galapagos utilized a fitness function to determine the goal it needed to solve, with the parametric variables' relationships defined as outputs (genomes) in this iterative process. The data recorder component in Grasshopper was employed to record the outputs of each generation, saving the data in Microsoft Excel for further analysis. The estimation of energy generation by PV layers involved determining the amount of radiation each layer could absorb. This information was then incorporated into a global formula to estimate the electricity generated by the photovoltaic systems. The dynamic PV facade was implemented on the south face of the office space with an 85% Window Wall Ratio (WWR) and spatial dimensions of 6 m (width), 7 m (depth), and 3 m (height) (Figure 4). The research process is outlined in three phases, as detailed below.

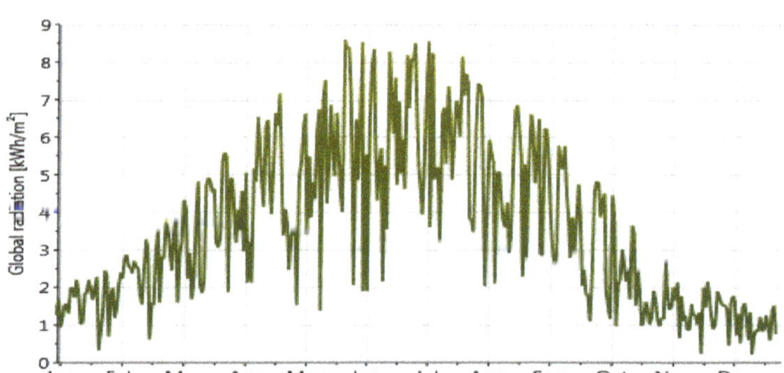

Figure 6. Daily global radiation in Montreal (Meteonorm 7 software).

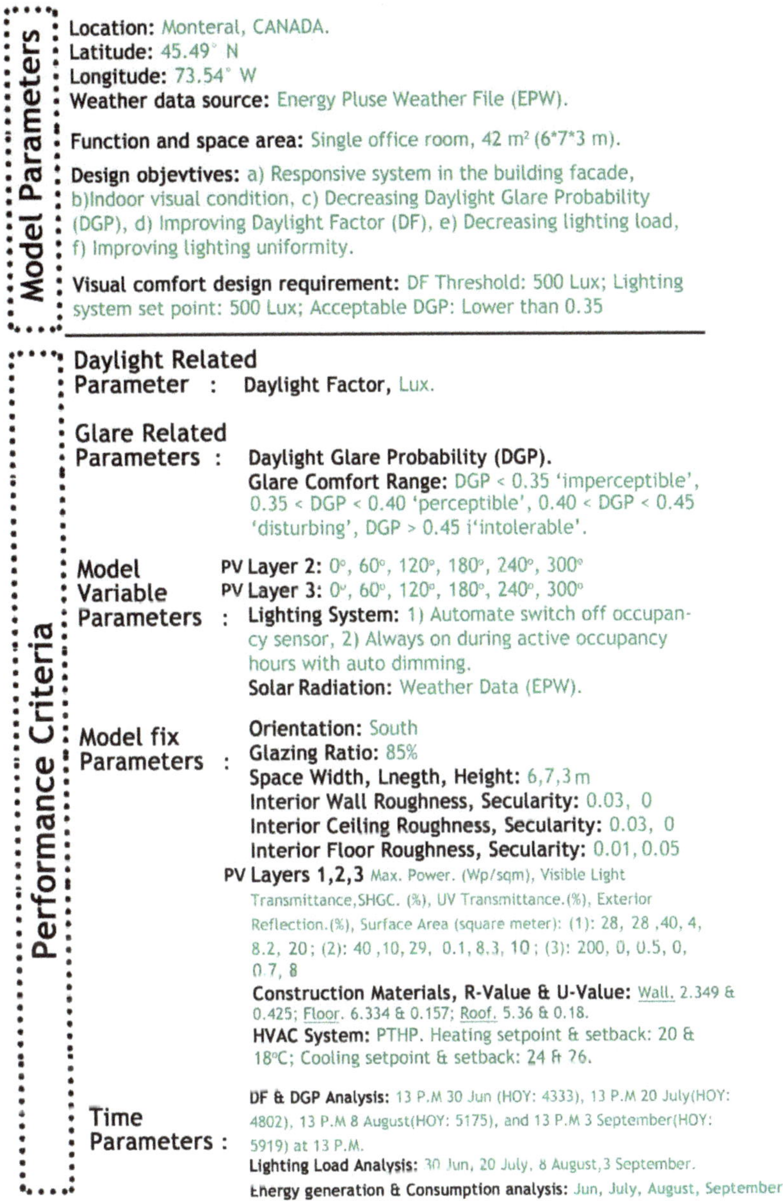

Figure 7. Model parameters and performance criteria in the simulation.

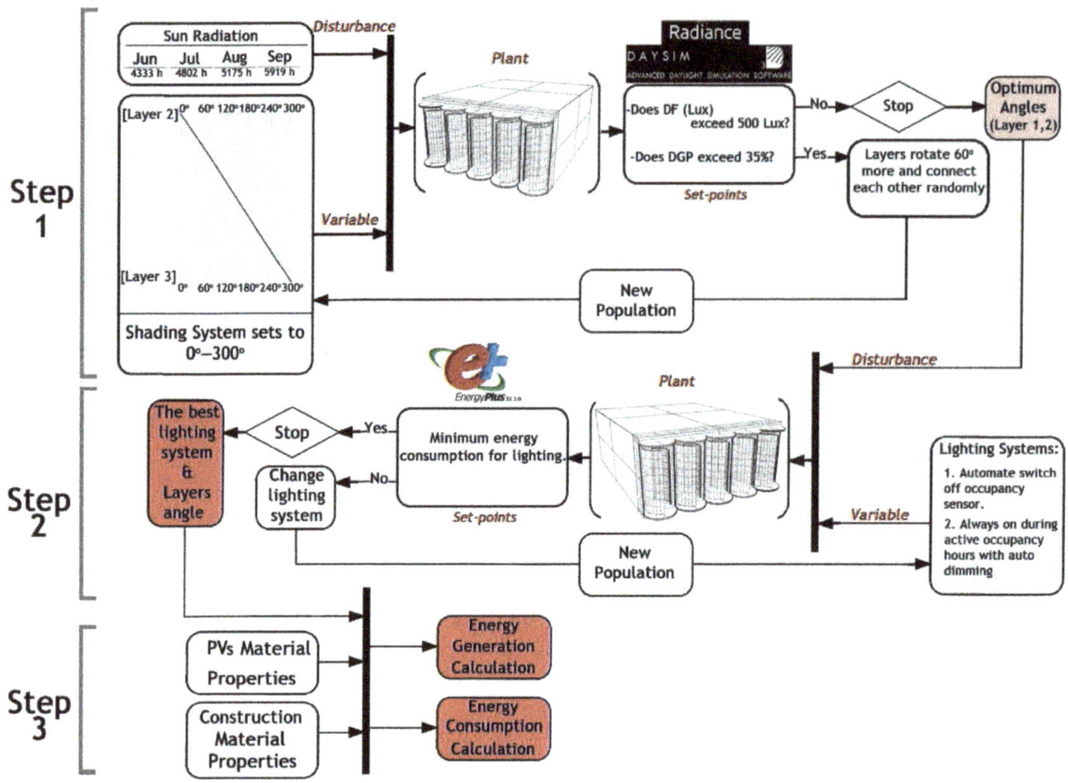

Figure 8. A visual representation of the constant feedback loop during the optimization processes.

- In the initial phase, the study investigated the dynamic behavior of the dynamic PV facade in controlling daylight uniformity and energy consumption at four different Hours of the Year (HOY). A parametrized prototype was developed, and critical DF and DGP indices for indoor visual comfort were identified through Grasshopper's optimization capabilities. The suggested glare comfort criteria for DGP included categories such as 'imperceptible', 'perceptible', 'disturbing', and 'intolerable'. The DF threshold for this analysis was set at 500 Lux. The Honeybee plug-in, integrated with Radiance and Daysim 3, facilitated the analysis of daylight performance and glare probability. An evolutionary solver function called Galapagos, a GA framework within Grasshopper, was utilized to find optimal solutions for dynamic PV facade system positions. Simulations were conducted for 1 P.M. on the 30th of June, 20th of July, 8th of August, and 3rd of September, corresponding to HOY: 4333, 4802, 5175, and 5919.
- The second phase involved determining the optimal positions of PV layers and two artificial lighting control systems to control DF and DGP, aiming to identify configurations that could achieve significant reductions in lighting load. The analysis of lighting load was performed using EnergyPlus 22.2.0 within the Honeybee plug-in. Daily simulations were conducted on the 30th of June, 20th of July, 8th of August, and 3rd of September.
- In the final phase, after identifying the optimum positions of dynamic PV facade layers and an efficient lighting control system, the study assessed the impact of the PV facade system on energy consumption and energy generation. Monthly simulations were conducted for June, July, August, and September, representing the sunniest times of

the year in Montreal, and selected using Meteonorm 7 software (global meteorological database for engineers, planners, and education). The assumption in this phase was that each month had an optimum PV layer position, and layers rotated monthly. It is worth noting that, in reality, the layers in the dynamic facade system can rotate on a smaller time scale (sub-hourly) through Lux sensors, actuators, and a Direct Digital Controller to continuously enhance visual comfort and reduce energy consumption.

3. Results and Discussion

This section presents the simulation results of the integrated control systems, namely the dynamic PV facade and artificial lighting control. The EnergyPlus Weather File (EPW) for Montreal, along with direct sunlight using the CIE sky, was utilized in this simulation process. The results and related data from all possible generations were categorized into three main sections for discussion:

- DF Analysis and DGP Analysis: This includes an analysis with a threshold of 500 Lux for DF and a threshold of DGP < 0.35 for DGP. The time scale for this analysis spans four different HOYs: 4333, 4802, 5175, and 5919.
- Artificial Lighting Control System Analysis: This section focuses on the analysis of the artificial lighting control system, with a threshold set for achieving the minimum lighting load. The time scale for this analysis covers four specific days.
- Electricity Energy Consumption and Generation: This involves the evaluation of electricity energy consumption in a simulated office room in Montreal and the generation by optimum positions of PV layers in the dynamic facade system. The time scale for this analysis spans four months.

The optimization involved 288 generations for DF and DGP analysis and 3456 generations for lighting load, employing an evolutionary solver known as a GA within Grasshopper. These categories provide a comprehensive overview of the performance and effectiveness of the integrated control systems in various aspects and time scales.

3.1. DF and DGP

In this step, the variations included the rotation of PV layers in six steps and different simulation periods (as shown in Figure 9, step 1). The optimization process generated extensive data and numerous generations, posing the challenge of finding the optimal trade-off between DF and DGP. To address this, data mining techniques were employed to extract actionable information from the optimization data obtained through the GA approach.

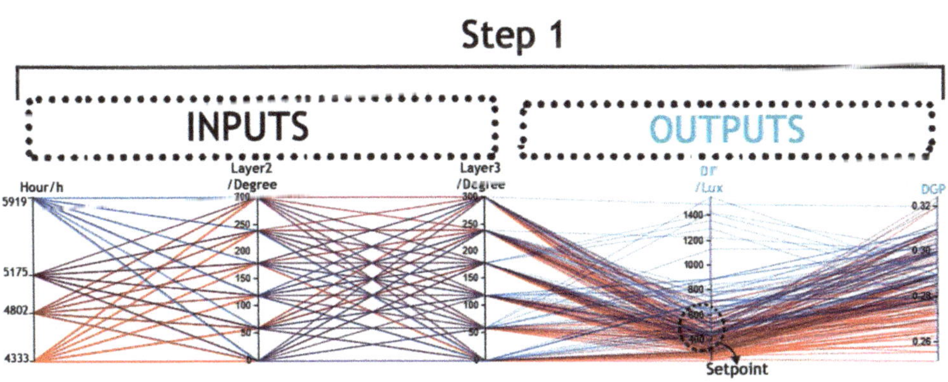

Figure 9. Optimization of Daylight Factor (DF) and Daylight Glare Probability (DGP) using Visual Data Mining Techniques..

A Decision Support System was enhanced using DM, with the online software Design Explorer 2 applied to analyze the multi-objective behavior of the prototype. DE facilitated a better understanding of the results, employing visual data mining techniques to display multivariate datasets in parallel coordinates. This approach transformed the search for relationships among variables into a 2D pattern.

The results from GA, along with inputs and outputs, were documented in Microsoft Excel spreadsheets. Figure 9 illustrates the visual relationships between multivariate inputs and outputs created by VDM to detect visual cues. In this analysis, a specific domain between 450 and 500 lux (representing acceptable DF) was chosen as a setpoint to identify correlations between DF domains and DGP, as shown in Figure 10.

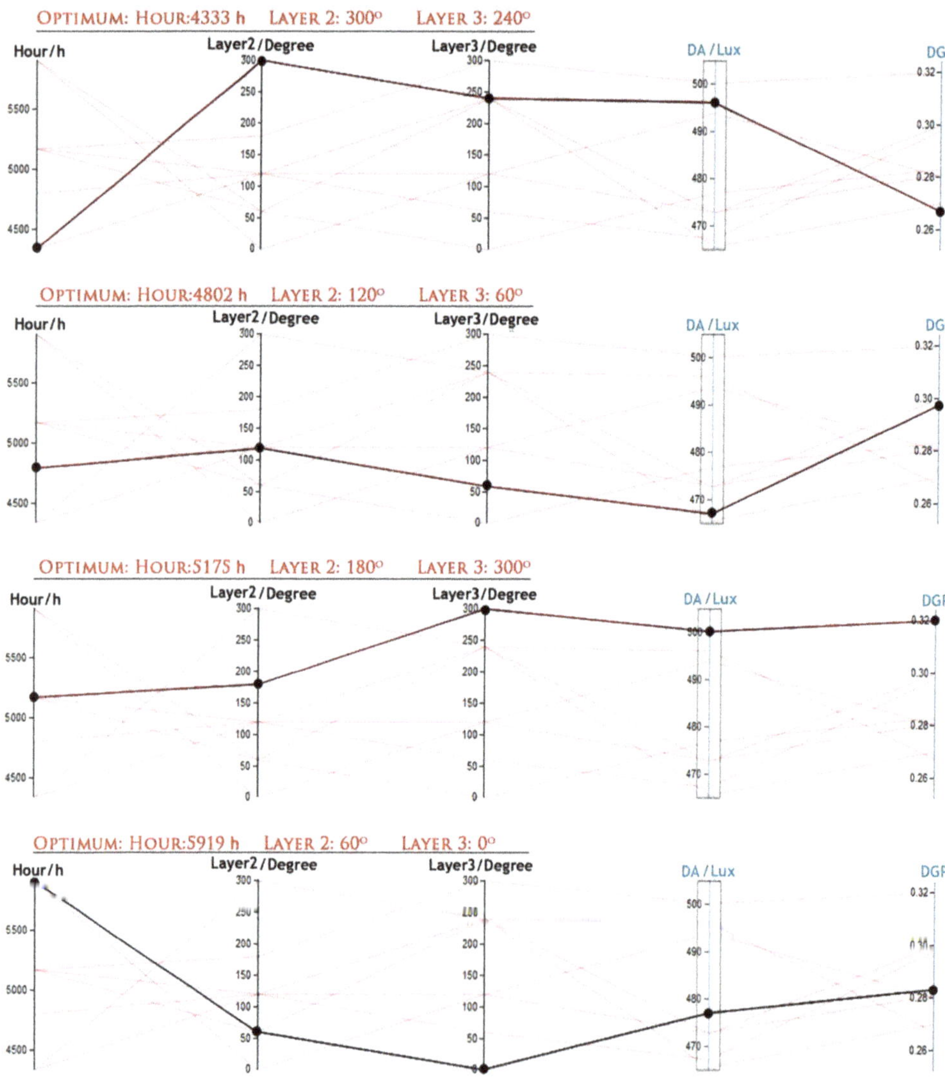

Figure 10. Detecting visual cues to find optimum layers of positions in DF and DGP optimization results.

In Figure 10, the optimal relationships between inputs and outputs to enhance visual comfort during four different HOYs were identified. The outcomes for each simulation period are as follows:

- June (HOY: 4333): Optimal angles for PV layers 2 and 3 were 300° and 240°, respectively. The outputs were 495 lux for DF, representing the average amount of lux over broad surfaces like a whole room, and 0.27 for DGP, falling within the imperceptible glare range.
- July (HOY: 4802): The optimal angle for PV layer 2 was 120°, and for layer 3, it was 60°. The results showed 468 lux for DF and 0.29 for DGP, with the Glare Comfort Range (GCR) categorized as imperceptible glare.
- August (HOY: 5175): An optimal angle of 180° was identified for PV layer 2, and for PV layer 3, 300° was optimal. DF was measured at 500 lux, with DGP at 0.32, falling within the imperceptible glare range.
- September (HOY: 5919): During this simulation period, PV layers 2 and 3 were set at 60° and 0°, respectively. The resulting outputs were DF: 478 lux and DGP: 0.28, with GCR falling within the range of imperceptible glare.

It is crucial to highlight that the dynamic PV facade effectively fulfills the essential building envelope functions concerning air, moisture, and heat transfer, encompassing vital features such as rainscreen protection, water-shedding mechanisms, air and vapor barriers, as well as insulation layers through PV layer 1. PV layer 1, characterized by high transparency, minimally affects daylighting within the space. Its design prioritizes natural ventilation capable of opening during favorable conditions and closing during winter months. However, it is worth noting that while the dynamic facade system enhances natural ventilation, the detailed investigation of its improvement lies beyond the scope of this paper.

The results of Daylight Autonomy (DA), DGP, and GCR calculations in an office room with a dynamic PV facade were compared with a baseline office room without the dynamic PV facade, as presented in Table 4. This comparison aimed to understand the impact of the dynamic PV facade on visual comfort. The table reveals that, in the baseline office room, there is visual discomfort in all four simulation periods, indicated by DA values exceeding 500 lux and DGP values surpassing 0.35. Additionally, the GCR in three HOYs is categorized as intolerable glare.

Table 4. The results of DA, DGP, and GCR with and without dynamic facade on an office room.

	Jun. HOY: 4333	July. HOY: 4802	Aug. HOY: 5175	Sep. HOY: 5919
Baseline (Lux)	8286	8691	13,876	14,099
Baseline with Dynamic Facade (Lux)	495	468	500	478
Baseline (DGP)	0.39	0.51	0.52	0.58
Baseline with Dynamic Facade (Lux)	0.27	0.29	0.32	0.28
Baseline (GCR)	Perceptible Glare	Intolerable Glare	Intolerable Glare	Intolerable Glare
Baseline with Dynamic Facade (Lux)	Imperceptible Glare	Imperceptible Glare	Imperceptible Glare	Imperceptible Glare

Figure 11 presents the grids used for grid-based daylight simulation in DF analysis and the monitor view for image-based simulation in DGP calculation. This comparison is made between the baseline office room and the office room equipped with optimum control systems of dynamic PV facade, as outlined in Table 4. The visual results in Figure 11 demonstrate that the use of PV materials in three layers enables sunlight penetration through diffuse transmission and effectively controls daylight to balance within the standard illuminance range.

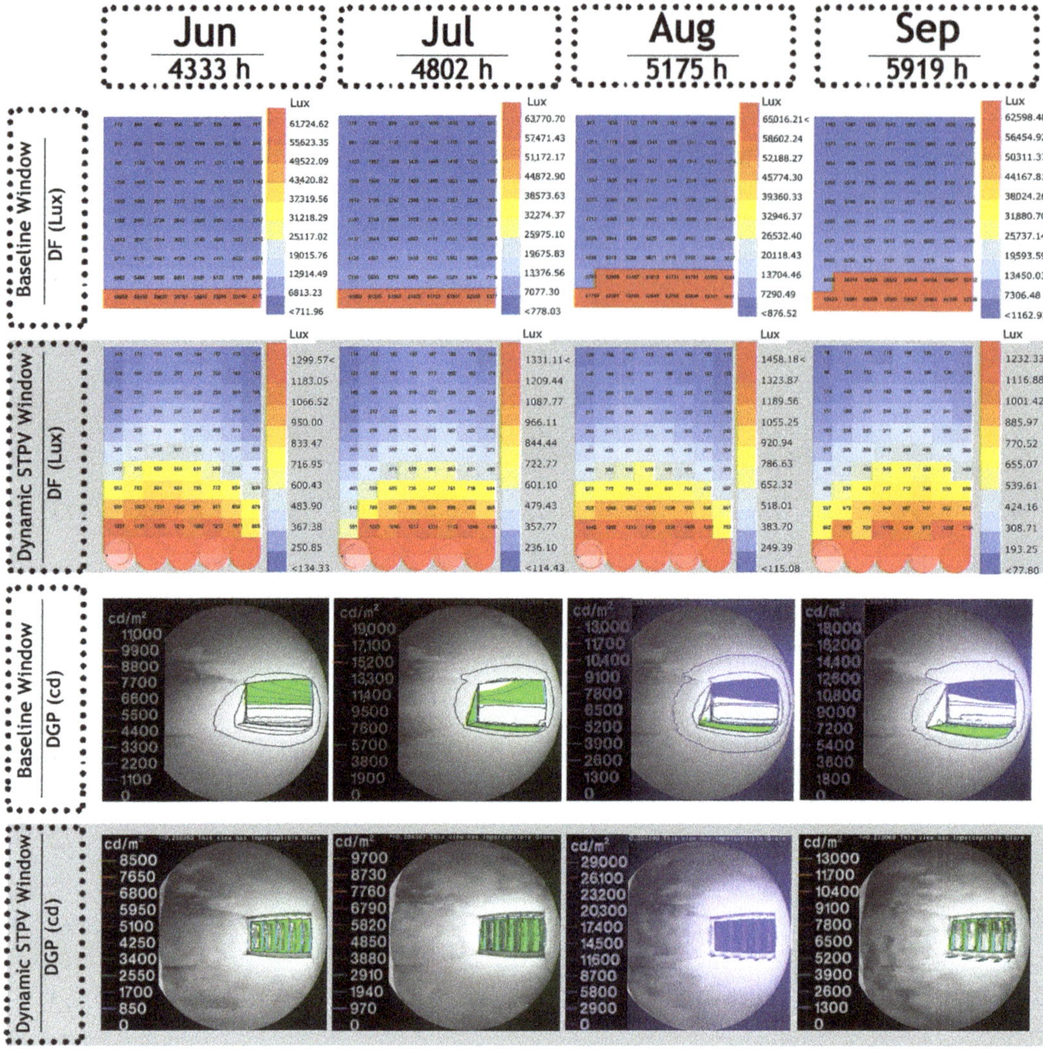

Figure 11. Above: (Top View) The ratio of the daylight illumination at points on the work plane in DF analysis. **Below**: (Perspective) The viewpoints chosen in the interior office room space and the radiance picture format in DGP analysis.

Despite the improvement in sunlight penetration, light quality, and uniformity achieved by adopting a dynamic PV system, the results indicate that the standard illuminance level cannot reach deeper into the space in the designated zone. Consequently, the implementation of artificial lighting was deemed necessary to achieve a suitable illuminance domain in all parts of the room.

Addressing a common concern regarding BIPV and dynamic facade systems potentially obstructing views outside, Figure 12 illustrates that, with the optimal degree of PV layers, users can still have a view outside. The white space represents areas covered with high-transparency PV, allowing users to see outside, while the black surface consists of PV opaque material with no transparency, and the grey region represents semi-transparent PV.

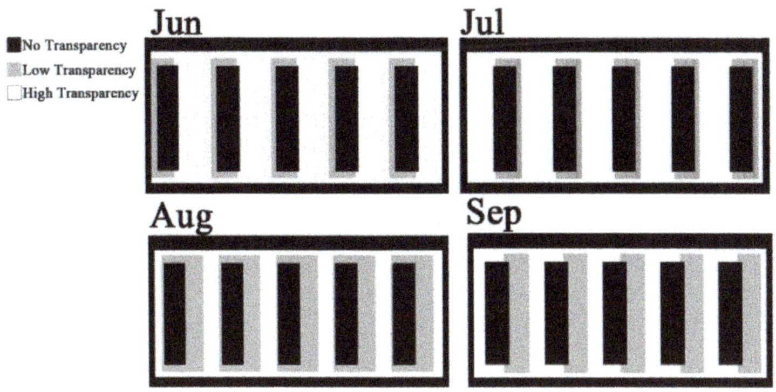

Figure 12. The analysis of the probability of the dynamic PV facade system in optimum angles to have a view outside.

3.2. Lighting Control System

In the subsequent phase following optimization, optimal positions were determined based on predefined set points in terms of DA and daylight DGP for four selected HOYs. This section delves into the evaluation of two distinct artificial lighting control systems: (1) an occupancy sensor system and (2) a dimming system. The optimization inputs encompassed four specific days of the year, the optimal positions of the dynamic PV layers in degrees, and the characteristics of the two lighting control systems.

The VDM results depicted in Figure 13 reveal that the lighting load associated with the occupancy sensor system surpasses that of the dimming system when integrated with the dynamic PV facade system. This outcome arises from the continuous occupation of the office room during working hours, coupled with the improved lighting uniformity facilitated by the dynamic facade system. Furthermore, Figure 14 provides an overview of the hourly energy consumption on the four simulation days for both lighting control systems. Notably, System 2, the dimming system, exhibits a noteworthy energy-saving performance, preserving 20.2% more energy in lighting on 30th June, 20.9% on 20th July, 10.71% on 4th August, and 12.8% on 3rd September. These energy savings contribute significantly to the reduction of peak loads and demand charges during operational hours.

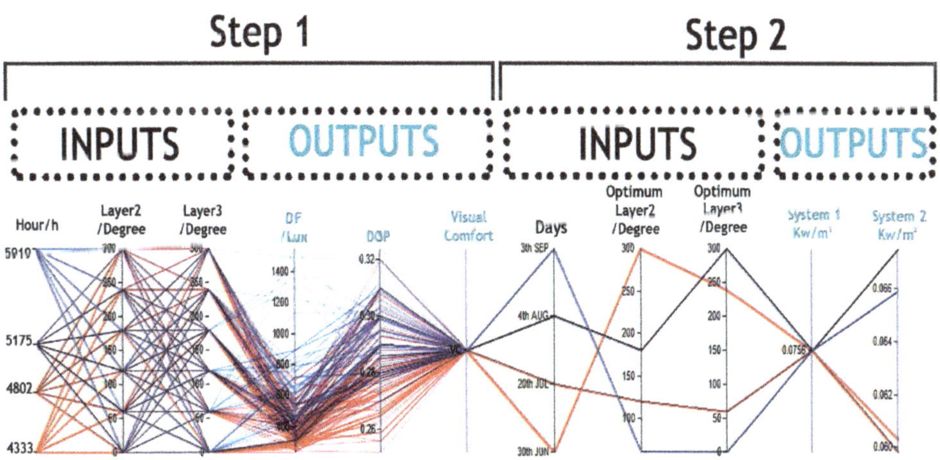

Figure 13. The visual data mining between the results of DF–DGP and lighting control system optimizations.

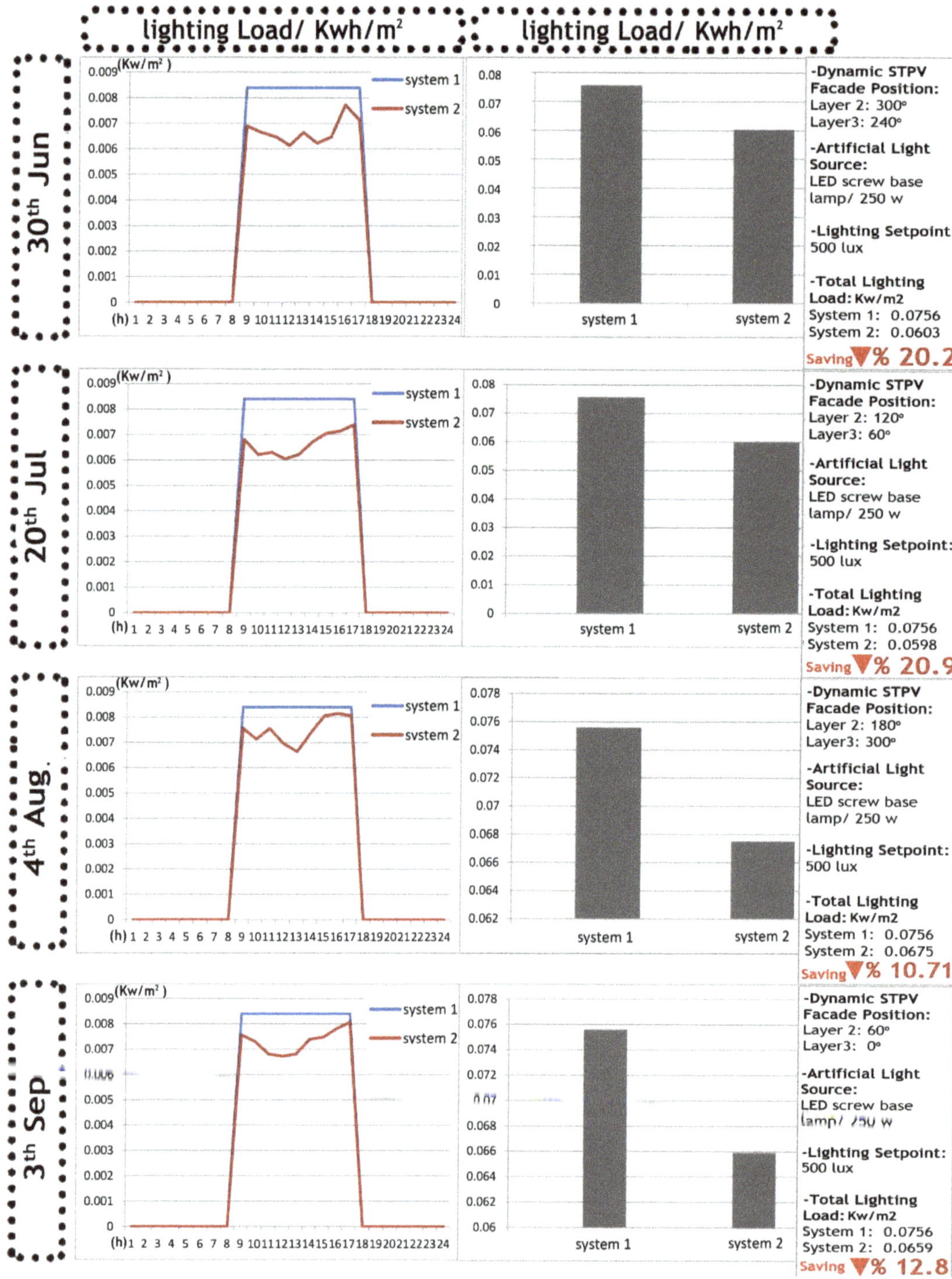

Figure 14. The hourly energy consumption in four simulation days in System 1 (occupancy sensor) and 2 (dimming lighting system).

3.3. Energy Generation

The energy generation in photovoltaic (PV) layers fluctuated based on their levels of transparency, specific PV properties (refer to Table 1), and the degree of layer orientation. To estimate PV energy generation, we employed Equation (4). Initially, the Daysim engine calculated the hourly radiation each layer could absorb at optimal positions.

Figure 15 displays the average hourly incident isolation data on PV layers set at optimal angles over a four-month period. The positions of the PV layers were optimized to enhance visual comfort, which means that the layers might not always be oriented to achieve maximum absorption and electricity generation. Table 5 shows the optimal degree of PV layers.

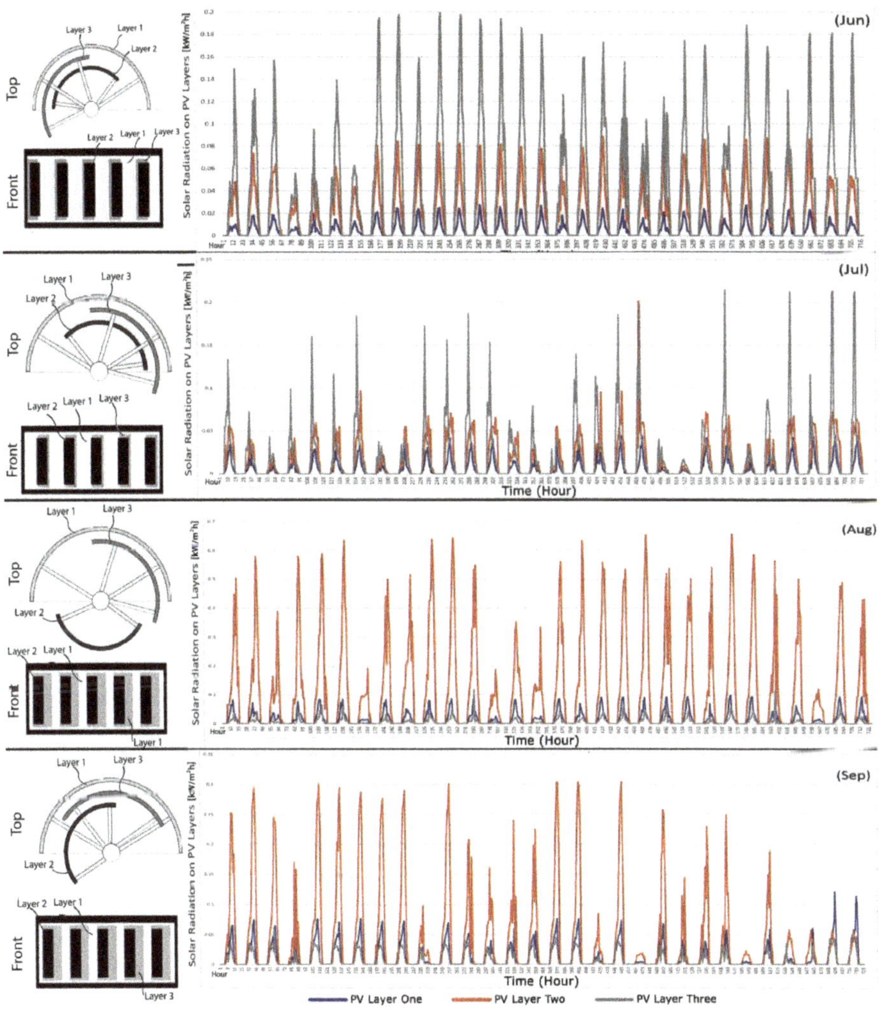

Figure 15. Left: The optimum PV layer positions. **Right**: Insolation incident on PV layers in optimum angles found in the last section of this research.

Table 5. The optimal degree of PV layer orientation.

Month	Optimal Angle, PV Layer 1 (Degree)	Optimal Angle, PV Layer 2 (Degree)	Optimal Angle, PV Layer 3 (Degree)
June	0	300	240
July	0	120	60
August	0	180	300
September	0	60	0

The degree of highly transparent PV layer 1 is zero in all positions in this study because it has minimal effect on daylighting within the space. Its design emphasizes natural ventilation, enabling it to open during favorable weather and close during winter months. However, it is important to note that while the dynamic facade system improves natural ventilation, a thorough examination of these enhancements is outside the scope of this paper. During June and July, PV layer 3 demonstrated higher absorption rates due to its position. Conversely, in August and September, the sun's angle resulted in PV layer 2 absorbing more radiation. These variations underscore the importance of considering both visual comfort and energy generation efficiency when positioning PV layers.

The total energy generation of PV layers and the energy consumption of this office room are shown in Figure 16. Typically, the total generation is used immediately by users, and storage is unnecessary. However, having an electric battery for weekends and holidays, when the office is unoccupied, can store electric generation during periods of low solar radiation or reduce peak load during the first hours of working time.

Figure 17 compares the energy consumption in the baseline office room and the office room with a dynamic PV facade system for four months. The dynamic PV facade reduces total energy consumption by 73% in June, 54% in July, 54.5% in August, and 52.55% in September. Significant savings are observed in heating and cooling, attributed to the dynamic facade system's ability to balance and control solar radiation during working hours.

In this context, the average total electricity generation of PV layers in June was 33.21 kWh, representing 12.57% of energy demand. In July, electricity generation was 36.53 kWh, equal to 16.23% of energy demand. For August, electricity generation was 18.82 kWh, accounting for 7.26% of energy demand, and in September, electricity generation was 18.15 kWh, constituting 7% of energy demand. Overall, compared to the baseline office room, the office room with a dynamic PV facade system and control strategy demonstrates energy savings of 76.4% in June, 61.45% in July, 57.8% in August, and 55.88% in September, showcasing the system's ability to enhance visual comfort, reduce energy consumption, and generate electric energy through PV layers' rotation and varying transparency levels.

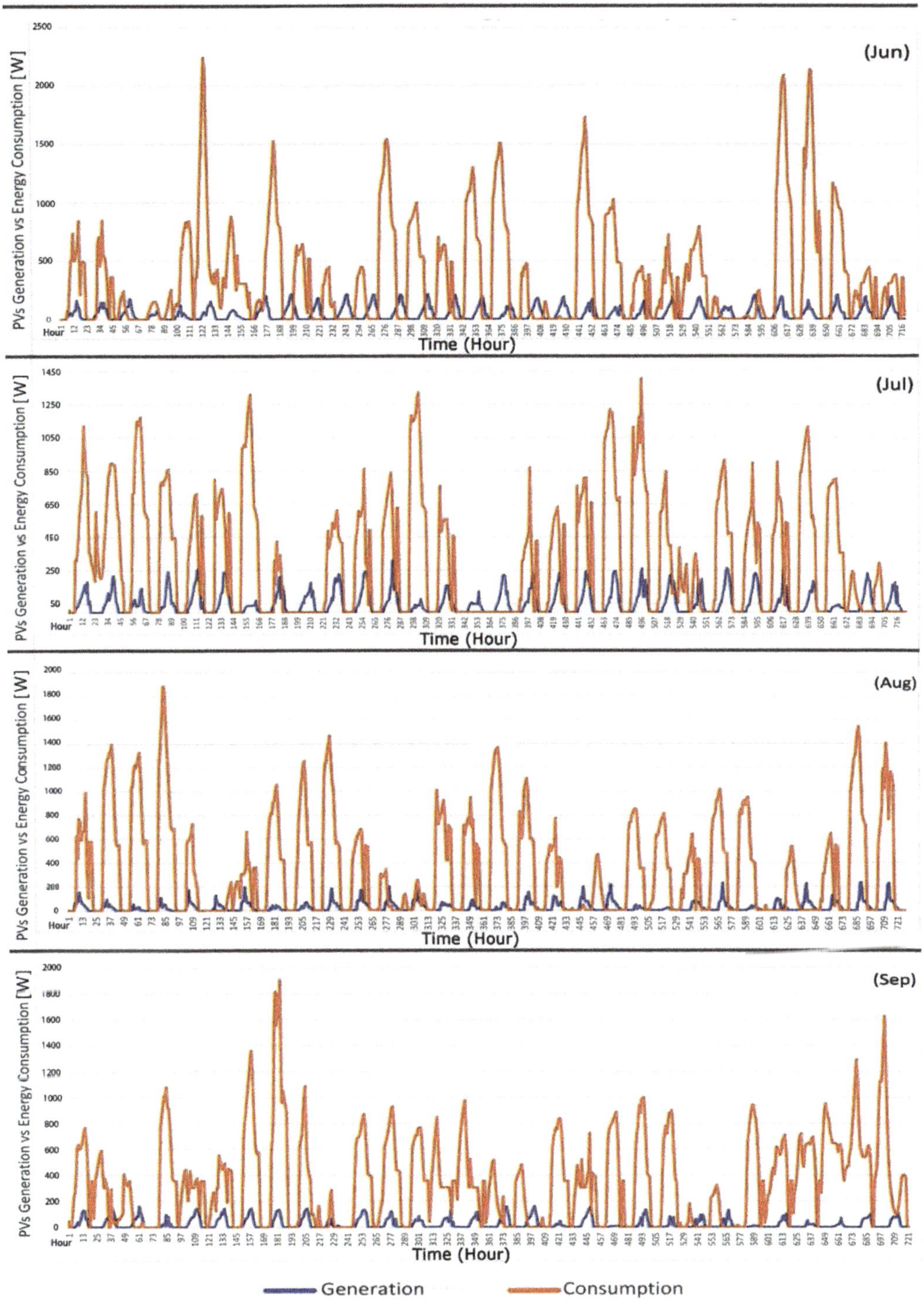

Figure 16. The hourly energy generation and consumption in four months.

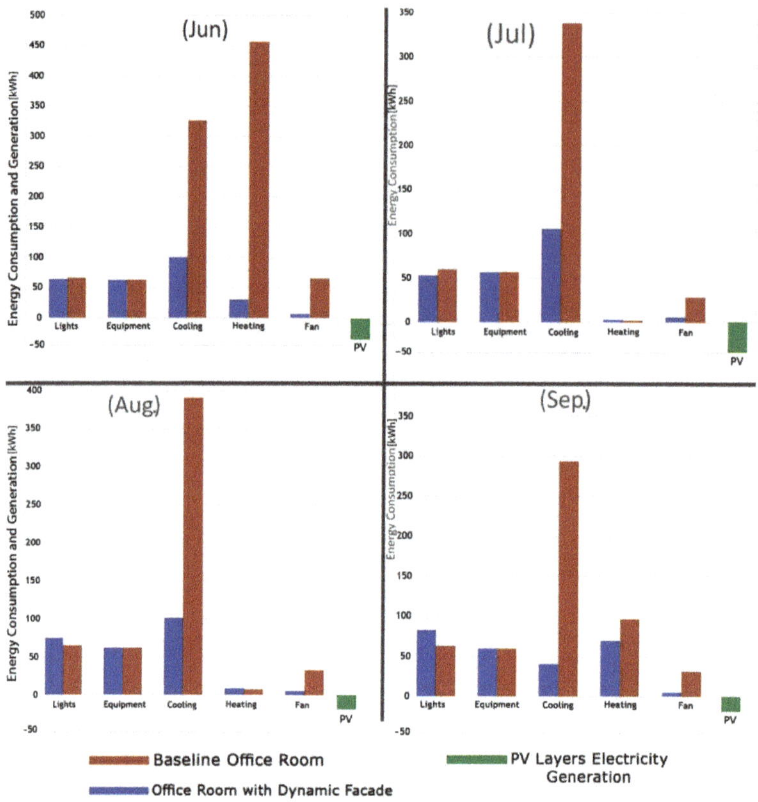

Figure 17. PV energy generation and energy consumption in the baseline office room and the office room with a dynamic PV facade system and control strategy.

4. Conclusions

Dynamic PV facades, inspired by biomimetics, have the potential to significantly improve energy efficiency, visual comfort, and electricity generation in office buildings. By mimicking natural systems that adapt to environmental changes, these facades optimize visual comfort, reduce glare, and minimize energy consumption in glazed office buildings. However, further work is needed to achieve a harmonious balance between these components.

The study shows that dynamic PV facades, demonstrated with a novel design concept, could play a crucial role in future architectural advancements by enhancing energy performance, improving indoor comfort, and contributing to sustainability. As illustrated in the simulation results, optimizing the positioning and transparency of PV layers is essential to maximize this potential. The integration of a dynamic PV facade system allows for precise regulation of daylight and solar radiation, leading to improved visual comfort, reduced energy consumption, and increased electricity generation. The design of the dynamic PV facade also incorporates features such as glare reduction, thermal comfort enhancement, and adaptable transparency levels, making it suitable for various climatic conditions and user needs. The dynamic PV façade significantly reduces energy consumption and enhances visual comfort. It achieves energy savings of 73% in June, 54% in July, 54.5% in August, and 52.55% in September, primarily by controlling solar radiation. Additionally, it generates electricity, covering 7% to 16.23% of energy demand during these months. Overall, the system demonstrates its effectiveness in reducing energy usage and producing electricity, showcasing an average energy savings of up to 76.4% compared to a baseline office room.

This research identifies several key areas for future exploration in dynamic PV facades. These include studying their performance in different climatic regions to assess adaptability and effectiveness, investigating long-term sustainability and practicality, and exploring seamless integration with building automation systems for enhanced efficiency and user experience. Additionally, researching advanced materials for PV layers, applying computational simulations and AI to optimize facade performance, and advancing control systems to improve responsiveness and energy management are crucial. Assessing cost-effectiveness and financial viability to promote broader adoption and evaluating maintenance requirements in large-scale projects are also essential.

In summary, the usefulness of dynamic PV facades lies in their ability to enhance sustainability and efficiency in building operations. They generate electricity, reduce the need for artificial lighting, and improve occupant comfort, thereby promoting innovative and sustainable building design practices rooted in the principles of biomimetics. Further research is vital for developing cost-effective and sustainable dynamic PV facades that meet environmental needs, enhance indoor comfort, and contribute to the overall energy efficiency of buildings.

Author Contributions: This paper represents the results of teamwork. Conceptualization, M.V.; methodology, investigation, resources, and data curation, M.V., A.K.A. and A.B.; writing—original draft preparation, M.V., A.K.A. and M.A.; validation, M.V.; writing—review and editing, M.V., A.K.A. and M.A. All authors have read and agreed to the published version of the manuscript.

Funding: Concordia University: bursary number N00111.

Institutional Review Board Statement: Not applicable.

Data Availability Statement: Data are contained within the article.

Conflicts of Interest: The authors declare no conflicts of interest.

References

1. European Parliament. "The Energy Performance of Buildings," Directive (EU) 2024/1275 of the European Parliament and of the Council on the Energy Performance of Buildings, Official Journal of the EuropeanUnion. 2024. Available online: https://eur-lex.europa.eu/legal-content/EN/TXT/?uri=CELEX:32024L1275 (accessed on 24 April 2024).
2. Soflaei, F.; Shokouhian, M.; Abraveshdar, H.; Alipour, A. The impact of courtyard design variants on shading performance in hot-arid climates of Iran. *J. Energy Build.* **2017**, *143*, 71–83. [CrossRef]
3. Waide, P.; Tanishima, S. *Light's Labour's Lost: Policies for Energy Efficient Lighting*; OECD/IEA: Paris, France, 2006.
4. USGBC. *LEED Reference Guide for Green Building Design and Construction*; US Green Building Council: Washington, DC, USA, 2009; Available online: https://www.usgbc.org/sites/default/files/LEED%202009%20RG%20BD+C-Supplement_GLOBAL_10_2014_Update.pdf (accessed on 7 June 2024).
5. Tabadekani, A.; Valinejad Shoubi, M.; Soflae, F.; Banihashemi, S. Integrated parametric design of adaptive facades for user's visual comfort. *Autom. Constr.* **2019**, *106*, 102857. [CrossRef]
6. Tzempelikos, A.; Athienitis, A.K. The impact of shading design and control on building cooling and lighting demand. *Sol. Energy* **2007**, *81*, 369–382. [CrossRef]
7. International Green Construction Code (IgCC). *Standard for the Design of High-Performance Green Buildings except Low-Rise Residential Buildings*; ANSI/ASHRAE/USGBC/IES 189.1; ASHRAE: Peachtree Corners, GA, USA, 2009.
8. Reynolds, J.; Rezgui, Y.; Kwan, A.; Piriou, S. A zone-level, building energy optimization combining an artificial neural network, a genetic algorithm, and model predictive control. *J. Energy* **2018**, *151*, 729–739. [CrossRef]
9. Valinejad Shoubi, M.; Ghanizadehgrayli, M.; Heidari, S. Modeling and fabrication of a kinetic solar energy-absorbing window as a green idea for sustainable future buildings. *J. Green Build.* **2018**, *13*, 145–162. [CrossRef]
10. Valinejad Shoubi, M.; Ghanizadehgrayli, M. *Multi-Functional Solar System for Generating Electricity, Useful Heat, and Ventilation in Building*; United States Patent and Trademark Office: Alexandria, VA, USA, 2019.
11. Valinejad Shoubi, M.; Valinejad Shoubi, M. Solar Wall System, The Sun-Centered Approach Toward Ecosystem. *J. Green Build.* **2013**, *8*, 39–54. [CrossRef]
12. Annunziata, E.; Frey, M.; Rizzi, F. Towards nearly zero-energy buildings: The state-of-art of national regulations in Europe. *J. Energy* **2013**, *57*, 125–133. [CrossRef]

13. Denton, J.C.; Stegou-Sagia, A.; Antonopoulos, K.; Angelopoulou, C.; Kotsiovelos, G. The impact of glazing on energy consumption and comfort. *Energy Convers. Manag.* **2007**, *48*, 2844–2852. [CrossRef]
14. Inanici, M.N.; Demirbilek, F.N. Thermal performance optimization of building aspect ratio and south window size in five cities having different climatic characteristics of Turkey. *J. Build. Environ.* **2000**, *35*, 41–52. [CrossRef]
15. Athienitis, A.K.; Tzempelikos, A. A Methodology for Simulation of Daylight Room Illuminance Distribution And Light Dimming For A Room With A Controlled Shading Device. *J. Sol. Energy* **2002**, *72*, 271–281. [CrossRef]
16. Tabadkani, A.; Banihashemi, S.; Hosseini, M.R. Daylighting and visual comfort of oriental sun responsive skins: A parametric analysis. *J. Build. Simul.* **2018**, *11*, 663–676. [CrossRef]
17. Bueno, B.; Cejudo-López, J.M.; Katsifaraki, A.; Wilson, H.R. A systematic workflow for retrofitting office façades with large window-to-wall ratios based on automatic control and building simulations. *J. Build. Environ.* **2018**, *132*, 104–113. [CrossRef]
18. CIBSE. *LG10 Daylighting-a Guide for Designers*; Society of Light and Lighting: London, UK, 2014; ISBN 9781906846480.
19. Rizala, Y.; Robandia, I.; Yuniarnoa, E.M. Daylight Factor Estimation Based on Data Sampling Using Distance Weighting. *Energy Procedia* **2016**, *100*, 54–64. [CrossRef]
20. Hafiz, D. Daylighting, Space, and Architecture: A Literature Review. *Enquiry* **2015**, *12*, 1–8. [CrossRef]
21. Kurnia, K.A.; Azizah, D.N.; Mangkuto, R.A.; Atmodipoero, R.T. Visual comfort assessment using high dynamic range images under daylight condition in the Main Library Building of Institut Teknologi Bandung Atmodipoero. Engineering Physics International Conference, EPIC 2016. *Procedia Eng.* **2017**, *170*, 234–239. [CrossRef]
22. Bellia, L.; Cesarano, A.; Iuliano, G.F.; Spada, G. Daylight glare: A review of discomfort indexes. In Proceedings of the Visual Quality and Energy Efficiency in Indoor Lighting: Today for Tomorrow, Rome, Italy, 31 March 2008.
23. Hopkinson, R.G. Glare from daylighting in buildings. *Appl. Ergon.* **1972**, *3*, 206–215. [CrossRef]
24. Park, C.-S. Occupant responsive optimal control of smart façade systems. In Proceedings of the 2003 Eighth International IBPSA Conference, Eindhoven, The Netherlands, 11–14 August 2003; pp. 1009–1016. Available online: https://publications.ibpsa.org/proceedings/bs/2003/papers/bs2003_1009_1016.pdf (accessed on 7 June 2024).
25. Attia, S.; Favoino, F.; Loonen, R.; Petrovski, A.; Monge-Barrio, A. Adaptive façades system assessment: An initial review. In Proceedings of the 10th Conference on Advanced Building Skins, Bern, Switserland, 3–4 November 2015; pp. 1265–1273.
26. Lee, E.S.; Selkowitz, S.E. *Design and Evaluation of Integrated Envelope and Lighting Control Strategies for Commercial Buildings*; LBL Publications: Berkeley, CA, USA, 1995; Available online: https://escholarship.org/uc/item/3xf9t5rf (accessed on 10 June 1994).
27. Morad, R.A.; Anca, D.G. Energy performance of daylight-linked automatic lighting control systems in large atrium spaces: Report on two field-monitored case studies. *Energy Build.* **2003**, *35*, 441–461. [CrossRef]
28. Zhou, J.; Tan, T.C.; Gao, C.P.; Valliappan, S.; Goh, A.; Seoh, A.; Babu, S.; Lamano, A.; Zhang, Z.; Sarvaiya, J.; et al. Development of a rotatable outdoor testbed and the testing of an integrated auto-dimming lighting and automated blind system in the tropics. In Proceedings of the World Sustainable Built Environment Conference (WSBE 2017): Innovative Processes and Methodologies to Transform SBE, Hong Kong, China, 5–7 June 2017; Available online: http://www.irbnet.de/daten/iconda/CIB_DC30359.pdf (accessed on 7 June 2024).
29. Carlucci, S.; Causone, F.; De Rosa, F.; Pagliano, L. A review of indices for assessing visual comfort with a view to their use in optimization processes to support building integrated design. *J. Renew. Sustain. Energy Rev.* **2015**, *47*, 1016–1033. [CrossRef]
30. Wienold, J.; Christoffersen, J. Towards a new daylight glare rating. In Proceedings of the 10th European Lighting Conference, Lux Europa 2005, Berlin, Germany, 19–21 September 2005; pp. 157–161. Available online: https://www.researchgate.net/publication/313608753_Towards_a_new_daylight_glare_rating (accessed on 7 June 2024).
31. Miyazaki, T.; Akisawa, A.; Kashiwagi, T. Energy savings of office buildings by the use of semi-transparent solar cells for windows. *Renew Energy* **2005**, *30*, 281–304. [CrossRef]
32. Namini, S.B.; Shakouri, M.; Tahmasebi, M.M.; Preece, C. Managerial sustainability assessment tool for Iran's buildings. *Proc. ICE-Eng. Sustain.* **2014**, *167*, 12–23. [CrossRef]
33. Banihashemi, S.; Tabadkani, A.; Hosseini, M.R. Integration of parametric design into modular coordination: A construction waste reduction workflow. *Autom. Constr.* **2018**, *88*, 1–12. [CrossRef]
34. Nagy, Z.; Svetozarevic, B.; Jayathissa, P.; Begle, M.; Hofer, J.; Lydon, G.; Willmann, A.; Schlueter, A. The adaptive solar facade: From concept to prototypes. *Front. Archit. Res.* **2016**, *5*, 143–156. [CrossRef]
35. Roudsari, M.S.; Pak, M.; Smith, A. Ladybug: A parametric environmental plugin for Grasshopper to help designers create an environmentally-conscious design. In Proceedings of the 13th International IBPSA Conference, Lyon, France, 26–28 August 2013; pp. 3128–3135. Available online: http://www.ibpsa.org/proceedings/bs2013/p_2499.pdf (accessed on 7 June 2024).
36. Sawicz, D. *Hobby Servo, Fundamental*; Princeton University: Princeton, NJ, USA, 2012; Available online: https://www.princeton.edu/~mae412/TEXT/NTRAK2002/292-302.pdf (accessed on 7 June 2024).
37. Soori, P.K.; Vishwas, M. Lighting control strategy for energy efficient office lighting system design. *Energy Build.* **2013**, *66*, 329–337. [CrossRef]
38. International Standard. Photovoltaic Devices–Part 1: Measurement of Photovoltaic Current-Voltage Characteristics. IEC 60904-1. 2006. Available online: https://standards.iteh.ai/catalog/standards/iec/2eaa7aa1-7f28-437b-b87f-a8c659d418c4/iec-60904-1-2006 (accessed on 12 September 2006).
39. Banihashemi, S.; Tabadkani, A.; Hosseini, M.R. Modular Coordination-based Generative Algorithm to Optimize Construction Waste. *Procedia Eng.* **2017**, *180*, 631–639. [CrossRef]

40. Baker, N. "Modelling and Analysis of Daylight, Solar Heat Gains and Thermal Losses to Inform the Early Stage of the Architectural Process." Degree Project in Building Services And Energy Systems. KTH Royal Institute of Technology, School of Architecture and the Built Environment. 2017. Available online: https://kth.diva-portal.org/smash/get/diva2:1181701/FULLTEXT01.pdf (accessed on 7 June 2024).
41. Reich, N.H.; Mueller, B.; Armbruster, A.; Van Sark, W.G.; Kiefer, K.; Reise, C. Performance ratio revisited: Is PR>90% realistic? *Prog. Photovolt. Res. Appl.* **2012**, *20*, 717–726. [CrossRef]
42. Reinders, H.M.E.; van Dijk, V.A.P.; Wiemken, E.; Turkenburg, W.C. Technical and economic analysis of grid-connected PV systems by means of simulation. *Prog. Photovolt. Res. Appl.* **1999**, *7*, 71–82. [CrossRef]
43. Swain, A. Solar Energy Generation Potential on National Highways. *Int. J. Energy Environ. Eng.* **2013**, *4*, 16. Available online: https://link.springer.com/article/10.1186/2251-6832-4-16 (accessed on 4 April 2013).

biomimetics

Article

Mimosa Kinetic Façade: Bio-Inspired Ventilation Leveraging the Mimosa Pudica Mechanism for Enhanced Indoor Air Quality

Sukhum Sankaewthong [1,2,*], Kazunori Miyata [1,*], Teerayut Horanont [2], Haoran Xie [1] and Jessada Karnjana [3]

1 Japan Advanced Institute of Science and Technology, Nomi 923-1211, Ishikawa, Japan; xie@jaist.ac.jp
2 School of Information, Computer and Communication Technology, Sirindhorn International Institute of Technology, Thammasat University, Bangkok 10200, Pathumthani, Thailand; teerayut.horanont@gmail.com
3 National Electronics and Computer Technology Centre (NECTEC), Bangkok 10400, Pathumthani, Thailand; jessada.karnjana@nectec.or.th
* Correspondence: sukhum.sanka@dome.tu.ac.th or sukhumphed@gmail.com (S.S.); miyata@jaist.ac.jp (K.M.)

Abstract: In light of pressing global health concerns, the significance of indoor air quality in densely populated structures has been emphasized. This research introduces the Mimosa kinetic façade, an innovative design inspired by the adaptive responsiveness of the Mimosa plant to environmental stimuli. Traditional static architectural façades often hinder natural ventilation, leading to diminished air quality with potential health and cognitive repercussions. The Mimosa kinetic façade addresses these challenges by enhancing effective airflow and facilitating the removal of airborne contaminants. This study evaluates the façade's impact on quality of life and its aesthetic contribution to architectural beauty, utilizing the biomimicry design spiral for a nature-inspired approach. Computational simulations and physical tests were conducted to assess the ventilation capacities of various façade systems, with a particular focus on settings in Bangkok, Thailand. The study revealed that kinetic façades, especially certain patterns, provided superior ventilation compared to static ones. Some patterns prioritized ventilation, while others optimized human comfort during extended stays. Notably, the most effective patterns of the kinetic façade inspired by the Mimosa demonstrated a high air velocity reaching up to 12 m/s, in contrast to the peak of 2.50 m/s in single-sided façades (traditional façades). This highlights the kinetic façade's potential to rapidly expel airborne particles from indoor spaces, outperforming traditional façades. The findings underscore the potential of specific kinetic façade patterns in enhancing indoor air quality and human comfort, indicating a promising future for kinetic façades in architectural design. This study aims to achieve an optimal balance between indoor air quality and human comfort, although challenges remain in perfecting this equilibrium.

Keywords: kinetic façade; biomimicry; air ventilation; *Mimosa pudica*; air change rate

1. Introduction

In the wake of recent global health challenges, the importance of understanding airborne transmission within densely populated structures such as office towers has been highlighted [1]. This heightened awareness has precipitated a paradigm shift, emphasizing the paramount importance of indoor air quality. Emerging from this exigency is the innovative Mimosa kinetic façade, a design inspired by the adaptive nature of the *Mimosa pudica* plant.

The *Mimosa pudica*, which is renowned for its outstanding responsiveness to environmental stimuli, offers a natural model for the adaptation of the kinetic façade. Traditional architectural features frequently obstruct natural ventilation because of their static façade [2,3]. This results in problems such as reduced air quality, which can worsen respiratory conditions and decrease mental abilities. Due to its dynamic adaptability, the Mimosa kinetic façade has the potential to completely transform this area by providing effective airflow, eliminating airborne contaminants, and keeping a consistent indoor temperature. In addition to its practical advantages, this façade provides a distinctive architectural component that gives buildings a unique aesthetic identity. Cross ventilation is emphasized in this design and is a crucial strategy for improving airflow, especially in areas with high occupancy. This reduces dependency on mechanical systems by purifying the air and assisting in temperature control [4,5].

This study underscores the pivotal role of a building's façade in enhancing the quality of life for its occupants. The primary emphasis is on how the effectiveness of the façade contributes to the wellness and wellbeing of those living within the space. As a subsequent outcome, an effective façade not only ensures the comfort and health of its inhabitants, but also bestows the building with architectural beauty. The project uses the biomimicry design spiral, a process drawn from nature, to imagine a façade that promotes cross ventilation [6,7]. By imitating natural processes, this methodical approach ensures innovative, efficient, and sustainable design outcomes.

This research addresses the critical need to mitigate airborne transmission in indoor environments. It underscores the importance of kinetic façades, particularly those inspired by the Mimosa plant. These façades not only improve indoor air quality but also contribute a unique aesthetic to buildings. Transitioning from innovative concepts to essential elements, these façades are increasingly recognized as key to ensuring safer indoor environments amid escalating airborne health challenges.

2. Literature Review

2.1. Related Papers

In this section, we synthesize the research gaps identified in various studies pertaining to both façade design and ventilation, with the objective of contextualizing the contributions of the Mimosa kinetic façade research. This synthesis is instrumental in bridging the identified gaps, thereby underpinning the significance of the proposed research.

Façade Design Research: In the realm of façade design and building architecture, numerous studies have pinpointed specific areas that necessitate further inquiry. The research on "Post-Occupancy Evaluation for Adaptive Façades [8]" underscores a notable absence of methodologies for assessing user interaction and satisfaction. Similarly, the study "Access to Daylight and View in Office Spaces [9]" acknowledges constraints arising from limited, potentially homogeneous participant groups, advocating for more inclusive and longitudinal investigations. A review on "Kinetic Façade Design for Visual and Thermal Comfort [7]" draws attention to the dearth of empirical data and the imperative for environmental impact evaluations. The paper titled "Performance Assessment of Adaptive Façade Systems [10]" identifies a lacuna in economic analysis and an incomplete understanding of the influence of user behavior on façade performance. Furthermore, the study "Adaptive Façade Systems [11–13]" emphasizes the paucity of information regarding market demands and the complexities entailed in integrating structural and energy aspects.

Moreover, research such as "Kinetic Facades in Hot, Dry Climates [14]" and "High Performance Building Façade Solutions [15]" exposes geographical focus limitations and the challenges of integrating these systems with existing building infrastructures, respectively. The paper "Adaptive Façade Definitions and Terms [16]" reveals an absence of comprehensive empirical data and a lack of focus on long-term sustainability. In "Solar Cooling Integrated Façades [17]", researchers note a scarcity of real-world data and economic analysis. The study "Manufacturing Adaptive Façades with Standard Products [18]" points to a narrow focus on specific case studies and a dearth of real-world performance data.

Research studies on "Active Dynamic Windows for Buildings [19]" and "Automated Dynamic Facades—User Satisfaction and Interaction [20]" both identify gaps in real-world performance data and the need for extensive studies on long-term user perceptions. The paper on "Innovative Responsive Façade Elements Using Building Performance Simulation [21]" discusses a theoretical focus that lacks practical application data. Studies such as "Dynamic Façade Shading Typologies [22]" and "Transparent Building Envelopes—Innovative Technologies [23]" highlight limitations in shading position considerations and characterizations of dynamic performance, respectively.

Additionally, "Performance Prediction with Responsive Building Elements (RBEs) [24]" and "Energy Performance of Multiple-Skin Facades [10]" emphasize a focus on specific case studies and geographical limitations. The article on "Responsive Carrier Component Envelope (RCCE) [25]" and the paper "Geometry in Shading Systems and Biological Role Models [6]" accentuate structural challenges and theoretical discussions devoid of empirical validation. Papers such as "Thermal Performance of Active Envelopes [26]" and "Kinetic Photovoltaic Architecture in Performative Design [27]" discuss issues related to practical feasibility and long-term durability. "Climate Adaptive Building Shells (CABS) [28]" and a second study on "Automated Dynamic Facades—User Interaction [20]" reveal gaps in economic analysis and context-specific findings.

Finally, the study "Biology and Architecture Hybridization in French Architectural Offices [29]" indicates geographic limitations and a qualitative emphasis. Collectively, these gaps across various studies underscore the need for a multidisciplinary approach integrating user-centric design, technological innovation, environmental sustainability, and economic feasibility, which are vital for advancing the field of façade design and building architecture.

Ventilation Research: The selected studies on ventilation in the context of COVID-19 offer valuable insights but also present several limitations. For example, the study on infection risks in a Shenzhen outpatient building is confined to a specific healthcare context [30] and overlooks variables such as ventilation systems and occupant behavior. Research focusing on ventilation and air disinfection in office buildings [31] is predominantly simulation-based, lacking a thorough assessment of cost, safety, and environmental impact. The study on double-glazed façades with louvers in Isfahan, Iran [32] is geographically specific and reliant on static simulations. Investigations into cough droplet dynamics in hospital isolation rooms [33] and the role of ventilation in a German nursing home during a COVID-19 outbreak [34] are limited by their dependence on computational models and a lack of definitive evidence, respectively. The study of aerosol control in gyms during the pandemic [35] is hindered by a small sample size and does not conclusively link aerosol reduction to a decreased transmission risk. The review focusing on the physical office workplace and mental health [36] emphasizes the necessity for more comprehensive research methodologies. Agent-based simulations for epidemic risk assessment [37] and ventilation recommendations for COVID-19 mitigation [38] lack real-world validation and specificity. Finally, strategies for improving indoor air quality in HVAC systems during the COVID-19 pandemic [39] and the impact of airborne fine particle pollution on natural ventilation in Asian megacities [40] are constrained by their lack of practical case studies and geographical focus. These studies highlight the importance of broader, more practical, and diverse research for enhancing the effectiveness of ventilation strategies in various settings during pandemics.

Mimosa Kinetic Façade: A Convergence of Research Gaps: The "Mimosa Kinetic Façade: Bio-Inspired Ventilation Leveraging the Mimosa Pudica Mechanism for Enhanced Indoor Air Quality" addresses several identified gaps in both façade design and ventilation research. Regarding façade design, this biomimetic kinetic façade, inspired by the *Mimosa pudica* plant, not only considers aesthetic and structural factors, but also dynamically responds to environmental stimuli to optimize natural ventilation. This directly addresses the need for empirical data, environmental impact assessments, the integration of structural and energy aspects, and real-world performance data. The design's responsiveness to environmental conditions positions the Mimosa kinetic façade as a user-centric and environmentally attuned solution.

In ventilation research, which is particularly relevant during the COVID-19 pandemic, the Mimosa kinetic façade introduces an innovative approach to natural ventilation, potentially mitigating airborne contaminants indoors. This addresses concerns about the lack of comprehensive assessments of ventilation systems and occupant behavior, the over-reliance on computational models, and the need for more practical and diverse research to enhance ventilation effectiveness. Utilizing the natural folding and unfolding mechanism of the *Mimosa pudica*, the façade design aims to improve air change rates and reduce CO_2 levels, which are critical for addressing airborne transmission risks.

Furthermore, this research article details both computational simulations and physical tests to evaluate the façade's impact on indoor air quality, responding to the call for more empirical data and the validation of theoretical models. This approach aligns with the demand for multidisciplinary research that not only considers user comfort and satisfaction, but also encompasses technological innovation and environmental sustainability.

In summary, this research article presents a holistic solution that fills identified research gaps by integrating responsive design elements, advancing empirical validation, and enhancing ventilation strategies, which are all crucial for the progression of façade design and the broader field of building architecture.

2.2. Ventilation Effect (Cross Ventilation)

The concept of ventilation, particularly cross ventilation, has long been recognized as an effective method for enhancing indoor air quality. Cross ventilation allows air to flow seamlessly from one side of a structure to another, replacing indoor air with fresher outdoor air as shown in Figure 1. This natural process not only disperses airborne contaminants such as dust and allergens, but also prevents the buildup of stagnant air, ensuring a healthier environment [41].

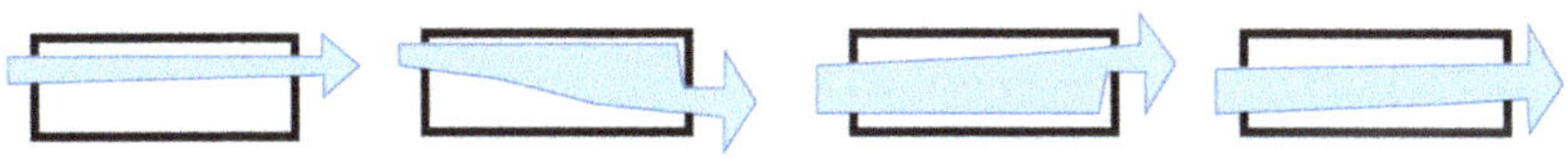

Figure 1. Impact of opening locations on cross-ventilation effect.

Drawing inspiration from nature, the research introduces a kinetic façade modeled after the *Mimosa pudica*, a plant known for its responsive movements. Just as the *Mimosa pudica* reacts to external stimuli by folding its leaves, the kinetic façade is designed to adapt to changing environmental conditions, optimizing the flow of air into the working space [2,3].

By integrating the principles of cross ventilation into this innovative façade design, several benefits are realized. The continuous flow of air is maintained, ensuring that airborne particles are effectively dispersed. As a passive ventilation method, the façade is energy efficient, reducing the reliance on mechanical systems that might recirculate contaminants. Furthermore, the design aids in maintaining lower indoor CO_2 levels, which is crucial for the wellbeing and productivity of occupants. The kinetic façade also addresses humidity concerns by allowing moist indoor air to be replaced with drier outdoor air, preventing mold growth and further enhancing air quality [41,42].

Beyond air quality, this kinetic façade, which is inspired by *Mimosa pudica* and utilizes cross ventilation, offers thermal benefits. It naturally cools the working space, reducing the need for energy-intensive air conditioning systems and providing a comfortable environment for occupants [43].

Enhancing Building Ventilation: The Synergy of EBC Annex 62 and the Mimosa Kinetic Façade

The integration of EBC Annex 62 principles into the cross-ventilation concept, particularly via the application of the Mimosa kinetic façade inspired by the *Mimosa pudica* plant, significantly enhances the efficacy of natural ventilation strategies in architectural design. Annex 62 underscores the critical role of ventilative cooling, a concept that is seamlessly actualized via the dynamic adaptability inherent in the Mimosa kinetic façade. This façade, mirroring the responsive movements of the *Mimosa pudica*, is adept at optimizing airflow in response to varying environmental conditions, a strategy that resonates with the core tenets of Annex 62. It dynamically modulates in order to enable cross ventilation, facilitating the unimpeded flow of fresh air throughout the building, thereby supplanting the indoor air with a cleaner, outdoor counterpart. This mechanism is in direct alignment with the objectives of Annex 62, which advocates for the utilization of natural ventilation as a means to enhance indoor air quality while concurrently reducing energy consumption [44].

Incorporating the VC tool, as delineated in Annex 62, allows for precision tuning of the façade's functionality, leveraging real-time climatic data and occupancy patterns. This tool is instrumental in evaluating the effectiveness of ventilative cooling, directing the façade's adjustments to optimize cross ventilation under favorable conditions. Such strategic operation ensures the maintenance of a comfortable indoor environment, characterized by diminished levels of airborne particles and CO_2, and effectively addresses issues of humidity and thermal comfort.

Moreover, the amalgamation of Annex 62 principles with the Mimosa kinetic façade fosters the development of energy-efficient buildings. This synergy significantly reduces the dependence on mechanical ventilation and air conditioning systems, thereby contributing to a reduction in energy consumption and operational expenses. This approach not only augments indoor air quality and occupant comfort, but also aligns with the sustainable building practices and energy efficiency standards emphasized in Annex 62.

In essence, the application of EBC Annex 62's ventilative cooling strategies via the Mimosa kinetic façade presents a sophisticated, biomimetic solution to the challenges of indoor air quality and thermal comfort in modern buildings. This integration culminates in edifices that are not only conducive to the health and comfort of occupants, but are also emblematic of energy efficiency and environmental stewardship.

2.3. Human Comfort

"Human comfort" is a multifaceted term that encapsulates the physical and psychological wellbeing of individuals in relation to their environmental conditions. It includes thermal comfort, influenced by temperature, humidity, and air movement; visual comfort, concerning the adequacy of lighting and glare reduction; acoustic comfort, focusing on ambient noise levels and their impact on activities; ergonomic comfort, relating to the physical interaction with the environment; and psychological comfort, which involves feelings of safety, privacy, and overall environmental satisfaction. This concept is integral to the fields of ergonomics, environmental psychology, and architectural design, with the objective of enhancing health, productivity, and wellbeing in both living and working spaces [45].

Central to this study is the aspect of thermal comfort, particularly the role of wind speed in indoor environments. Research indicates that a wind speed range of 1 to 5 m/s is considered comfortable, balancing air movement to avoid feelings of stagnation while preventing discomfort from drafts or excessive cooling. Below 1 m/s, air can feel stuffy and lead to increased perceived temperatures and discomfort, while speeds above 5 m/s can be uncomfortable or even dangerous, causing drafts, cooling, and disturbances. Therefore, maintaining indoor air movement akin to an outdoor wind speed of 1 to 5 m/s is crucial for optimal human comfort [46] as shown in Table 1. Maintaining a balance between indoor environmental quality and energy efficiency is crucial in the design of heating, ventilation, and air conditioning (HVAC) systems, as well as in planning natural ventilation in buildings. This balance is essential for creating environments that promote health and wellbeing. In the context of Bangkok, Thailand, this consideration becomes even more significant due to the city's tropical monsoon climate. Bangkok is characterized by year-round warm temperatures, typically ranging from 25 °C to 35 °C, and high humidity levels, often exceeding 70%. The city undergoes a rainy season from May to October, influenced by the southwest monsoon, which brings heavy rainfall and stronger winds. Conversely, the dry season, from November to April, is marked by lighter winds. These climatic factors, including moderate wind speeds averaging 2 to 5 m per second, play a significant role in influencing the design and effectiveness of kinetic façades in the region. Such façades are integral to achieving the optimal balance in indoor climate control, ensuring that buildings are both comfortable and energy efficient.

Table 1. The relationship between wind speed and human comfort.

Wind Speed Range (m/s)	Human Body Feeling
<1.0	Breezeless
1.0~5.0	Comfortable
5.0~10.0	Uncomfortable, movement affected
10.0~15.0	Very uncomfortable, movement greatly affected
15.0~20.0	Intolerable
>20.0	Dangerous

2.4. Indoor Air Quality (IAQ)

Indoor air quality (IAQ) is crucial in enclosed spaces such as homes and offices, particularly during health crises when traditional ventilation methods may be inadequate [38,39]. This study explores the development of advanced façades that modulate ventilation rates based on room size and occupancy, with a focus on maintaining CO_2 levels below 700 ppm to optimize air quality [47] as shown in Table 2. These façades, which blend contemporary design with sustainable architecture, either enhance natural ventilation, integrate with HVAC systems, or dynamically adjust as kinetic façades using real-time data.

Table 2. Indoor air quality index.

IAQ Index	
CO_2 (ppm)	Hazard Level
<700	Good
800	Moderate
1100	Poor
1500	Unhealthy
2000	Very Unhealthy
3000	Hazardous
>5000	Extreme

A critical measure for evaluating these façades is the air change rate (ACR), quantified in air changes per hour (ACH). The ACR is essential for assessing the efficacy of ventilation in diluting airborne particles, including viruses such as COVID-19. For example, an ACH of 1 indicates complete air replacement within an hour. The American Society of Heating, Refrigerating, and Air-Conditioning Engineers (ASHRAE) recommends a minimum outdoor air ventilation rate of 5 cubic feet per minute per person, or 0.06 per square foot for offices [48,49], with residential buildings typically requiring an ACR of 0.35. However, office spaces may necessitate an ACH of 2–3, particularly for mitigating airborne diseases [50].

The study involves measuring the ACR of kinetic façades using methods such as direct airflow rate measurements and computational fluid dynamics simulations. Comparing the achieved ACR against ASHRAE standards is crucial. If kinetic façades yield higher ACR values, enhanced ventilation is indicated, potentially reducing airborne contaminants and significantly improving IAQ. This research underscores the importance of integrating an innovative façade design in building architecture to promote healthier indoor environments.

3. Methodology

3.1. Research Methodology Overview

Figure 2 illustrates the research methodology used for this study on kinetic façades inspired by the *Mimosa pudica* plant. It is structured into six distinct parts:

1. Biomimicry Method: This involves creating the kinetic façade by emulating the physical movement and behavior of the *Mimosa pudica*. The façade's design is derived using biomimicry principles.
2. Computational Design Method: This focuses on façade optimization using Rhino 7 for 3D modeling and Grasshopper for environmental analysis (Climate Studio), as well as air velocity and CO_2 concentration analysis (Eddy 3D and OpenFOAM (version2212)). This step also includes optimizing façade patterns.
3. Validation: This uses a simulated box space (378 cm × 480 cm × 250 cm) to validate different façade types, including the Mimosa-pudica-inspired kinetic façade, assessing their impact on ventilation and indoor CO_2 levels.
   The room dimensions of 378 cm × 480 cm × 250 cm, totaling 27.216 sq.m, were chosen for the simulation for several reasons. Firstly, these dimensions closely resemble the common room sizes found in real-world settings, such as offices, enabling the simulation to realistically depict practical scenarios. Secondly, the ample size of the room is crucial for evaluating the effectiveness of various façade designs in enhancing ventilation and reducing indoor CO_2 levels. Thirdly, these specific dimensions are optimized for computational modeling, achieving a balance between realistic space representation and computational resource efficiency. Fourthly, the room is spacious enough to clearly demonstrate the effectiveness of different façade designs in improving ventilation, especially the Mimosa-pudica-inspired kinetic façade. Lastly, the insights obtained from simulations conducted in this room will provide a solid baseline for future comparative analyses with different room dimensions, contributing to a comprehensive understanding of the impact of façade designs on ventilation. This meticulous selection of room dimensions ensures that the simulation results will be both highly relevant and applicable, thereby enhancing the overall validity and utility of the study.
4. Wind Analysis: This examines how air velocity affects human comfort and the air change rate (ACR) to evaluate the kinetic façade's effectiveness in reducing airborne contaminants.
5. Initial Summary: Post-simulation, a prototype is developed, incorporating materials and mechanisms such as capacitive touch sensors, servo motors, and an Arduino board. The prototype undergoes validation in a wind tunnel, focusing on the CO_2 rate, air velocity, airborne particle concentration, and ventilation time.
6. Comparison: This compares simulation results with real-world test outcomes to conclude the kinetic façade's efficacy in enhancing air ventilation and indoor air quality.

This methodology combines computational and experimental approaches to assess the potential of a biomimicry-inspired kinetic façade in improving indoor environmental conditions.

3.2. Variable

This research on the kinetic façade, inspired by the *Mimosa pudica* concept, comprises computational simulations and actual testing. The simulations explore a wide range of variables, while the actual tests prioritize façade selections from simulations in order to compare them with static façade cases. Both methods aim to assess the façade's efficacy in enhancing ventilation and indoor air quality. The variables for the two sections are shown in Table 3:

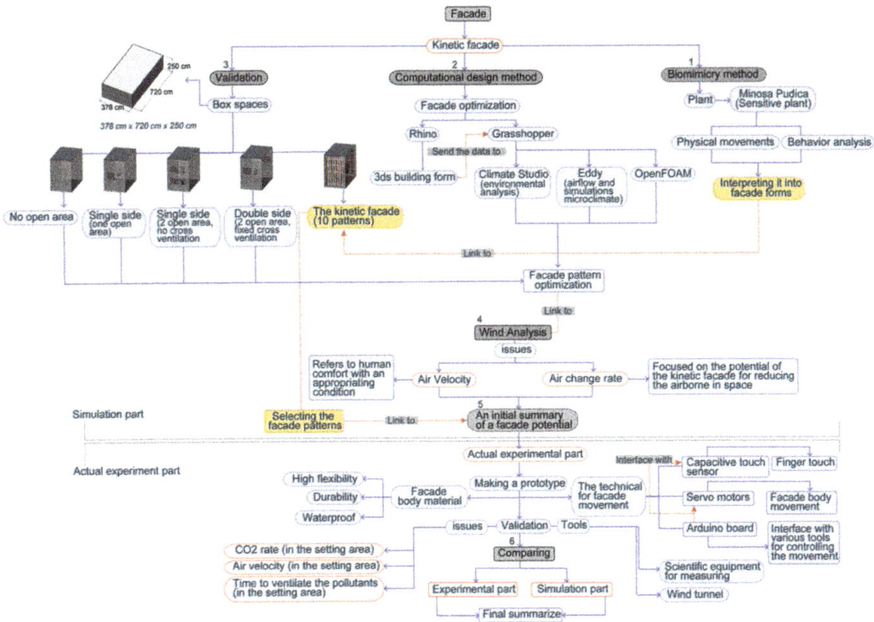

Figure 2. Flowchart illustrating the research methodology employed in this study.

Table 3. Variables used in computational simulation (all façade types) and actual testing (*Mimosa pudica* façade concept) for assessing the potential of kinetic façades.

Simulation Variable	Actual Experiment Variable
1. Independent variable	1. Independent variable
1.1 Façade types	1.1 Façade patterns
1.2 Box size: 378 cm × 720 cm × 250 cm	
2. Dependent variable	2. Dependent variable
2.1 Air velocity: The speed of indoor air movement influencing human comfort and airborne contaminant reduction.	2.1 CO_2 rate (in the setting area)
2.2 Air change rate (ACR): The rate of outdoor air replacement in a workspace, vital for indoor air quality and reducing health risks.	2. Air velocity (in the setting area)
	3. Potential for airborne particle ventilation (in the setting area)
3. Control variable	3. Control variable
3.1 The number of occupants	3.1 The level of airborne particles
3.2 External Conditions: This research considers the impact of outdoor factors such as temperature, humidity, and wind on the performance of kinetic façades, specifically in the context of Bangkok, Thailand's weather conditions.	3.2 The actual experiment was conducted in a controlled environment in Bangkok, Thailand.
3.3 Building type: Office or workspaces used to assess the kinetic façade's potential.	

3.3. Simulation Part: The Façade Type for Simulation

The simulation of the kinetic façade inspired by the *Mimosa pudica* is an essential step before moving to prototype testing, serving multiple key functions. It provides a preliminary evaluation of the façade's effectiveness in reducing airborne contaminants, allowing for optimization and risk mitigation. Additionally, it saves time and costs by allowing for digital refinements, leading to better-informed decisions regarding the design and performance. The research considers various aspects of airflow within the workspace, dividing the section plane into six positions, as shown in Figure 3, and focusing on elements such as airflow velocity and air change rate (ACR) for a comprehensive understanding of the façade's potential performance.

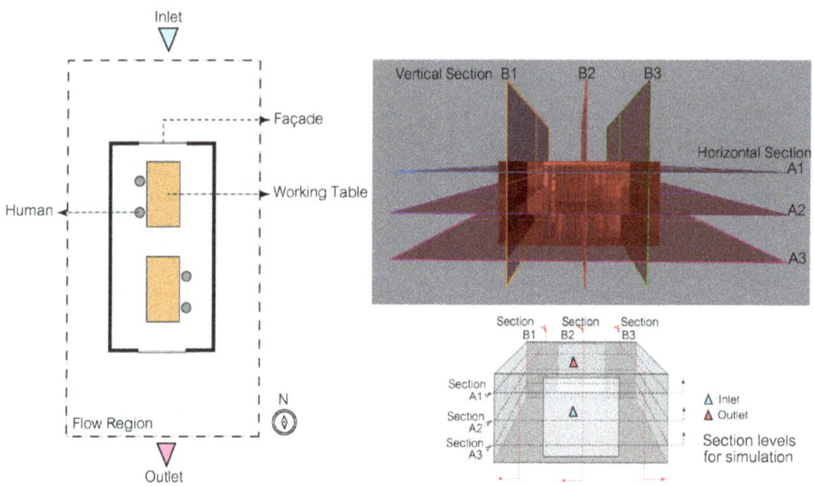

Figure 3. Determined plan and section for simulation.

In this study, two façade designs were compared: a standard version, as depicted in Figure 4 (A. Base cases), and another design inspired by the Mimosa plant that showcases ten different patterns, as also shown in Figure 4 (B. Kinetic façade cases). The relationship between the façade's physical appearance and its code is defined by the kinetic façade codes. These codes, used in the Grasshopper software , dictate how the kinetic façade module opens, influencing the cross-ventilation effect. Different codes result in varied façade patterns, as reflected in their physical appearance. The research aimed to assess the kinetic façade's influence on air circulation. Emphasizing cross ventilation, where cool air enters one side and warm air exits the opposite, is essential for air quality. Façade patterns on the north and south should be staggered, not grid-aligned, to enhance airflow. The design must also account for factors such as wind direction, sun orientation, and weather to ensure optimal ventilation and a comfortable workspace.

3.4. Actual Experiment Section: The Static vs. Kinetic Façades

- The static façades (A–D)—base cases: These represent traditional or conventional façade designs as shown in Figure 5. Testing these designs provides a baseline against which the performance of kinetic façades can be compared. It is essential to understand how much improvement or difference kinetic façades offer over traditional designs.
- The kinetic façades (E–I)—bio-inspired façades: These are dynamic façades whose designs were selected based on simulations as shown in Figure 5. The idea is to evaluate if the simulated potential of these façades translates effectively into real-world scenarios.

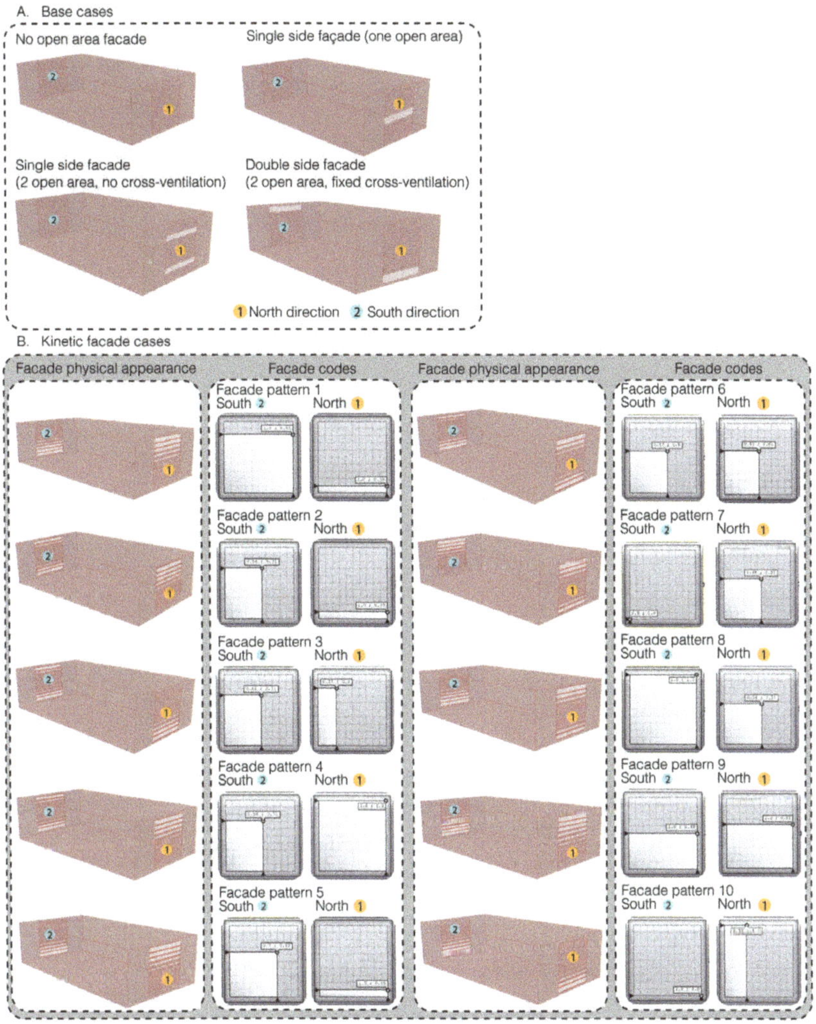

Figure 4. Comparative visuals of static façades ((**A**) Base cases) and bio-inspired façades ((**B**) Kinetic façade cases): this illustration contrasts established façade types with ten distinct kinetic patterns, each inspired by the adaptive *Mimosa pudica* plant.

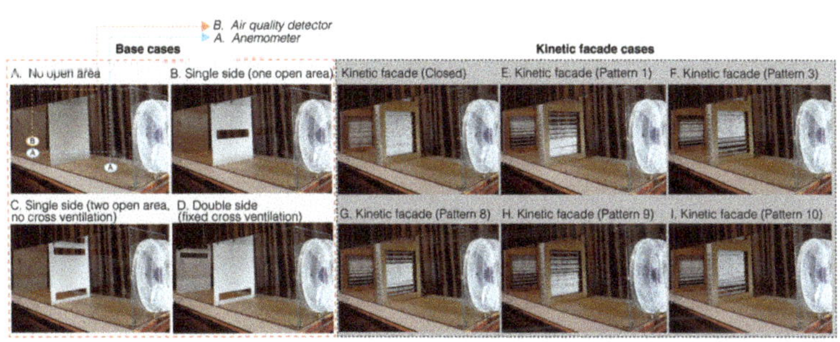

Figure 5. Experimental patterns of the static and kinetic façades.

4. Biomimicry Method

4.1. Applying the Biomimicry Approach to the Concept of a Kinetic Façade

Biomimicry draws inspiration from nature to develop sustainable solutions for challenges that humans face [51]. It seeks to harmonize our environment with the natural world, fostering a symbiotic relationship between humans and nature [52]. Carl Hastrich pioneered the biomimicry design spiral in 2005 [51,53], outlining a systematic six-step approach to transform nature's innovations into sustainable human designs. As Figure 6 shows, this methodology is exemplified by the kinetic façade concept, inspired by the *Mimosa pudica* plant. This innovative design aims to enhance indoor workspaces by applying the principles of biomimicry, translating nature's systems into practical human applications via a structured six-step process [53,54].

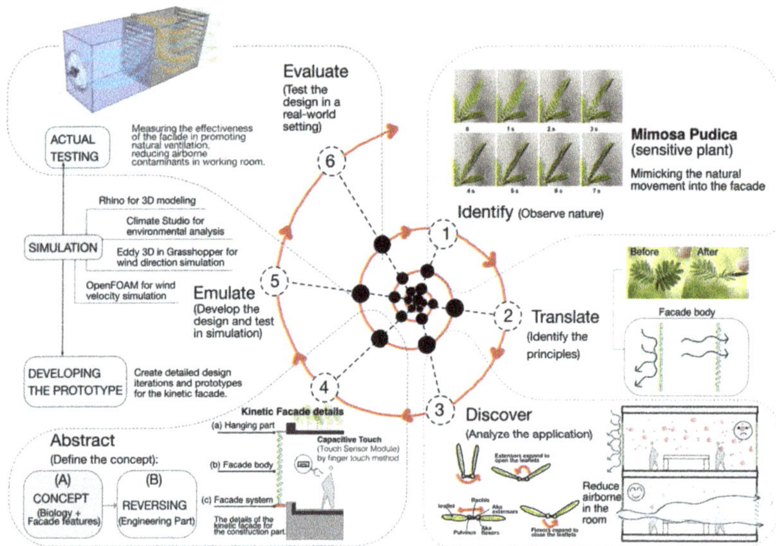

Figure 6. The biomimicry approach to the concept of a kinetic façade.

1. Identify (observe nature): Study the *Mimosa pudica* plant, the unique folding and unfolding behavior of its leaves, and the mechanisms behind its responsive movement to external stimuli, focusing on their relevance to types of office buildings.
2. Translate (identify the principles): Understand how the plant's behavior can be translated into a kinetic façade system with three distinct elements (hanging part, façade body, and façade system) that reacts to environmental conditions (e.g., air quality, temperature, and humidity) to optimize natural ventilation and improve indoor air quality in working spaces.
3. Discover (analyze the application): Investigate the potential applications of the Mimosa-pudica-inspired kinetic façade in office buildings, with a focus on enhancing ventilation and reducing airborne contaminants such as CO_2 and COVID-19 in working rooms.
4. Abstract (define the concept): Develop a conceptual design for the kinetic façade, integrating the principles of the *Mimosa pudica* plant, the identified applications, and the three façade elements (hanging part, façade body, and façade system) in a functional, aesthetically pleasing, and structurally sound manner.

5. Emulate (develop the design and test in simulation): Create detailed design iterations and prototypes of the kinetic façade using computational methods, such as Rhino for 3D modeling, Climate Studio for environmental analysis, Eddy 3D in Grasshopper for wind direction simulation, and OpenFOAM for wind velocity simulation, focusing on the performance in office building environments.
6. Evaluate (test the design in a real-world setting): After the simulation phase, construct a physical prototype or scaled model of the kinetic façade under controlled environmental conditions in a working space (office building type) to assess its performance in a real-world scenario. This step should include measuring the effectiveness of the façade in promoting natural ventilation, reducing airborne contaminants, and improving overall indoor air quality in working rooms.

By following the six-step biomimicry method, the behavior of the *Mimosa pudica* can be translated into a kinetic façade system that promotes healthier and more sustainable indoor environments in workspaces.

4.2. The Process of Mimicking the Mimosa Pudica Movement

Translating the movement of the *Mimosa pudica* into a digital form involves a series of multiple stages. Initially, the study began by meticulously observing the movements of the *Mimosa pudica* plant as it responded to stimuli. For the sake of clarity, this observation was divided into several sections. A 3D digital model of the *Mimosa pudica* was created using Rhino, emphasizing its moving parts. A parametric model was created in Grasshopper using this digital representation to match the movements of the plant as shown in Figure 7. In order to control the folding and unfolding of the digital model and mimic the behavior of the real plant, a sequence was built in Grasshopper. Sensing devices such as capacitive touch were added into the model to increase realism, and actuators were integrated to control movement as shown in Figure 8. This allowed the façade to mimic the behaviors of the *Mimosa pudica*. The digital model was rigorously tested and improved in the last step to maximize its responsiveness and effectiveness.

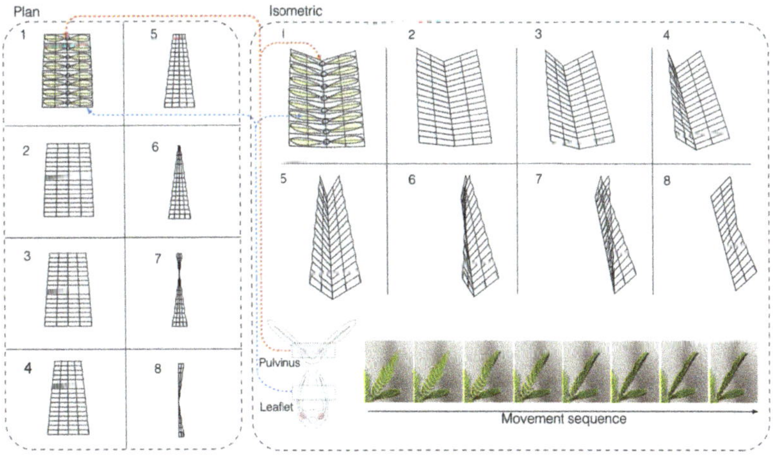

Figure 7. Sequential illustration of the *Mimosa pudica*'s movement in eight steps. A parametric model was created in Grasshopper, using this digital representation to match the movements of the plant as shown in Figure 7. The numbers from 1 to 8 represent the leaflet movement at each step.

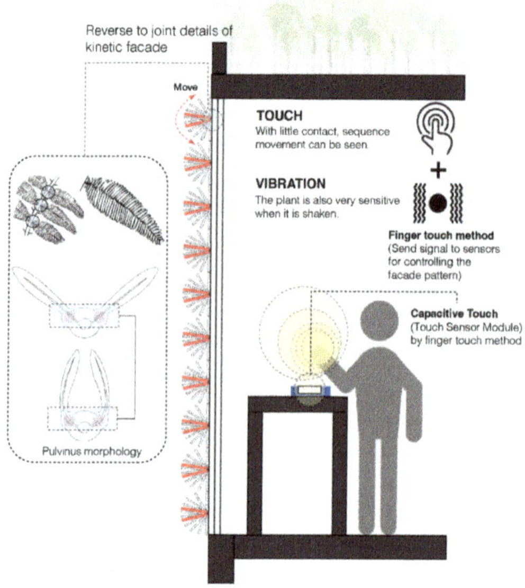

Figure 8. Details of the kinetic façade interpreted from the *Mimosa pudica*.

5. Experiment and Results

5.1. Simulation Part: The Results of Air Velocity in the Space

During the simulation phase, the façade's effectiveness was assessed using two primary metrics. Firstly, air velocity, which pertains to the speed of indoor air circulation, plays a pivotal role in both human comfort and the reduction in airborne contaminants. Secondly, the air change rate (ACR) measures the frequency with which outdoor air replaces indoor air in a workspace, which is a vital metric for ensuring optimal indoor air quality and minimizing health risks. Through these simulations, the goal is to identify the most promising façade patterns for subsequent validation in actual experimental procedures.

The accurate measurement of air velocity is crucial for ensuring human comfort in both indoor and outdoor environments. Adequate air movement is essential for maintaining good indoor air quality and thermal comfort, while excessive wind speeds can lead to discomfort. Proper design and assessment of ventilation systems play a crucial role in maintaining optimal air velocity levels for human comfort as shown in Table 1. In the present study, the focus was on evaluating the efficacy of different façade types in this regard, and simulation was employed as the primary method used to assess their performance.

In this study, the air velocity ranges associated with various façade designs were evaluated to determine their efficacy in ventilating airborne particles in indoor spaces and maintaining human comfort. The assessment took into account the climatic conditions in Bangkok, Thailand.

The results are summarized as follows:

1. Figure 9 depicts a space with no open areas, resulting in a complete absence of air ventilation and making the space unsuitable for human habitation. In contrast, Figure 9 also presents spaces with single-sided openings and two open areas (with and without cross ventilation), where air velocities ranged from 0.00 to 2.50 m/s. While these configurations ensured comfort for occupants, they were not effective in ventilating airborne particles.
2. Figures 10 and 11 illustrate kinetic façades inspired by *Mimosa pudica* (patterns 1, 2, 4, 5, 6, 7, 8, and 9); they demonstrated the ability to ventilate airborne particles with air

velocity ranges between 0.00 and 12.00 m/s. Nevertheless, these designs may cause discomfort for occupants over prolonged periods.

3. Figures 10 and 11 illustrate Mimosa-inspired kinetic façades with patterns 3 and 10, which provided occupant comfort (air velocity ranges of 0.00–3.50 m/s and 0.00–3.00 m/s, respectively). However, these designs showed limited efficacy in ventilating airborne particles.

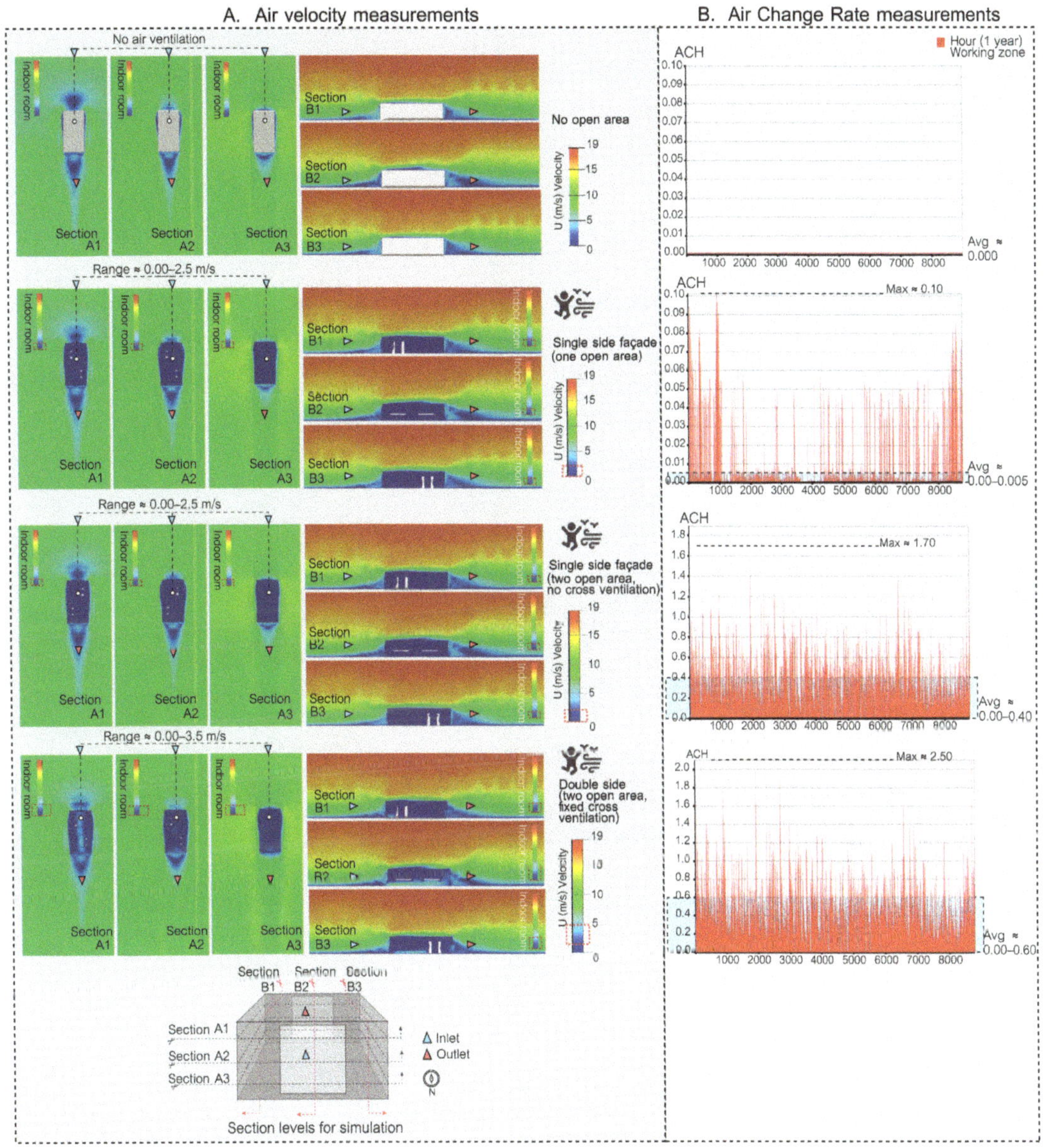

Figure 9. Air velocity and air change rate measurements in a space with base cases (no open areas), single-sided façade (one open area), single-sided façade (two open areas, no cross ventilation), and double-sided façade (two open areas, fixed cross ventilation).

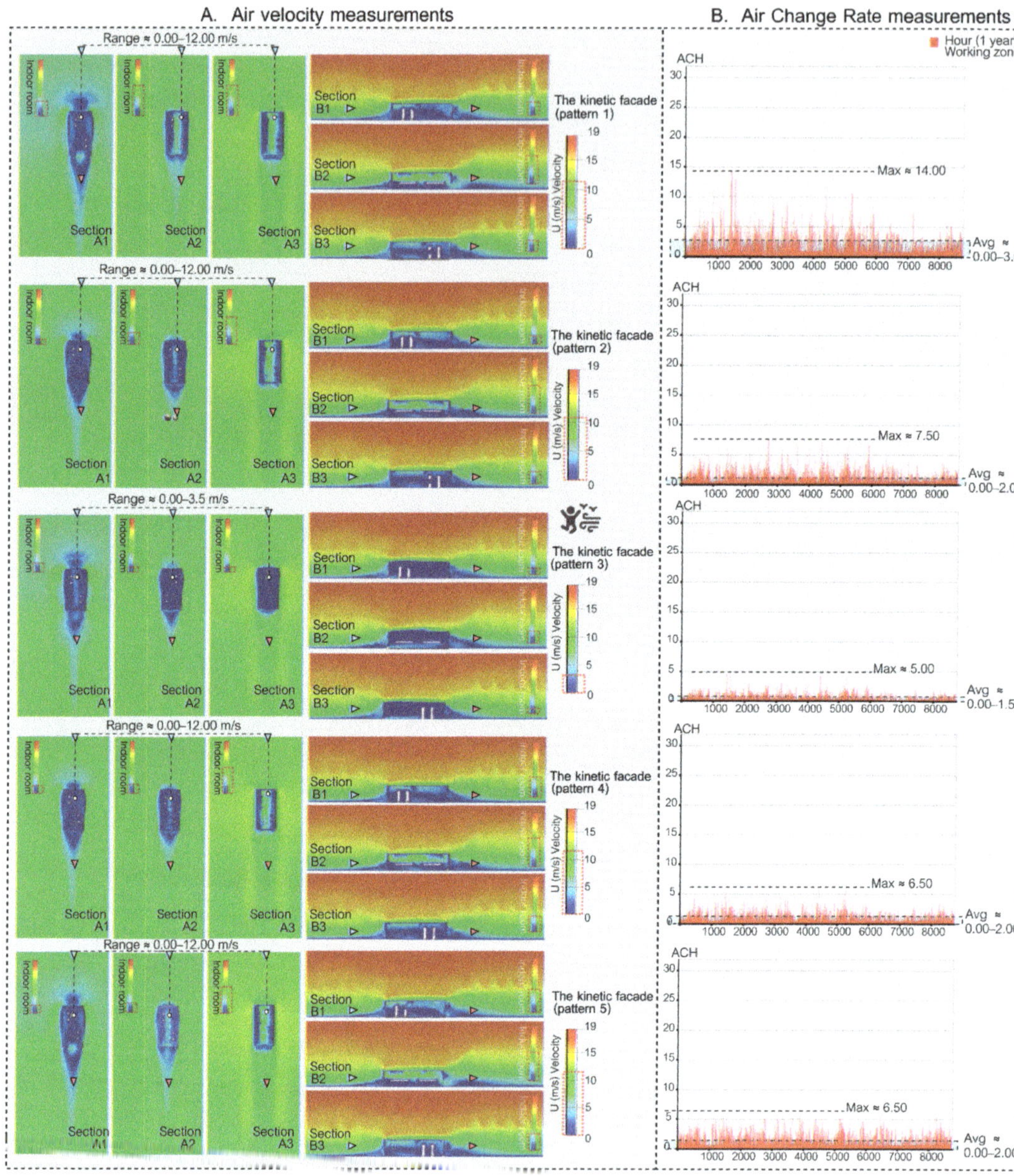

Figure 10. Air velocity and air change rate measurements in a space featuring Mimosa-inspired façade patterns 1, 2, 3, 4, and 5.

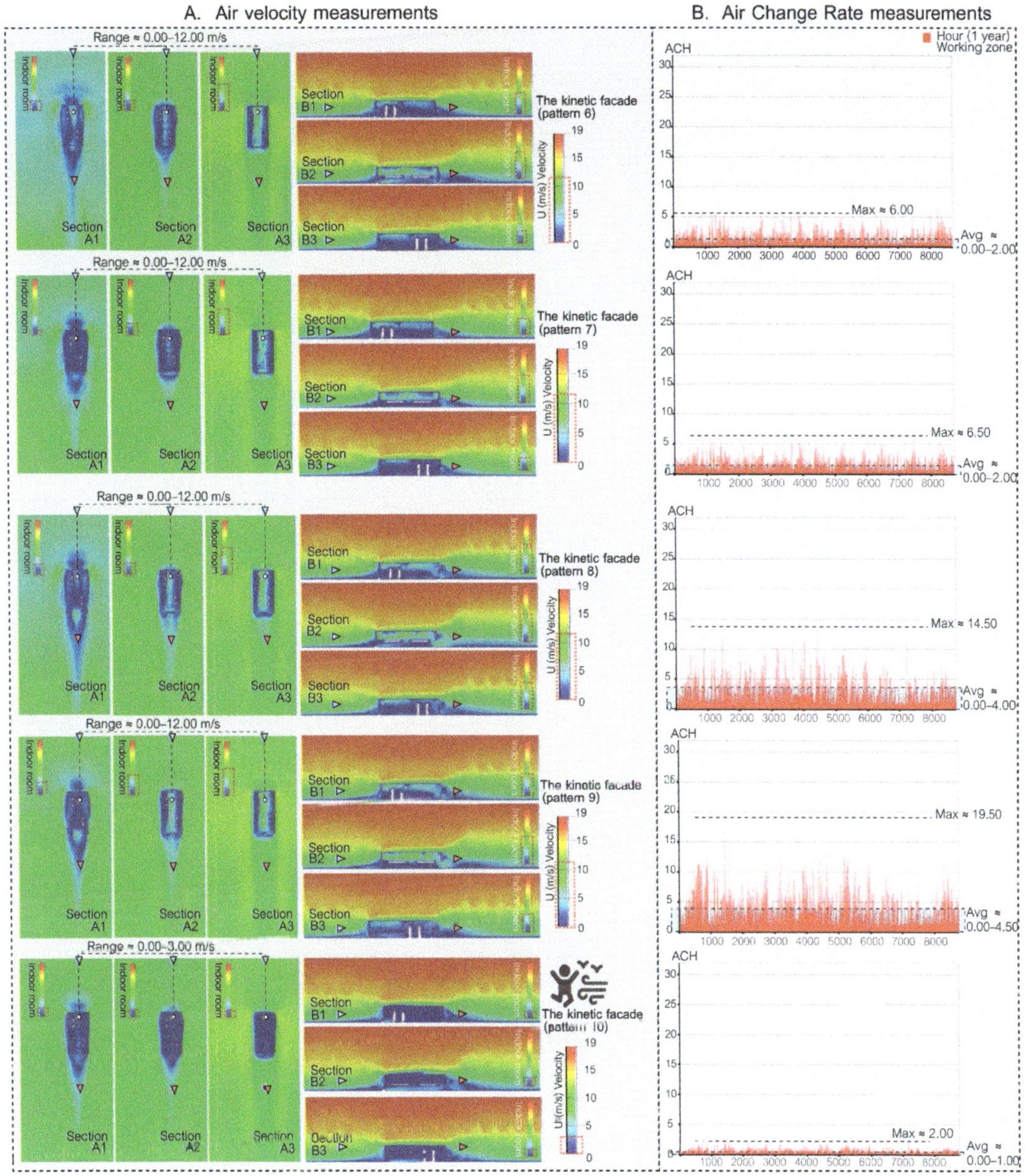

Figure 11. Air velocity and air change rate measurements in a space featuring Mimosa-inspired façade patterns 6, 7, 8, 9, and 10

As depicted in Table 4, air velocity measurements play a significant role in assessing human comfort. It can be deduced that achieving an optimal façade design necessitates the integration of elements from both effective and less effective ventilation patterns, thereby striking a balance between human comfort and airborne particle ventilation. Consequently, in order to select an appropriate pattern for the façade prototype, the research will focus on five specific patterns that exhibit distinct characteristics. Patterns 1, 8, and 9 of the kinetic façade are notably proficient in ventilating airborne particles within the space. On the other hand, patterns 3 and 10 of the kinetic façade are deemed suitable for facilitating prolonged human habitation within the enclosed environment. By meticulously considering the strengths and weaknesses of these patterns, a well-informed decision can be made regarding the most suitable choice for the façade prototype.

Table 4. Summary of the measurement of air velocity with regard to human comfort (Table 1).

Façade Types	Air Velocity Range (m/s)	Airborne Ventilation	Human Comfort
No open area	no air ventilation	NO	NO
Single-sided façade (one open area)	0.00–2.50	NO	YES
Single-sided façade (two open areas, no cross ventilation)	0.00–2.50	NO	YES
Double-sided (two open areas, fixed cross ventilation)	0.00–3.50	NO	YES
Mimosa-inspired façade (pattern 1)	0.00–12.00	YES	NO
Mimosa-inspired façade (pattern 2)	0.00–12.00	YES	NO
Mimosa-inspired façade (pattern 3)	0.00–3.50	NO	YES
Mimosa-inspired façade (pattern 4)	0.00–12.00	YES	NO
Mimosa-inspired façade (pattern 5)	0.00–12.00	YES	NO
Mimosa-inspired façade (pattern 6)	0.00–12.00	YES	NO
Mimosa-inspired façade (pattern 7)	0.00–12.00	YES	NO
Mimosa-inspired façade (pattern 8)	0.00–12.00	YES	NO
Mimosa-inspired façade (pattern 9)	0.00–12.00	YES	NO
Mimosa-inspired façade (pattern 10)	0.00–3.00	NO	YES

5.2. The Result of Air Change Rate in the Space

The air change rate per hour (ACH) serves as a vital metric for evaluating the potential of various façade designs in terms of air ventilation, which is crucial for mitigating airborne contaminants in enclosed spaces. This study conducted simulations over a full year, accounting for 8760 hourly data points, in a small office setting with a meeting room function in Bangkok, Thailand.

The American Society of Heating, Refrigerating, and Air-Conditioning Engineers (ASHRAE) recommends a minimum ACH of 2.00–3.00 for office spaces, while COVID-19-related recommendations for hospital areas vary depending on the function. In this research, a minimum ACH of 2.00 was set as a criterion for reducing airborne contaminants in the space using passive ventilation methods.

The ACH values for each distinct façade design are as follows.

Figure 9 outlines four distinct base case ventilation scenarios for comparison with the kinetic façade, each marked by different ACH values. The "no open area case" indicates that there is no air ventilation in the indoor space. In the "single side (one open area)" scenario, ventilation slightly improves, evidenced by a maximum ACH of 0.10 and an average ranging from 0.000 to 0.005. The scenario featuring a single side with two open areas but without cross ventilation exhibits a notable increase in ventilation, with a peak ACH of 1.70 and an average spanning 0.00 to 0.40. The most effective ventilation is observed in the double-sided case with two open areas and fixed cross ventilation, where the ACH peaks at 2.50 and the average varies between 0.00 and 0.60.

Figures 10 and 11 showcase the enhanced ventilation performance of kinetic façade designs inspired by the Mimosa plant, featuring patterns 1 to 10. Patterns 1 to 7, as depicted in Figure 10, achieve maximum air change rate per hour (ACH) values ranging from 5.00 to 14.00, with average ACH values between 0.00 and 3.00. Figure 11 highlights pattern 8's outstanding ventilation, with a peak ACH of 14.50 and an average ranging from 0.00 to 4.00. It also presents pattern 9's exceptional ventilation capabilities, reaching a maximum ACH of 19.50 and an average between 0.00 and 4.50, while pattern 10 exhibits satisfactory performance with a maximum ACH of 2.00 and an average from 0.00 to 1.00. In conclusion, kinetic façades inspired by the Mimosa plant (patterns 1–10) show a higher potential for air ventilation and effectively reduce airborne particles in office spaces using passive ventilation methods. In particular, patterns 8 and 9 exceed the ASHRAE's minimum criteria for office spaces. The study highlights the advantages of these innovative façade designs in enhancing air change rates and promoting healthier indoor environments for workspaces.

In Bangkok's hot and humid climate, managing indoor air quality and comfort is challenging, particularly when the air change rate (ACR) exceeds 15, due to increased demands for cooling and dehumidification. However, innovative kinetic façade designs inspired by the Mimosa plant, especially pattern 9, show promise in effectively managing these challenges. While occasionally reaching a maximum ACH of 19.50, pattern 9 generally maintains an average ACH well within the comfortable range of 0.00 to 4.50. This performance suggests that, despite brief periods of a higher ACR, these façades can align with Bangkok's climatic needs for a comfortable and energy-efficient indoor environment. Further enhancing this potential, a holistic façade prototype emerges by amalgamating the strengths of various Mimosa-inspired patterns. This prototype combines the comfort-centric attributes of patterns 3 and 10 with the superior airborne particle ventilation capabilities of patterns 1, 8, and 9. Such a unified façade design aims to create spaces that are not just habitable but offer a comprehensive living experience, balancing enhanced air velocities for occupant comfort with an optimal ACH for maintaining air quality. This approach contributes to a future where architectural innovation harmoniously blends human comfort with effective air quality management.

5.3. Actual Experiment Section: Evaluation of Façade Types Based on Ventilation Potential: A Study Using the Indoor Air Quality Index (IAQI)

This study emphasizes the value of real-world application and places the priority on actual evaluations over explicit simulations. The rate of airborne particle ventilation, air velocity for comfort, and CO_2 content as an indicator of air quality are important testing criteria. Prototypes are tested in controlled environments to acquire a real understanding of how well the façade performs. This practical learning approach helps design modifications and strikes a balance between innovative ideas and practical implementation.

In the research, smoke fog oil was provided as a substitute for airborne particles. Before testing the ventilation capacities, this oil saturated the experimental area to a concentration of 6000 ppm. This decision was crucial because, despite being harmless, smoke fog oil mimics the dispersion and settlement tendencies of airborne viruses. Recent global health crises have heightened worries regarding airborne transmission, making this scenario all the more vital. By evaluating the behavior of smoke fog oil under the effect of the kinetic façade, it is possible to gain insight into the potential behavior of authentic airborne particles in similar environments.

The study broadened its examination to assess the impact of diverse façade designs on indoor CO_2 reduction. CO_2 levels, monitored in a controlled environment for 30 min, are depicted in Figure 12, with outcomes categorized as per the indoor air quality index in Table 2. An observational analysis highlighted the airborne transmission within 30 min after the opening of the façade using a kinetic façade (pattern 9). Figure 12 distinctly shows that, from 0 to 5 min (prior to façade opening), smoke density remains high indoors. Following the opening of the façade when employing kinetic façade pattern 9, there is a significant smoke reduction after 5 min, with the smoke nearly vanishing from 20 to 25 min.

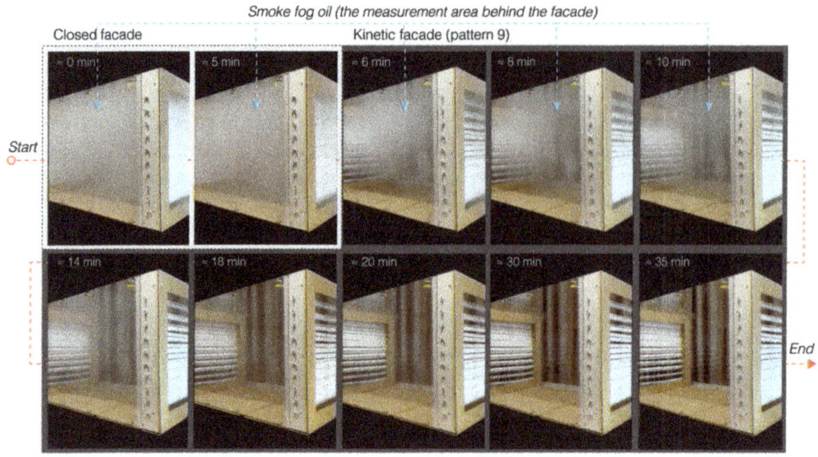

Figure 12. Observational analysis of airborne transmission within 30 min following façade opening utilizing kinetic façade (pattern 9).

The selected timeframe was determined based on multiple considerations: it reflects the usual duration that an individual might spend in a confined area; it permits several air exchange cycles typical of most ventilation systems; it is consistent with the duration adopted in comparable studies; and it provides a sufficient window for detecting notable CO_2 variations while still being concise enough to retain experimental oversight. Essentially, this timeframe ensures that the study's findings are both scientifically accurate and applicable to real-world situations.

Table 5 presents the rankings of façade types based on their CO_2 concentration levels at the end of a 30 min evaluation period. Additionally, a comparative analysis of CO_2 levels over time for different façade types is illustrated in Figure 13.

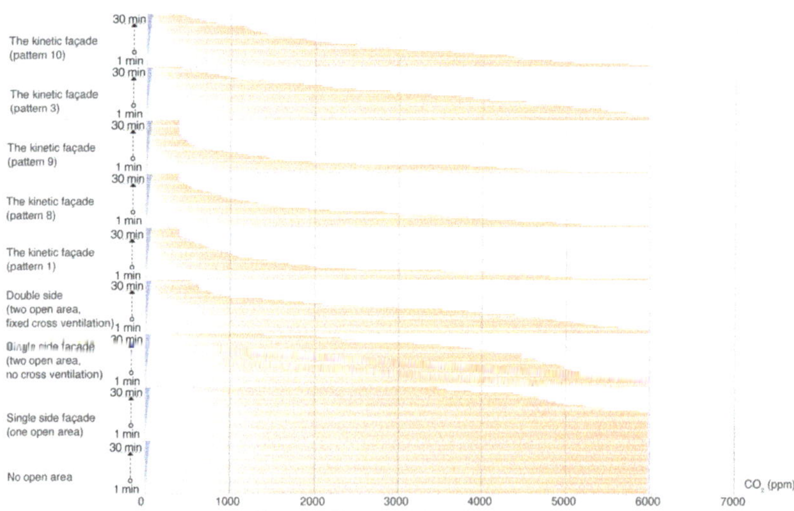

Figure 13. Comparative analysis of CO_2 levels over time for different façade types.

Table 5. Rankings of façade types based on CO_2 concentration levels at the conclusion of a 30 min evaluation period, as detailed in Table 2.

Façade Types	CO_2 Concentration (ppm)	IAQI Rating	Interpretation
The kinetic façade (patterns 1, 8, and 9)	400	Good	This façade type exhibited the best performance, achieving the lowest CO_2 concentration and securing a "Good" rating on the IAQI.
The kinetic façade (pattern 3)	440	Good	This façade type exhibited the best performance, achieving the lowest CO_2 concentration and securing a "Good" rating on the IAQI.
The kinetic façade (pattern 10)	450	Good	This façade type had a marginally higher CO_2 concentration than pattern 3 but remained within the "Good" IAQI rating.
Double-sided façade (two open areas with fixed cross ventilation)	550	Good	Despite its effective ventilation design, it recorded a higher CO_2 concentration than the kinetic façades.
Single-sided façade (two open areas, no cross ventilation)	1410	Unhealthy	The absence of cross ventilation hindered its performance, resulting in an "Unhealthy" IAQI rating.
Single-sided façade (one open area)	3500	Hazardous	Its design, with only one open area, limited its ventilation potential, leading to a "Hazardous" IAQI rating.
Façade without open area	6000	Extreme	This façade type demonstrated the poorest performance, with the highest CO_2 concentration, resulting in an "Extreme" IAQI rating.

The kinetic façades, regardless of their specific patterns, emerged as the top performers in terms of ventilating airborne particles within the stipulated 30 min, all achieving a commendable "Good" IAQI rating. In contrast, the façade with no open area was markedly less effective, falling into the "Extreme" IAQI category. The study underscores the significance of open areas and cross ventilation in façade design for optimal indoor air quality.

Analysis of Air Velocity Concentrations across Various Façade Types over Time

The graph in Figure 14 illustrates the observed air velocities (m/s) over a span of 40 min for each façade type, subjected to standardized testing conditions.

The research employed a 40 min testing duration for air velocity to ensure consistent measurements across façade designs. This timeframe captures the façades' steady behavior, reduces the impact of short-term disturbances such as gusts, and reflects real-world conditions. By reaching a quasi-steady state, especially under thermal influences, the study offers a balanced, accurate, and practical insight into the façades' performance in indoor environments.

The study revealed that façades without open areas showed no air movement, registering a consistent velocity of 0 m/s. Similarly, single-sided façades, regardless of the number of open areas, failed to facilitate any significant air movement. Double-sided façades with fixed cross ventilation showed modest improvements, with velocities reaching up to 1.05 m/s. However, kinetic façades showcased the most promise. Pattern 1 peaked at 2.21 m/s, while pattern 3 achieved velocities of up to 1.05 m/s. Pattern 8 closely followed pattern 1, reaching 1.98 m/s. Impressively, pattern 9 emerged as the top performer, with velocities ranging between 2.21 m/s and 2.56 m/s. In contrast, pattern 10 lagged behind other kinetic designs, peaking at just 0.8 m/s. Overall, the research accentuates the pivotal role of architectural design in influencing ventilation efficiency, with kinetic façades, especially pattern 9, demonstrating superior potential in optimizing air movement within interior spaces.

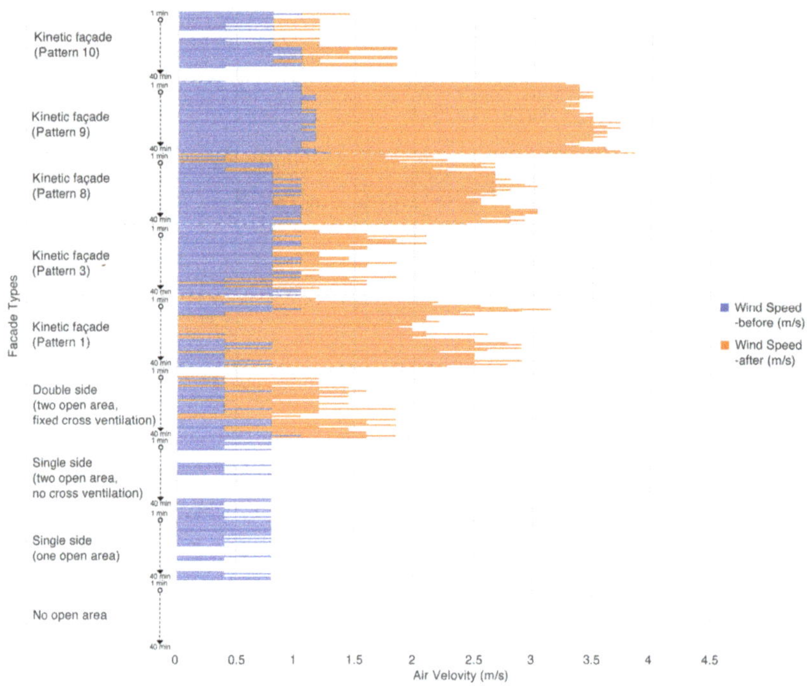

Figure 14. Comparative analysis of air velocity across different façade types.

In a comparative analysis of various façade types, it becomes evident that the design and structure of a façade significantly influence indoor air quality and ventilation efficiency. Static façades, such as those with no open areas or limited open areas, demonstrate restricted ventilation capabilities. For instance, façades with no open areas exhibit a complete absence of ventilation, leading to an "Extreme" IAQI rating due to elevated CO_2 concentrations. Similarly, façades with a single open area, despite occasional peaks in wind speed, ultimately succumb to poor air quality, as indicated by their "Hazardous" IAQI rating. Conversely, kinetic façades, irrespective of their specific patterns, consistently outperform their static counterparts. These kinetic façades not only achieve "Good" IAQI ratings but also maintain efficient ventilation, even at varying wind speeds. Notably, certain kinetic façade patterns, such as pattern 9, demonstrate exceptional ventilation efficiency, with wind speeds reaching up to 2.56 m/s, while maintaining consistent CO_2 concentrations comparable to other kinetic façades. In conclusion, the kinetic façades undeniably present a more promising potential in terms of ventilation and air quality when juxtaposed with static façades. Their dynamic design, coupled with multiple open areas, ensures optimal indoor air quality, underscoring their superiority in architectural design for sustainable and healthy living environments.

6. Discussion

The results indicate that kinetic façades, particularly patterns 1, 8, and 9, markedly surpass static façades (base cases) in ventilation efficiency. This significant finding underscores the practical applicability of these innovative designs, extending beyond theoretical expectations. Furthermore, patterns 3 and 10, though less effective in air purification, are identified as more suitable for prolonged human occupancy, emphasizing the necessity of balancing ventilation effectiveness with human comfort in façade design.

The comparative analysis, as illustrated in Figure 15, further reinforces these findings. The study's comprehensive investigation, which combined computational simulations with physical tests, highlighted the superior ventilation capabilities of kinetic façades, especially patterns 1, 8, and 9. These patterns not only showed theoretical promise but also demonstrated remarkable practical utility in ventilating spaces. In contrast, static façades, particularly those with limited or single openings, showed significantly lower ventilation efficacy. This stark difference underscores the transformative potential of kinetic façades in architectural design.

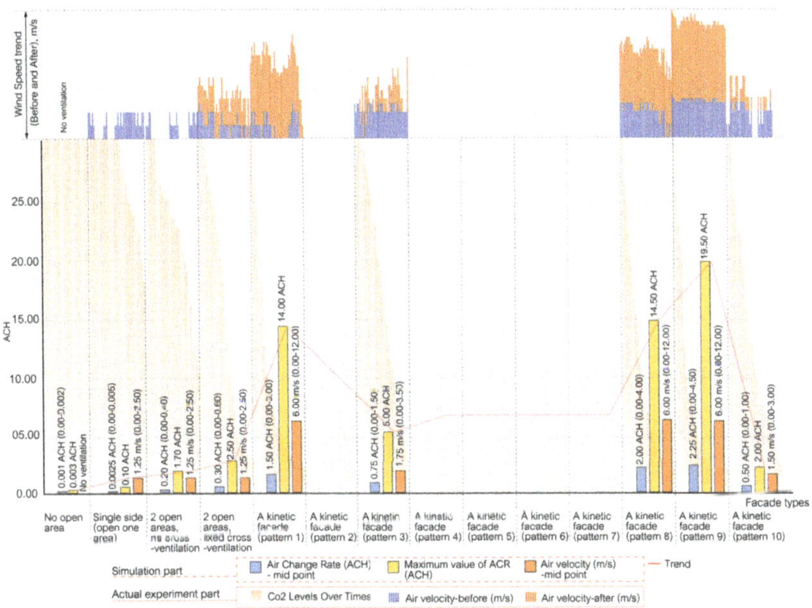

Figure 15. Comparative analysis of façade types: a synthesis of simulation and actual experiment results.

The study also revealed the nuanced roles of different façade patterns. While patterns 1, 8, and 9 excelled in ventilation, patterns 3 and 10 carved out a niche in enhancing long-term human comfort, despite their lower effectiveness in air purification. This finding is crucial, as it highlights the importance of a multifaceted approach to façade design: one that considers both air quality and occupant comfort. The benefits of double-sided façades, particularly in terms of cross-ventilation efficiency, were also evident. However, patterns 2, 4, 5, 6, and 7 were less effective, primarily due to their limited impact on air exchange rates and velocities.

Contextually, the study aligns with the expanding research on dynamic, responsive building designs. The kinetic façades, inspired by the Mimosa plant, represent a notable advancement in this field, introducing a unique method for managing indoor air quality and human comfort. This is particularly relevant in the context of urban environments such as Bangkok, where the challenges of heat and humidity are persistent, thereby contributing to the discourse on sustainable building practices.

The implications of these findings are substantial for architectural design, especially in hot and humid climates. The demonstrated capability of kinetic façades in efficiently ventilating airborne particles while optimizing human comfort could significantly influence future building codes and standards, advocating for the integration of dynamic façade systems to enhance indoor environmental quality.

However, this study is not without limitations. Its focus on Bangkok's specific climatic conditions may restrict the applicability of its findings to other climates. Moreover, the reliance on simulations and controlled experiments, while valuable, might not entirely reflect the complexities encountered in real-world building environments. These aspects highlight areas for further research and consideration in the application of kinetic façade systems.

7. Conclusions and Future Studies

The research findings suggest that kinetic façades have the potential to significantly improve indoor air quality and optimize human comfort. However, there is still work to be carried out in achieving a harmonious balance between these components.

The research suggests that kinetic façades could play a crucial role in future architectural advancements by enhancing indoor air quality, improving human comfort, and increasing productivity. As depicted in Figure 16, maximizing this potential necessitates the integration of the most effective patterns into the kinetic façade design. One such pattern is the Mimosa kinetic façade, which is characterized by its touch-based adjustability. This innovative approach allows for the regulation of airflow and ventilation, leading to improved indoor environmental conditions. The design of the Mimosa kinetic façade also includes features such as the modulation of airflow, effective management of natural light, mitigation of solar heat gain, and provision of shading. Additionally, the façade's design is adaptable to different climates, allowing users to modify patterns based on their comfort needs and environmental conditions.

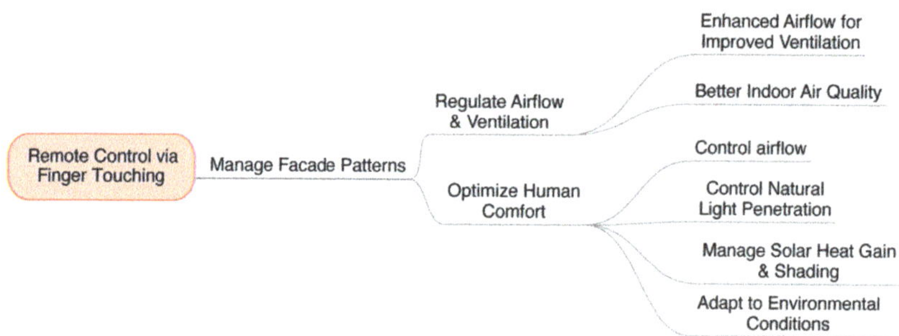

Figure 16. The guidelines for utilizing the Mimosa kinetic façade.

This research identifies several key areas for future exploration in kinetic façades. These include studying their performance in various climates to assess adaptability and effectiveness, investigating long-term sustainability and practicality, and exploring seamless integration with smart technologies for enhanced efficiency and user experience. Additionally, researching other responsive organisms for innovative façade design insights, applying computational fluid dynamics and AI to improve façade movement predictions, and advancing material science and engineering fields to boost the façades' adaptive capabilities are crucial. Assessing cost effectiveness and financial viability to promote broader adoption and evaluating practicality, economic feasibility, and maintenance in large-scale projects, focusing on airflow regulation and human comfort, are also essential. In summary, further research is vital for developing cost-effective, sustainable kinetic façades that meet environmental needs and enhance indoor human wellbeing.

Author Contributions: This paper represents the results of teamwork. Conceptualization, S.S., K.M. and T.H.; methodology, investigation, resources, and data curation, S.S., K.M., T.H., H.X. and J.K.; writing—original draft preparation, S.S.; validation, S.S.; writing—review and editing, S.S., K.M., T.H., H.X. and J.K. All authors have read and agreed to the published version of the manuscript.

Funding: This study was encouraged by the Japan Advanced Institute of Science and Technology (JAIST) Research Fund and the research fund of Thailand's National Electronics and Computer Technology Centre (NECTEC) and Sirindhorn International Institute of Technology, Thammasat University: JAIST-G-S1Y19/002.

Institutional Review Board Statement: Not applicable.

Data Availability Statement: Data are contained within the article.

Conflicts of Interest: The authors declare no conflict of interest.

References

1. Awada, M.; Becerik-Gerber, B.; White, E.; Hoque, S.; O'Neill, Z.; Pedrielli, G.; Wen, J.; Wu, T. Occupant health in buildings: Impact of the COVID-19 pandemic on the opinions of building professionals and implications on research. *Build. Environ.* **2022**, *207*, 108440. [CrossRef]
2. Hagihara, T.; Toyota, M. Mechanical Signaling in the Sensitive Plant *Mimosa pudica* L. *Plants* **2020**, *9*, 587 . [CrossRef]
3. Patil, H.S.; Vaijapurkar, S. Study of the Geometry and Folding Pattern of Leaves of Mimosa pudica. *J. Bionic Eng.* **2007**, *4*, 19–23. [CrossRef]
4. Alonso, M.J.; Liu, P.; Marman, S.F.; JØrgensen, R.B.; Mathisen, H.M. Holistic methodology to reduce energy use and improve indoor air quality for demand-controlled ventilation. *Energy Build.* **2023**, *279*, 112692. [CrossRef]
5. Wang, H.; Li, X. Visualization and Measurement of Indoor Airflow by Color Sequence Enhanced Particle Streak Velocimetry. In *Handbook of Indoor Air Quality*; Zhang, Y., Hopke, P.K., Mandin, C., Eds.; Springer: Singapore, 2022; pp. 569–608. [CrossRef]
6. Gosztonyi, S. The role of geometry for adaptability: Comparison of shading systems and biological role models. *J. Facade Des. Eng.* **2018**, *6*, 163–174. [CrossRef]
7. Hosseini, S.M.; Mohammadi, M.; Rosemann, A.; Schröder, T.; Lichtenberg, J. A morphological approach for kinetic façade design process to improve visual and thermal comfort: Review. *Build. Environ.* **2019**, *153*, 186–204. [CrossRef]
8. Attia, S.; Navarro, A.L.; Juaristi, M.; Monge-Barrio, A.; Gosztonyi, S.; Al-Doughmi, Z. Post-occupancy evaluation for adaptive façades. *J. Facade Des. Eng.* **2018**, *6*, 001–009. [CrossRef]
9. Jamrozik, A.; Clements, N.; Hasan, S.S.; Zhao, J.; Zhang, R.; Campanella, C.; Loftness, V.; Porter, P.; Ly, S.; Wang, S.; et al. Access to daylight and view in an office improves cognitive performance and satisfaction and reduces eyestrain: A controlled crossover study. *Build. Environ.* **2019**, *165*, 106379. [CrossRef]
10. Saelens, D.; Carmeliet, J.; Hens, H. Energy Performance Assessment of Multiple-Skin Facades. *HVAC R Res.* **2011**, *9*, 167–185. [CrossRef]
11. Attia, S.; Favoino, F.; Loonen, R.C.G.M.; Petrovski, A.; Monge-Barrio, A. Adaptive façades system assessment: An initial review. In Proceedings of the 10th Conference on Advanced Building Skins, Bern, Switzerland, 3–4 November 2015; Economic Forum: Cologny, Switzerland , 2015; pp. 1275–1283.
12. Struck, C.; Almeida, M.; Silva, S.; Mateus, R.; Lemarchand, P.; Petrovski, A.; Rabenseifer, R.; Wansdronk, R.; Wellershoff, F.; Wit, J. Adaptive facade systems—Review of performance requirements, design approaches, use cases and market needs. In Proceedings of the 10th Energy Forum on Advanced Building Skins, Bern, Switzerland, 3–4 November 2015. [CrossRef]
13. Attia, S.; Bilir, S.; Safy, T.; Struck, C.; Loonen, R.; Goia, F. Current trends and future challenges in the performance assessment of adaptive façade systems. *Energy Build.* **2018**, *179*, 165–182. [CrossRef]
14. Bacha, C.B.; Bourbia, F. Effect of kinetic facades on energy efficiency in office buildings -hot dry climates. In Proceedings of the 11th Conference on Advanced Building Skins, Bern, Switzerland, 10–11 October 2016; Volume 1, pp. 458–468.
15. Lee, E.S.; Selkowitz, S.E.; DiBartolomeo, D.L.; Klems, J.H.; Clear, R.D.; Konis, K.S.; Hitchcock, R.; Yazdanian, Y.; Mitchell, R., Konstantoglou, M. High Performance Building Facade Solutions: PIER Final Project Report. California Energy Commission Public Interest Energy Research Program. December 2009; pp. 27–50. Available online: https://eta-publications.lbl.gov/sites/default/files/lbnl-4583e.pdf (accessed on 5 October 2023).
16. Romano, R.; Aelenei, L.; Aelenei, D.; Mazzucchelli, E.S. What is an adaptive façade? Analysis of recent terms and definitions from an international perspective. *J. Facade Des. Eng.* **2018**, *6*, 065–076. [CrossRef]
17. Prieto, A.; Knaack, U.; Klein, T.; Auer, T. Possibilities and constraints for the widespread application of solar cooling integrated façades. *J. Facade Des. Eng.* **2018**, *6*, 10–18. [CrossRef]
18. Basarir, B.; Altun, M.C. A redesign procedure to manufacture adaptive façades with standard products. *J. Facade Des. Eng.* **2018**, *6*, 77–100. [CrossRef]
19. Casini, M. Active dynamic windows for buildings: A review. *Renew. Energy* **2018**, *119*, 923–934. [CrossRef]
20. Bakker, L.G.; van Oeffelen, E.C.M.H.; Loonen, R.C.G.M.; Hensen, J.L.M. User satisfaction and interaction with automated dynamic facades: A pilot study. *Build. Environ.* **2014**, *78*, 44–52. [CrossRef]
21. de Klijn-Chevalerias, M.L.; Loonen, R.C.; Zarzycka, A.; de Witte, D.; Sarakinioti, M.V.; Hensen, J.L. Assisting the development of innovative responsive façade elements using building performance simulation. *Simul. Ser.* **2017**, *49*, 202–209. [CrossRef]

22. Elzeyadi, I. The impacts of dynamic façade shading typologies on building energy performance and occupant's multi-comfort. *Archit. Sci. Rev.* **2017**, *60*, 316–324. [CrossRef]
23. Paper, C.; Pierleoni, A.; Spa, F.; Politecnico, V.S.; Politecnico, L.B. Innovative technologies for transparent building envelopes: Experimental assessment of energy and thermal comfort data to facilitate the decision-making process. In Proceedings of the CIGOS Innovatiion and Construction, CIGOS Innovatiion and Construction, Paris, France, 11–12 May 2015; pp. 1–9
24. Loonen, R.C.G.M.; Hoes, P.; Hensen, J.L.M. Performance prediction of buildings with responsive building elements challenges and solutions. In Proceedings of the 2014 Building Simulation and Optimization Conference (BSO14), London, UK, 23–24 June 2014; pp. 1–8.
25. Chang, T.W.; Huang, H.Y.; Datta, S. Design and fabrication of a responsive carrier component envelope. *Buildings* **2019**, *9*, 84. [CrossRef]
26. Saelens, D.; Hens, H. Evaluating the Thermal Performance of active envelopes. In Proceedings of the Buildings VIII: Thermal Performance of Exterior Envelopes of Whole Buildings, Trondheim, Norway, 17–19 June 2002; pp. 1–10.
27. Jayathissa, P.; Caranovic, S.; Hofer, J.; Nagy, Z.; Schlueter, A. Performative design environment for kinetic photovoltaic architecture. *Autom. Constr.* **2018**, *93*, 339–347. [CrossRef]
28. Loonen, R.C.G.M.; Trčka, M.; Cóstola, D.; Hensen, J.L.M. Climate adaptive building shells: State-of-the-art and future challenges. *Renew. Sustain. Energy Rev.* **2013**, *25*, 483–493. [CrossRef]
29. Chayaamor-Heil, N.; Vitalis, L. Biology and architecture: An ongoing hybridization of scientific knowledge and design practice by six architectural offices in France. *Front. Archit. Res.* **2021**, *10*, 240–262. [CrossRef]
30. Li, C.; Tang, H. Study on ventilation rates and assessment of infection risks of COVID-19 in an outpatient building. *J. Build. Eng.* **2021**, *42*, 103090. [CrossRef]
31. Srivastava, S.; Zhao, X.; Manay, A.; Chen, Q. Effective ventilation and air disinfection system for reducing coronavirus disease 2019 (COVID-19) infection risk in office buildings. *Sustain. Cities Soc.* **2021**, *75*, 103408. [CrossRef] [PubMed]
32. Pourshab, N.; Tehrani, M.; Toghraie, D.; Rostami, S. Application of double glazed façades with horizontal and vertical louvers to increase natural air flow in office buildings. *Energy* **2020**, *200*, 117486. [CrossRef]
33. Dao, H.T.; Kim, K.S. Behavior of cough droplets emitted from Covid-19 patient in hospital isolation room with different ventilation configurations. *Build. Environ.* **2022**, *209*, 108649. [CrossRef] [PubMed]
34. Hurraß, J.; Golmohammadi, R.; Bujok, S.; Bork, M.; Thelen, F.; Wagner, P.; Exner, D.; Schönfeld, C.; Hornei, B.; Kampf, G.; et al. Explosive COVID-19 outbreak in a German nursing home and the possible role of the air ventilation system. *J. Hosp. Infect.* **2022**, *130*, 34–43. [CrossRef] [PubMed]
35. Blocken, B.; van Druenen, T.; Ricci, A.; Kang, L.; van Hooff, T.; Qin, P.; Xia, L.; Ruiz, C.A.; Arts, J.H.; Diepens, J.F.; et al. Ventilation and air cleaning to limit aerosol particle concentrations in a gym during the COVID-19 pandemic. *Build. Environ.* **2021**, *193*, 107659. [CrossRef] [PubMed]
36. Bergefurt, L.; Weijs-Perrée, M.; Appel-Meulenbroek, R.; Arentze, T. The physical office workplace as a resource for mental health—A systematic scoping review. *Build. Environ.* **2022**, *207*, 108505 . [CrossRef]
37. Harweg, T.; Bachmann, D.; Weichert, F. Agent-based simulation of pedestrian dynamics for exposure time estimation in epidemic risk assessment. *J. Public Health* **2021**, *31*, 221–228. [CrossRef] [PubMed]
38. Elsaid, A.M.; Ahmed, M.S. Indoor Air Quality Strategies for Air-Conditioning and Ventilation Systems with the Spread of the Global Coronavirus (COVID-19) Epidemic: Improvements and Recommendations. *Environ. Res.* **2021**, *199*, 111314. [CrossRef]
39. Agarwal, N.; Meena, C.S.; Raj, B.P.; Saini, L.; Kumar, A.; Gopalakrishnan, N.; Kumar, A.; Balam, N.B.; Alam, T.; Kapoor, N.R.; et al. Indoor air quality improvement in COVID-19 pandemic: Review. *Sustain. Cities Soc.* **2021**, *70*, 102942. [CrossRef] [PubMed]
40. Martins, N.R.; da Graça, G.C. Effects of airborne fine particle pollution on the usability of natural ventilation in office buildings in three megacities in Asia. *Renew. Energy* **2018**, *117*, 357–373. [CrossRef]
41. Fordham, M. Natural ventilation. *Renew. Energy* **2000**, *19*, 17–37. [CrossRef]
42. Aflaki, A.; Mahyuddin, N.; Mahmoud, Z.A.C.; Baharum, M.R. A review on natural ventilation applications through building façade components and ventilation openings in tropical climates. *Energy Build.* **2015**, *101*, 153–162. [CrossRef]
43. Sankaewthong, S.; Horanont, T.; Miyata, K.; Karnjana, J.; Busayarat, C.; Xie, H. Using a Biomimicry Approach in the Design of a Kinetic Façade to Regulate the Amount of Daylight Entering a Working Space. *Buildings* **2022**, *12*, 2089. [CrossRef]
44. Holzer, P.; Psomas, T. Ventilative Cooling Sourcebook: Energy in Buildings and Communities Programme. Department of Civil Engineering, Aalborg University, Thomas Manns Vej 23, 9220, Aalborg Ø, Denmark. EBC Bookshop: March 2018; p.156. Available online: https://www.aivc.org/resource/ventilative-cooling-source-book (accessed on 5 October 2023).
45. Su, X.; Yuan, Y.; Wang, Z.; Liu, W.; Lan, L.; Lian, Z. Human thermal comfort in non-uniform thermal environments: A review. *Energy Built Environ.* **2023**. [CrossRef]
46. Hou, Y. Effect of wind speed on human thermal sensation and thermal comfort. *AIP Conf. Proc.* **2018**, *1971*, 30012. [CrossRef]
47. Olesen, B.W. International standards for the indoor environment. *Indoor Air Suppl.* **2004**, *14*, 18–26. [CrossRef]
48. Maddalena, R.; Mendell, M.J.; Eliseeva, K.; Chan, W.R.; Sullivan, D.P.; Russell, M.; Satish, U.; Fisk, W.J. Effects of ventilation rate per person and per floor area on perceived air quality, sick building syndrome symptoms, and decision-making. *Indoor Air* **2015**, *25*, 362–370. [CrossRef]
49. Persily, A.K. Rationale for Ventilation Rate Requirements in ASHRAE Standard 62-1999 and in Potential Revisions to the Standard. In *Air Quality and Comfort in Airliner Cabins*; ASTM International: West Conshohocken, PA, USA, 2000. [CrossRef]

50. Mendell, M.J.; Apte, M.G. *Balancing Energy Conservation and Occupant Needs in Ventilation Rate Standards for "Big Box" Stores and Other Commercial Buildings in California: Issues Related to the ASHRAE 62.1 Indoor Air Quality Procedure*; Technical Report; Lawrence Berkeley National Lab. (LBNL): Berkeley, CA, USA, 2010 . . [CrossRef]
51. Speck, O.; Speck, D.; Horn, R.; Gantner, J.; Sedlbauer, K.P. Biomimetic bio-inspired biomorph sustainable? An attempt to classify and clarify biology-derived technical developments. *Bioinspir. Biomim.* **2017**, *12*, 011004 . [CrossRef]
52. Silverstein, D.; Samuel, P.; Decarlo, N. Biomimicry. In *The Innovator's Toolkit* ; Wiley: Hoboken, NJ, USA, 2011; pp. 153–158. [CrossRef]
53. Hapsari, F.N.; Purwaningsih, R.; Azzahra, F.; Sari, D.P. Velcro Product Design with Biomimicry Approaches. *IOP Conf. Ser. Earth Environ. Sci.* **2022**, *1111*, 012057. [CrossRef]
54. Visser, W.; Benyus, J.M. Biomimicry. In *The Top 50 Sustainability Books*; Visser, W., Ed.; Routledge: London, UK, 2013; pp. 104–107. [CrossRef]

biomimetics

MDPI

Article

Strategies for the Design and Construction of Nature-Inspired & Living Laboratory (NILL 1.0)™ Buildings

Mariam AlAli [1,*], Salwa Beheiry [2] and Serter Atabay [2]

[1] Engineering Systems Management, American University of Sharjah, Sharjah 26666, United Arab Emirates
[2] Department of Civil Engineering, American University of Sharjah, Sharjah 26666, United Arab Emirates
* Correspondence: g00093066@aus.edu

Abstract: This article explores the growing prominence of nature-inspired design philosophies in the context of sustainability and human well-being within the built environment and focuses on their application within laboratory buildings. Biomimicry and biophilic design are highlighted as key nature-inspired design approaches, with biomimicry drawing inspiration from nature for innovations and biophilic design promoting human health through enhancing the connection with the surrounding natural elements. This paper further discusses living building strategy as an emerging method for creating dynamic and adaptable spaces by prioritizing user experience through co-creation and focusing on sustainable and regenerative structures. The potential of integrating these approaches is emphasized using laboratory buildings as an example, with nature-inspired and living laboratories serving as models for future built environments that promote both environmental responsibility and a positive human experience. Accordingly, this work aims to investigate the design and construction of laboratory buildings based on nature-inspired design strategies and the living building concept. Moreover, the paper discusses the application of biomimicry and living building concepts within laboratory buildings as a novel contribution to the body of knowledge, and concludes by proposing the Nature-inspired & Living Laboratory (NILL 1.0)™ Building Assessment index to serve as a guideline for the design and construction of laboratory buildings using nature as an inspiration and the analogy of human body systems.

Keywords: nature-inspired; living; laboratory; biophilic; construction; strategies; buildings; biomimicry; green; sustainable

1. Introduction

Nature-inspired design is becoming more popular in a world where sustainability and wellbeing are increasingly prioritized. Biomimicry and biophilic design are two methods that fall under the umbrella of nature-inspired design [1]. By directly imitating natural forms, processes, and ecosystems, biomimicry attempts to address problems facing humanity and produces innovations in construction methods, building materials, and energy systems [2,3]. Additionally, biophilic design, much like green and sustainable design that emerged in the 1970s, aims to improve human comfort, productivity, and health by bringing natural elements into the built environment [4]. These design philosophies offer numerous advantages by utilizing the power of nature, encouraging environmental responsibility as well as an enhanced human experience in the built environment [5].

Furthermore, biophilic design has been identified by Kellert and Wilson in 1993 as "the deliberate attempt to translate an understanding of the inherent human affinity to affiliate with natural systems and processes", known as biophilia [6]. This definition was further developed by Kellert in 2011, who emphasized that "the positive experience of natural systems and processes in our buildings and constructed landscapes remains critical to human performance and well-being" [7]. On the other hand, biomimicry, a promising research field providing possible nature-inspired solutions for design problems, has been coined as a term by Janine Benyus in 1997 through merging the Greek words "bios", meaning "life", and "mimesis", meaning imitation [8]. Over the span of 3.8 billion years, natural processes have developed technologies comparable to, or surpassing, those created by humans, using sustainable and efficient methods [9]. Therefore, biomimicry, an interdisciplinary scientific field, holds promise for delivering sustainable solutions through collaboration among biologists, physicists, chemists, engineers, and architects [10].

Moreover, creative methods for designing dynamic, flexible spaces are emerging as the built environment changes to meet the demands of the 21st century. Two examples of these developments are living laboratories and living building techniques. By encouraging co-creation and in-the-moment testing of design features, living labs put the user experience first [11]. With the help of this user-centered approach, researchers, users, and designers can work together to create inventive and useful spaces. On the other hand, living building strategies concentrate on developing structures that are sustainable, regenerative, and functional [12]. These tactics seek to reduce their negative effects on the environment, maximize resource efficiency, and even improve the local ecosystem [13]. Through the integration of these approaches, living laboratories can be designed to serve as models for innovative design as well as showcases for sustainable and user-centric built environments.

Accordingly, this paper aims to explore the approaches/practices employed in the design and construction of laboratory buildings based on the concepts of Living Lab and Living Building and nature-inspired design through biophilic design and biomimicry. The aim is achieved through setting multiple objectives, which include the review of peer-reviewed journals published through the international database SCOPUS from 2000 until 2024, highlighting the major research contributions within the area of the topic of interest, and, finally, providing a set of strategies to design and construct Nature-inspired & Living Laboratory (NILL) Buildings through the Nature-inspired Living Laboratory (NILL 1.0)^{TM} Assessment Tool.

2. Materials and Methods

2.1. Data Collection

This article explored the application of nature-inspired strategies, including biomimicry and biophilic design, as well as the living lab approach and living building concept within the context of laboratory buildings. The data were primarily collected from a large assortment of scholarly articles acquired from SCOPUS database and Google Scholar. The examined publications covered the time period between 2000 and 2024 and used the following code for SCOPUS search: ("Biomimicry" OR "Living Building Challenge" OR "LBC" OR "biophilia" OR "biophilic design" OR "sustainable construction") AND ("lab*" AND "building"). The reviewed publications were selected based on the following criteria: range of publication between years 2000 and 2024, relevance of title and abstract to the addressed topic, and the content being relevant to architecture, construction, and engineering aspects of laboratory buildings, while ensuring that the publication is in the final stage and written in the English language. Moreover, the exclusion criterion strictly omitted any publication not belonging to a SCOPUS-indexed scholar. However, due to the lack of scholarly articles discussing the application of the previously mentioned concepts within laboratory buildings, additional resources, such as conference papers and official websites, were reviewed through search queries in Google Scholar.

2.2. Data Processing and Text Mining

Titles and abstracts were scanned for relevance to the selected topics of biomimicry, biophilic design, living lab approach, and living building within laboratory buildings in order to facilitate text mining and the review process of the selected papers. Papers were then categorized according to their respective fields. Using Microsoft Excel (Microsoft Office Home and Student 2019 version 2304, Redmond, WA, USA), themes and key ideas were taken from the papers and entered into tables. The main themes that were addressed while examining the articles were applications or case studies concerning biomimicry, biophilic design, living laboratory, living building, and sustainable building. Those themes were used to categorize the articles that are further discussed and summarized in the Section 3. Where data from the articles that appeared in SCOPUS were not enough to proceed with the next phase of the methodology, additional resources such as Google Scholar were used to capture relevant articles using the same keywords mentioned in Section 2.1.

2.3. Index Creation

The reviewed literature was assessed for the extraction of relevant features/indicators that can be compiled within an index to assess laboratory buildings by combining key features inspired by biomimicry, biophilic design, living building, and sustainable features. The indicators were then categorized into relevant groups/constructs under one common theme that collectively combines similar aspects. The main approach of designing the index was based on the analogy of the human body, which consists of different systems performing specific functions. Accordingly, each system was applied within the context of a laboratory building based on its primary function and how it can be applied or utilized for inspiration to create a similar system in a laboratory building. Therefore, each system is considered as a separate category or "construct" within the index. Furthermore, the indicators of each category/system were chosen based on their suitability to achieve the overall purpose of the category/system. Overall, this conceptual guideline is proposed as a first stage of multi-stage research that will include a future validation for the entire index and all its attributes through subject matter experts from the industry and subjected to further quantitative analysis and possible future publication.

3. Results

3.1. SCOPUS Search

The results of the initial search through SCOPUS database using the keywords mentioned in the methodology showed a low number of publications during the period of 2000 until 2024, ranging between 1 and 13 publications per year, as shown in Figure 1. However, the topic under investigation is still relatively new and not thoroughly investigated. Overall, the accumulative total of the SCOPUS-indexed articles was 89 articles, of which a selection is examined and further discussed in this section.

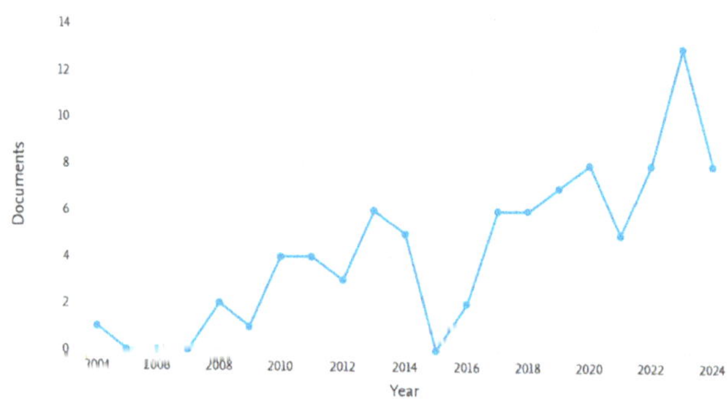

Figure 1. Number of annual publications around the topic of interest during the period of 2000–2024.

Further analysis into the publications found on SCOPUS suggests that the topic under investigation has received a relatively equal percentage of interest amongst researchers in the fields of Energy and Social Sciences, with the highest interest being in Engineering, followed by Environmental Sciences, as shown in Figure 2, which further supports the notion that there is a need to consolidate the acquired knowledge in all three fields in a comprehensive approach to effectively address the topic of interest.

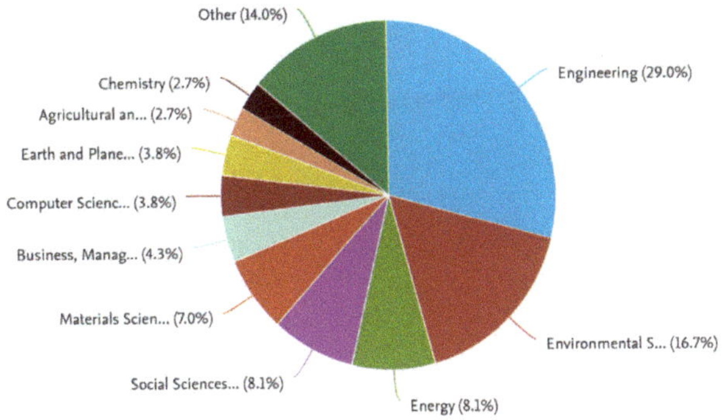

Figure 2. Research interest from different fields concerning the topic of interest during the period of 2000–2024.

3.2. VOS Viewer Analysis

Additional analysis into the trending keywords using VOS Viewer relied on the criterion of selecting keywords that appeared two times or more within the 89 selected articles. As a result, 21 publications showed several links between multiple terminologies emerging from the topic of interest, such as biophilia, living building challenge, architectural design, sustainability, sustainable construction, living buildings, energy efficiency, thermal comfort, and biophilic design, amongst others, as shown in Figure 3. Accordingly, this analysis further supports the notion that the consolidation of biomimicry, biophilic design, living building, and sustainable features has started to capture the interest of researchers, with potential for further enhancements and developments in the future.

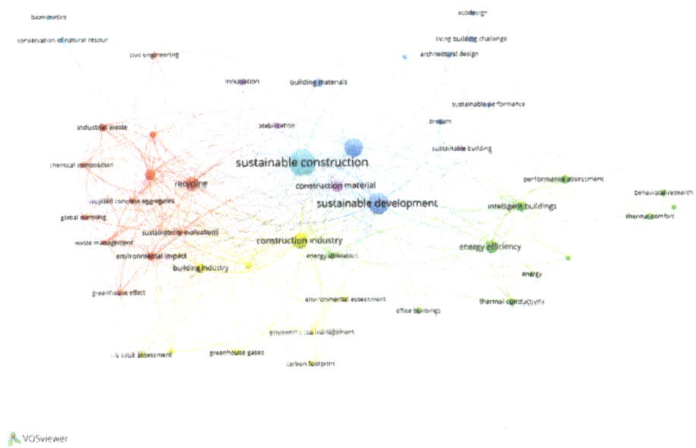

Figure 3. Author keywords mentioned at least two times within the publications during the period of 2000–2024.

3.3. Summary of Articles Findings

Further examination of the extracted 21 articles through evaluating the abstract content, purpose, methodology, and results reduced the number of articles to 12 articles, with their main findings summarized in Table 1.

Table 1. Summary of major findings from the final 12 articles acquired through SCOPUS.

Title	Authors	Year	Main Findings	Theme
Biophilia in the workplace: A pilot project for a living wall using an interactive parametric design approach [14]	Assem A.; Hassan D.K.	2024	This study explored how biophilic design, specifically integrating a living wall, can enhance workplace aesthetics and well-being. It employed a parametric design approach to optimize the integration process, focusing on creating varied green wall forms inspired by natural concepts. Interactive elements were incorporated to enhance user experience and perception of the living wall's dynamics, aiming to create a versatile ambiance conducive to diverse workplace activities. Results indicated successful implementation of these design strategies, showcasing improved user engagement and functionality within workplace environments.	Biophilic Design and Living Building
Conceptualization of Biomimicry in Engineering Context among Undergraduate and High School Students: An International Interdisciplinary Exploration [15]	Yeter I.H.; Tan V.S.Q.; Le Ferrand H.	2023	The study found that engineering students grasped the bottom-up approach of biomimetics more readily, using traditional engineering tools to apply biological knowledge for engineering solutions, while perceiving the top-down approach, which identifies technical problems and applies natural solutions, as vaguer. It suggests that combining both approaches in teaching biomimicry, along with hands-on learning, could effectively enhance student comprehension of these concepts.	Biomimicry
Short-term effects of natural view and daylight from windows on thermal perception, health, and energy-saving potential [16]	Jiang Y.; Li N.; Yongga A.; Yan W.	2022	Visual windows enhanced thermal comfort, potentially reducing HVAC energy use; physiological measures were more sensitive than subjective questionnaires in assessing their impact on occupant health, indicating positive effects on well-being by alleviating symptoms of sick building syndrome and reducing stress and fatigue indicators.	Sustainable Building and Biophilic Design
Clay 3D printing as a bio-design research tool: development of photosynthetic living building components [17]	Crawford A.; In-na P.; Caldwell G.; Armstrong R.; Bridgens B.	2022	The study utilized digital fabrication to embed living microalgae in ceramic building components, examining how design factors like geometry and firing temperature affect algae growth. It highlighted the importance of managing evaporation and moisture levels for optimal performance and proposed digital manufacturing as a method to develop viable, integrated systems for living building applications.	Living Building
Photosynthetic textile biocomposites: Using laboratory testing and digital fabrication to develop flexible living building materials [18]	Stefanova A.; In-Na P.; Caldwell G.S.; Bridgens B.; Armstrong R.	2021	The study explored the development of 3D-printed biocomposites containing the microalgae Chlorella vulgaris, which can be integrated into building materials to help sequester CO_2, demonstrating their effectiveness in supporting algae growth despite occasional challenges in cell distribution and fluctuations. Kappa-carrageenan, full-strength BG11 nutrient medium, and Auro Clay Paint were promising when used with cotton and polyester textiles, despite occasional cell distribution challenges and fluctuations.	Living Building
The impact of a view from a window on thermal comfort, emotion, and cognitive performance [19]	Ko W.H.; Schiavon S.; Zhang H.; Graham L.T.; Brager G.; Mauss I.; Lin Y.-W.	2020	The study discovered that providing office occupants with a window view resulted in slight yet noteworthy enhancements in their thermal comfort, positive emotions, and specific cognitive functions such as working memory and concentration, in comparison to those in windowless environments. Having a window could potentially contribute to energy savings as it makes occupants more tolerant of minor thermal comfort variations. Windows were also found to boost occupants' psychological well-being by amplifying positive emotions and minimizing negative ones.	Biophilic Design

Table 1. *Cont.*

Title	Authors	Year	Main Findings	Theme
Building on the inherent strengths of green space environments: Promoting trust, democracy, and resilience among ethnically diverse groups [20]	Hoffman A.	2020	This quasi-experimental study explored how participating in community service activities within green space environments, such as community gardens and urban forestry programs, influenced individuals' perceptions of community service-learning programs and democratic processes associated with green space development. Sixteen volunteers shared their subjective experiences, noting increased appreciation for living things, a stronger connection with plants and animals, and a heightened sense of belonging to nature. Interviews highlighted how exposure to these environments shaped participants' views on the importance of nature in urban settings and their personal sense of connection to both community and the natural world.	Biophilia
Leed gold but not equal: Two case study buildings [21]	Baja F.D.F.; Bajracharya S.; Freeman M.A.; Gray A.J.; Haglund B.T.; Kuipers H.R.; Opatola O.R.	2019	The study aimed to resolve issues of glare, thermal discomfort, and excessive brightness in the Education building's west- and south-facing study spaces due to highly reflective materials. Researchers proposed an integrated shading solution with vertical fins and elongated light shelves to mitigate direct sunlight and enhance visual and thermal comfort. Additionally, the paper compared two LEED Gold-rated buildings at the University of Idaho, highlighting varying ecological performance despite similar certifications, and offered recommendations to address comfort issues in the Education building.	Sustainable Building and Biophilic Design
The R.W. Kern center as a living laboratory: Connecting campus sustainability goals with the educational mission at Hampshire college, Amherst, MA [22]	Cianfrani C.M.; Hews S.; Tor J.; Jewhurst J.J.; Shillington C.; Raymond M.	2018	The R.W. Kern Center at Hampshire College exemplified how sustainable design can educate future sustainability leaders, transforming the campus to prioritize pedestrians over cars and showcasing features like optimized building orientation, insulation, natural ventilation, and educational displays as a living laboratory for students.	Sustainable Building and Living Laboratory
Modelling to drive design: Honing the SU + RE house through performance simulations [23]	May E.	2018	Digital simulation technologies transformed architectural design by enabling direct study and manipulation of energy, water, air, heat, and sound flows impacting building occupants. The SU + RE House exemplified this advancement, showcasing how integration of data and environmental analysis techniques could create genuinely sustainable and resilient buildings. Led by Ed May from Stevens Institute of Technology and BLDGtyp, the project marked a significant shift towards designing structures that prioritize environmental performance and occupant comfort	Biophilic Design

Focusing on the importance of biomimicry education in sustainability [15], Yeter et al. conducted workshops in Singapore to explore how students from two distinct groups—local high school students and undergraduate engineering students from the United States—conceptualized biomimicry approaches. The study concluded that students struggled with the top-down method, which involves identifying human problems and seeking solutions from nature. This highlights the need for curriculum development that strengthens students' ability to identify the unique principles that make natural objects function effectively and fosters interdisciplinary knowledge for a more comprehensive grasp of biomimicry.

As for the application of biophilic design, the SCOPUS search revealed a study by Jiang et al. that investigated the impact of windows on occupants in buildings [16], finding that having a window led to increased comfort and tolerance of thermal conditions (potentially saving energy), reduced stress and fatigue (based on physiological measurements), and offered energy-saving potential through daylight, though it did not significantly affect perceived lighting, as users were satisfied with both levels of natural and artificial lighting. Overall, integrating windows as a biophilic design feature to enhance user experience and comfort level was evaluated and showed potential for further exploration and consideration in building design.

Concerning the use of living building materials in the construction of buildings, a study by Crawford et al. explored the potential of microalgae, a type of microscopic organism, as a sustainable building material [17]. Researchers investigated the use of 3D-printed ceramic structures embedded with microalgae to create "living" building components. The study evaluated how different designs affected the survival and growth of the algae. The findings suggest that integrating these micro-ecologies into buildings could contribute to sustainability efforts and mitigate anthropogenic carbon dioxide emissions.

The use of living building components, such as living walls, has also been discussed by Assem and Hassan [14]. This study investigated the use of living walls, plant installations integrated into walls, to improve employee well-being and combat sick building syndrome in workplaces. Researchers employed a parametric design approach to optimize the design of these living walls, considering factors like aesthetics, user interaction, and functionality within the workspace. The results, though not explicitly detailed, likely demonstrate the effectiveness of this approach in creating visually appealing living walls that enhance user experience and create unique ambiances suitable for various workplace activities.

The application of the Living Building Challenge (LBC) has also been discussed in a previous study by Cianfrani et al. [22]. This study examined the R.W. Kern Center, Hampshire College's first new building in 40 years, designed with a focus on forward-thinking sustainability. The multi-purpose facility aims to be self-sufficient in energy generation, water management, and waste processing. It prioritizes nontoxic materials, local sourcing, and biophilic design principles to promote human well-being and natural beauty. Beyond functionality, the building serves as a gateway to the campus, attracting prospective students and fostering a sense of community. The R.W. Kern Center achieved Living Building Challenge certification in 2018, demonstrating its commitment to a holistic approach to sustainability.

Energy consumption patterns was another theme discussed within the reported studies. One study by Yeter et al. investigated the effectiveness of three energy-saving initiatives in a university office/laboratory building designed with sustainable features, such as natural ventilation in an open-plan office area and biomass to meet heating requirements, at the University of Cambridge [24]. The building employed voluntary design frameworks (BREEAM) and a bespoke post-construction strategy (Cambridge Work Plan) alongside mandatory EU reporting requirements. Eventually, the study revealed a significant discrepancy between the building's actual energy consumption (140% higher than estimated needs) and its design projections. Furthermore, the three initiatives implemented fell short in actively reducing this energy performance gap. As such, the study recommended alternative approaches that prioritize monitoring operational performance and ensure realistic energy estimations during the design stages, including building energy management strategies focused on operational performance, such as the Living Building Challenge [25]. As such, LBC can drive buildings to achieve net-zero energy consumption using renewable sources, thus integrating biomimicry to mimic nature's efficiency and resilience. It can also promote biophilic design by enhancing occupant well-being through natural elements like light and vegetation, while reducing energy needs. Thus, LBC can support harmonizing environmental responsibility with human comfort and health. Other suggested tools for utilization included the Chartered Institution of Building Services Engineers' Technical Memorandum 54, a technical guideline that aids building professionals during building design to create more accurate energy models by transforming low-energy designs into buildings that meet designated energy targets and offering precise instructions for thoroughly and accurately assessing operational energy consumption during the design phase [26].

Overall, the result of analyzing the SCOPUS-indexed articles concerning the topic of interest revealed a general lack of application of biomimicry, biophilic design, and living building strategies within laboratory buildings. Further suggestions are made by the authors on the applicability of those concepts and detailed in the Discussion section.

4. Discussion

4.1. SCOPUS-Indexed Publications

Building on the findings in the previously mentioned study of biomimicry application in education [15], the study's outcomes can be applied to bridge this gap by emphasizing the importance of understanding nature's "unique principles". This could involve studying phenomena like termite mounds for inspiration in passive cooling systems or mimicking spider silk's strength-to-weight ratio for lightweight structures. Additionally, fostering interdisciplinary knowledge by combining biology, engineering, and architecture can lead to broader biomimicry inspiration. By addressing these knowledge gaps, professionals can be equipped with the skills to translate biomimicry into practical applications for laboratories. This can ultimately lead to the design of more sustainable, adaptable, and user-centric lab environments that promote scientific progress.

Regarding the application of biophilic design in the built environment that was previously discussed [16], the reported findings support the notion of using biophilic design principles, which incorporate elements of nature like natural light and views to enhance occupant well-being and sustainability in laboratory buildings. To explain, incorporating windows with natural views can create a more comfortable and healthier workplace for lab personnel, as natural light and views can potentially lead to increased comfort in cooler temperatures and reducing energy consumption for cooling. Additionally, biophilic design principles promote a connection with nature, which has been shown to decrease stress and fatigue in occupants, leading to increased productivity and better job performance. Thus, windows can serve as a valuable design feature to improve the access to nature through the building envelope, while other alternatives may also provide a similar experience when windows are not present in the building, such as a roof glass ceiling. By integrating these elements, architects and designers can create lab environments that prioritize both human health and sustainability.

Concerning the use of living building materials in the construction of buildings using microalgae, as discussed previously [17], this holds promise for laboratory facilities. To illustrate, laboratories often require specific environmental conditions, and these living materials could potentially help regulate temperature or air quality within the space. Additionally, the controlled environment of a laboratory would be ideal for studying the interaction between microalgae and different materials. Researchers could leverage this setting to further explore the potential benefits and limitations of these living materials for broader use in architecture.

The concept of living walls in building setups, as previously discussed [14], has potential for application in laboratory facilities. To illustrate, laboratories can often be sterile and lack natural elements. Therefore, living walls could introduce a connection to nature, potentially improving employee well-being and reducing stress, which can be crucial for scientific research. Additionally, the study's focus on user interaction with the living wall could be particularly relevant in a lab setting. Interactive elements could provide employees with opportunities to engage with nature during breaks or stressful moments, further promoting well-being.

Moreover, the LBC certification may be assessed for suitability to apply in laboratory buildings, taking buildings such as R.W. Kern Center as an example to follow [22]. To explain, the sustainable design principles employed at the R.W. Kern Center hold promise for application in laboratory buildings, since laboratories often require significant energy and water resources. Therefore, implementing similar self-sufficiency measures could minimize environmental impact. Additionally, focusing on nontoxic materials and biophilic design could promote the health and well-being of laboratory personnel, who are considered a crucial factor in scientific research. By adopting these principles, laboratories can contribute to a more sustainable future while creating healthy and inspiring work environments for scientific discovery.

Additionally, the findings of the study on energy consumption at the University of Cambridge [24] hold significant value for laboratory buildings, which are known for high energy consumption. The identified shortcomings of focusing solely on design-stage initiatives highlight the need for a multi-pronged approach. Laboratories can benefit from incorporating operational performance monitoring into their energy management strategies. Additionally, collaborating with design professionals who utilize tools like the Chartered Institution of Building Services Engineers' Technical Memorandum 54 can ensure the creation of realistic energy models at the design stage. By implementing these recommendations, laboratory buildings can bridge the energy performance gap and achieve true sustainability goals.

Overall, the potential of applying nature-inspired design strategies and the living building concept within laboratory buildings is still a relatively new field, with significant potential for further work and advancements. The authors discuss some strategies detailed in the following subsections.

4.2. *Commandments of Biomimicry and Laboratory Buildings*

Pioneering advocate for biomimicry, Janine Benyus, established the core principles, often referred to as "commandments", that provide a framework for applying biomimicry to problem solving across various disciplines [8]. These principles hold significant influence within the built environment, shaping sustainable and innovative design approaches. By encouraging the study of nature's models, ecosystems, and processes as a source of inspiration, biomimicry translates into real-world applications such as mimicking natural ventilation systems for passive cooling or emulating spider silk for lightweight building materials. Furthermore, these principles advocate for using nature as a benchmark for measuring the environmental impact and as a mentor for fostering adaptable designs. Buildings designed with biomimicry in mind can minimize their environmental footprint by considering energy use and resource consumption throughout their lifecycle. Additionally, these principles can inspire the creation of structures that evolve over time, similar to how organisms adapt to their environment. Finally, there is an emphasis on the importance of respecting the interconnectedness of nature. This translates to buildings designed with locally sourced materials and a consideration for the impact on the surrounding ecosystem. By adhering to these principles, architects and engineers can create buildings that are more sustainable, resilient, and user-centric, ultimately promoting a built environment that thrives in harmony with the natural world.

As the current literature does not discuss the application of Benyus's commandments of biomimicry to laboratory buildings, the authors suggest the following strategies (Table 2), where each commandment is applied within the context of a laboratory building to serve as a general guidance for designers, contractors, laboratory managers, or laboratory building users/occupants.

Table 2. Strategies for the application of Benyus's commandments of biomimicry to laboratory buildings.

Commandments of Biomimicry	Application to Laboratory Buildings
Use Waste as a Resource:	Design laboratories that incorporate waste-to-resource systems, utilizing lab waste for energy production or recycling materials within the facility.
Diversify and Co-operate to Fully Use the Habitat:	Create laboratory environments that mimic the diversity and co-operation found in natural ecosystems. Design spaces that accommodate various research activities and encourage collaboration among scientists.
Gather and Use Energy Efficiently:	Implement energy-efficient technologies and systems in laboratories, such as renewable energy sources, smart energy management, and energy recovery systems.
Optimize Rather Than Maximize:	Focus on optimizing laboratory processes and spaces, avoiding excessive resource use and square footage. Prioritize efficiency and functionality over unnecessary expansion.
Use Materials Sparingly:	Design laboratories with a focus on minimal material use, incorporating sustainable and low-impact materials. Prioritize durability and recyclability in material selection.
Don't Foul Their Nests:	Ensure that laboratory activities, waste disposal, and emissions are managed in an environmentally responsible manner, minimizing negative impacts on the surrounding ecosystem.
Don't Draw Down Resources:	Design laboratories with a commitment to sustainable resource management, avoiding the depletion of natural resources and promoting circular economy practices.
Remain in Balance with the Biosphere:	Align laboratory design with the local ecosystem, considering factors like water usage, biodiversity, and ecological balance. Implement landscaping that supports local flora and fauna.
Run on Information:	Incorporate smart technologies and information systems in laboratories for efficient data collection, analysis, and communication. Embrace data-driven decision making for sustainability.
Shop Locally:	Source laboratory materials and equipment locally whenever possible to reduce the carbon footprint associated with transportation. Support local businesses and contribute to the regional economy.

4.3. Biomimicry Life Principles and Laboratory Buildings

The organization Biomimicry 3.8 acts as a champion for the field of biomimicry, drawing inspiration from the vast library of biological strategies gleaned from Earth's 3.8-billion-year evolutionary history [27]. This knowledge base informs the development of the Biomimicry Life Principles, which constitute a core element of Biomimicry 3.8's methodology. These principles function as a blueprint for sustainable and efficient design, guiding designers and engineers to emulate nature's time-tested patterns and processes. Furthermore, Biomimicry 3.8 offers tools such as the DesignLens: Life's Principles to facilitate the application of these principles across a wide range of design challenges [28]. In essence, the Biomimicry Life Principles serve as the foundational pillar upon which Biomimicry 3.8 constructs its approach. By harnessing nature's wisdom through these principles, Biomimicry 3.8 fosters the development of ecologically sustainable and innovative solutions. This ensures that human creations are well adapted to the Earth's ecosystem and contribute positively to it.

As the current literature shows lack of publications discussing the application of Biomimicry Life Principles in laboratory buildings, the authors present several strategies (Table 3) where each principle is translated into a practical application during the design, construction, and operation of a laboratory facility.

Table 3. Strategies for the application of biomimicry life principles to laboratory buildings.

Biomimicry Life Principles	Application to Laboratory Buildings
Evolve To Survive	Design labs to be adaptable and flexible to accommodate changing research needs, technologies, and environmental conditions. Utilize modular and adaptable infrastructure to facilitate future expansion and modifications.
Replicate Strategies that Work	Emulate proven natural designs and processes to enhance lab functionality and efficiency. Implement biophilic design elements to improve occupant well-being and productivity.
Integrate the Unexpected	Incorporate unexpected elements and unconventional approaches to foster innovation and creativity within the lab environment. Encourage cross-disciplinary collaboration and embrace serendipitous discoveries.
Reshuffle Information	Analyze and utilize data effectively to optimize lab operations, resource management, and energy consumption. Implement smart sensors and control systems to gather real-time data and make informed decisions.
Adapt To Changing Conditions	Design labs to respond to changing environmental conditions, such as fluctuating temperatures, sunlight intensity, and occupancy levels. Utilize passive design strategies to minimize energy consumption and ensure thermal comfort.
Incorporate Diversity	Create diverse lab spaces that cater to different research needs and encourage collaboration among scientists. Foster a culture of inclusivity and diversity of thought to maximize the potential of the lab environment.
Maintain Integrity through Self-Renewal	Design labs with self-healing and self-repairing mechanisms to minimize maintenance requirements and extend the lifespan of the building. Utilize renewable materials and sustainable practices to ensure long-term functionality.
Embody Resilience	Design labs to withstand natural hazards, climate change, and other disruptions. Implement robust structural systems, redundant power sources, and disaster preparedness plans to ensure continuous operation.
Be Locally Attuned and Responsive	Design labs to harmonize with the local climate, ecology, and cultural context. Utilize locally sourced materials, adapt to site conditions, and respect the surrounding environment.
Leverage Cyclic Processes	Integrate cyclical processes, such as rainwater harvesting, greywater recycling, and natural ventilation, to reduce resource consumption and minimize environmental impact.
Use Readily Available Materials and Energy	Utilize sustainable and locally sourced materials with low environmental impact. Prioritize renewable energy sources and optimize energy efficiency throughout the building.
Use Feedback Loops	Implement feedback loops to continuously monitor and improve lab performance. Utilize data analytics to identify areas for optimization and implement corrective measures.
Cultivate Cooperative Relationships	Encourage collaboration and knowledge sharing among lab users, researchers, and the surrounding community. Foster partnerships with local universities, research institutions, and businesses.
Integrate Development with Growth	Design labs that can accommodate future growth and expansion without compromising sustainability or functionality. Utilize modular and adaptable infrastructure to facilitate seamless expansion.
Self-Organize	Design labs with self-organizing capabilities, such as intelligent lighting systems and automated climate control, to optimize resource utilization and occupant comfort.
Build from the Bottom Up	Adopt a bottom-up approach to lab design, involving users, researchers, and community members in the planning and decision-making process.
Combine Modular and Nested Components	Utilize modular and nested components to create flexible and adaptable lab spaces that can be easily reconfigured and expanded as needs evolve.
Be Resource Efficient (Material and Energy)	Minimize material consumption and energy use throughout the design, construction, and operation of the lab.Implement sustainable practices and prioritize renewable energy sources.
Use Low Energy Processes	Utilize low-energy processes and technologies to minimize the environmental impact of lab operations. Employ energy-efficient equipment, appliances, and HVAC systems.
Use Multi-Functional Design	Design lab spaces with multi-functional capabilities to reduce the need for additional infrastructure and maximize resource utilization.
Recycle All Materials	Implement comprehensive recycling and waste management strategies to divert materials from landfills and promote circularity. Utilize waste streams as energy sources whenever possible.
Fit Form to Function	Design lab spaces that prioritize functionality and efficiency over unnecessary aesthetics. Follow biophilic design principles to create a harmonious and stimulating environment.
Use Life-friendly Chemistry	Implement green chemistry principles to minimize the use of hazardous chemicals and promote sustainable alternatives. Utilize nontoxic and biodegradable materials whenever possible.

Table 3. *Cont.*

Biomimicry Life Principles	Application to Laboratory Buildings
Break Down Products into Benign Constituents	Design products and processes that break down into benign constituents at the end of their lifecycle, minimizing environmental impact.
Build Selectively with a Small Subset of Elements	Utilize a limited number of well-understood and sustainable materials to reduce complexity and facilitate recycling and reuse.
Do Chemistry in Water	Implement water-based chemistry whenever possible to minimize the use of harmful solvents and reduce environmental impact. Utilize water as a reaction medium and solvent.

4.4. Living Building Challenge (LBC) and Laboratory Buildings

The Living Building Challenge Imperatives offer a thorough framework for creating living laboratories inspired by nature that aim for a regenerative impact rather than just sustainability [29]. These Seven Imperatives [30], which are divided into seven interconnected petals (Place, Water, Energy, Health & Happiness, Materials, Equity, and Beauty), have a major impact on all phases of a laboratory's lifecycle.

To further illustrate this possible application, the "Place" petal has "Imperative 01: Site Ecology" that necessitates a detailed examination of the surrounding environment, which may have an impact on the design stage by prescribing a building footprint that reduces disturbance and blends in with the natural features already in place. Then, using biomimicry features (motivated by the "Beauty" petal) like naturally occurring ventilation systems modelled after animal respiratory systems can be considered, which lessens the need for mechanical equipment and advances the "Energy" petal. Moreover, to lessen its impact on the environment, construction (the "Materials" petal) may use recycled or readily renewable materials that can be found locally (Imperative 05: Biobased & Recycled Content). Rainwater harvesting systems in line with the "Water" petal's "Imperative 09: Water Capture & Treatment" can considerably lessen dependency on municipal supplies during operation and management. Meanwhile, occupant well-being tactics from the "Health & Happiness" petal, such as plenty of natural light and better indoor air quality, can boost user satisfaction and productivity. These are just a few instances that demonstrate how Living Building Challenge Imperatives, when carefully implemented, can turn labs into functional, naturally inspired living ecosystems that also improve the surrounding area and the health of their occupants. As such, the authors propose the following tactics (Table 4) where each imperative of the LBC's petals can be applied in a laboratory building, along with the predicted implications.

Table 4. Strategies for the application of Living Building Challenge (LBC) to laboratory buildings.

Living Building Challenge		Application to Laboratories	Impact on Laboratories
Petal	Imperative		
Place	Limits to Growth Place	Design laboratories that respect ecological limits, considering local ecosystems, biodiversity, and resource availability.	Ensures the laboratory's impact aligns with the natural capacity of the surrounding environment.
	Urban Agriculture	Incorporate green spaces and possibly rooftop gardens to promote urban agriculture within the laboratory setting.	Enhances the laboratory's connection to nature, provides greenery for researchers, and contributes to local food production.
	Habitat Exchange	Implement measures to enhance and protect local ecosystems, possibly through partnerships with conservation organizations.	Demonstrates a commitment to preserving and enhancing the natural habitats surrounding the laboratory.
	Human-Powered Living	Promote alternative transportation methods, such as cycling or walking, and design spaces that encourage physical activity.	Aligns with a nature-inspired approach by encouraging a healthy and active lifestyle for laboratory occupants.
Water	Net Positive Water	Design water-efficient laboratories with water capture and reuse systems, ensuring a positive impact on the local water balance.	Demonstrates responsibility in water usage, mirroring natural systems' efficiency.

Table 4. *Cont.*

Living Building Challenge		Application to Laboratories	Impact on Laboratories
Petal	**Imperative**		
Energy	Net Positive Energy	Implement renewable energy sources and energy-efficient design to achieve net-positive energy consumption.	Aligns with the sustainability aspect of a nature-inspired approach by reducing energy demand.
Health and Happiness	Civilized Environment	Design laboratories with a focus on promoting social well-being, comfort, and a sense of community.	Creates a workspace that aligns with the harmonious and civilized aspects of nature.
	Healthy Interior Environment	Prioritize air quality, lighting, and acoustics to create a healthy and comfortable indoor environment.	Ensures researchers work in spaces that support their well-being, similar to the health-promoting aspects of nature.
	Biophilic Environment	Integrate nature-inspired design elements, such as natural light, greenery, and biomimicry, to create a biophilic laboratory.	Enhances the connection between researchers and the natural world, fostering a positive and inspired work environment.
Materials	Red List	Avoid the use of materials on the Red List, prioritizing healthier and environmentally responsible choices.	Aligns with the nature-inspired principle of using materials in harmony with the environment.
	Embodied Carbon Footprint	Minimize the embodied carbon footprint of construction materials, choosing low-impact options.	Reflects a commitment to reducing the overall carbon impact, aligning with sustainable practices found in nature.
	Responsible Industry	Source materials from responsible and sustainable suppliers and promote ethical practices within the laboratory.	Demonstrates a commitment to responsible and sustainable industry practices.
	Living Economy Sourcing	Support local economies and choose materials and services that contribute to a living economy.	Aligns with the nature-inspired principle of interconnectedness and community support.
	Net Positive Waste	Minimize waste generation and implement recycling and composting systems to achieve net-positive waste.	Mirrors the efficiency and waste reduction found in natural ecosystems.
Equity	Human Scale and Humane Places	Design laboratories with a human-centric approach, focusing on comfort, accessibility, and a sense of place.	Aligns with the nature-inspired principle of creating spaces that resonate with human well-being.
	Universal Access to Nature and Place	Ensure that all laboratory occupants have access to natural elements, whether through views, green spaces, or biophilic design.	Fosters inclusivity and promotes a connection to nature for everyone in the laboratory.
	Equitable Investment	Prioritize equitable investment in laboratory facilities, ensuring fair distribution of resources and benefits.	Reflects a commitment to fairness and equity, similar to the balanced relationships in nature.
	Just Organization	Implement just and equitable policies within the laboratory organization, considering the well-being and fairness of all occupants.	Creates a work environment that aligns with the principles of justice found in nature.
Beauty	Beauty and Spirit	Design laboratories with aesthetic appeal, incorporating natural elements and inspiring spaces.	Aligns with the beauty and inspiration found in the natural world.
	Inspiration and Education	Design spaces that inspire creativity and provide educational opportunities for researchers and visitors.	Creates a laboratory environment that encourages learning and innovation, mirroring the inspiration found in nature.

4.5. Comparative Overview of Biomimicry and LBC in Laboratory Buildings

A significant alignment exists between the philosophies of biomimicry and the Living Building Challenge (LBC), both advocating for a built environment that fosters harmony with nature. Biomimicry draws inspiration from biological systems, encouraging the design of buildings that emulate nature's efficiency and resilience in areas like resource management and structural integrity. The LBC, on the other hand, establishes a rigorous framework for sustainable construction, pushing for buildings to function as self-sufficient ecosystems with minimal environmental impact. This synergy between biomimicry and the LBC presents intriguing possibilities for the future of architecture, particularly within the domain of laboratory buildings. By integrating biomimicry principles into LBC design strategies, laboratories have the potential to evolve beyond their role as purely functional research spaces, transforming into models of environmental responsibility. One can envision buildings that utilize natural ventilation strategies inspired by termite mounds, or self-cleaning facades mimicking the lotus leaf. These biomimetic elements, when coupled with the LBC's emphasis on renewable energy and water conservation, could lead to the creation of highly sustainable and functional laboratory buildings. This convergence of biomimicry and the LBC holds the potential to transform the built environment, fostering a future where human innovation coexists seamlessly with a healthy and thriving natural world. A detailed comparison between biomimicry and LBC is further discussed in Table 5.

Table 5. Comparison between the petals and imperatives of the Living Building Challenge (LBC) and the possible application within laboratory buildings.

Biomimicry Principle	Living Building Challenge Petal: Imperative	Similarities	Difference	Application to Laboratories
Use Waste as a Resource	Materials: Net Positive Waste	Both emphasize minimizing waste generation and finding valuable uses for waste materials.	Biomimicry emphasizes emulating nature's ability to transform waste into valuable resources, while the Living Building Challenge focuses on quantifying waste reduction and diversion.	Implement composting systems, establish waste segregation and recycling programs, and explore opportunities to reuse or repurpose lab materials.
Diversify and Cooperate to Fully Use the Habitat	Place: Ecology of Place	Both emphasize creating diverse and interconnected ecosystems that support a variety of life.	Biomimicry focuses on emulating nature's complex ecosystems, while the Living Building Challenge emphasizes restoring and enhancing biodiversity on the lab site.	Integrate native landscaping, create habitats for wildlife, and promote interactions between different species within the lab's environment.
Gather and Use Energy Efficiently	Energy: Net Positive Energy	Both emphasize reducing energy consumption and utilizing renewable energy sources.	Biomimicry focuses on emulating nature's ability to harness energy efficiently from natural sources, while the Living Building Challenge focuses on quantifying energy production and consumption.	Implement energy-efficient appliances and lighting, optimize HVAC systems, and utilize renewable energy sources such as solar panels or geothermal systems.
Optimize Rather Than Maximize	Materials: Responsible Materials	Both emphasize using materials efficiently and responsibly.	Biomimicry focuses on emulating nature's ability to achieve functionality with minimal material use, while the Living Building Challenge emphasizes using nontoxic, renewable, and locally sourced materials.	Select materials with low environmental impact, prioritize reusable and recyclable materials, and minimize material consumption throughout the lab's design and construction.
Use Materials Sparingly	Materials: Red List	Both emphasize minimizing the use of harmful materials and reducing the environmental impact of materials.	Biomimicry focuses on emulating nature's use of nontoxic and biodegradable materials, while the Living Building Challenge emphasizes avoiding red list materials and quantifying embodied carbon emissions.	Eliminate the use of hazardous materials, prioritize sustainable and bio-based alternatives, and consider the lifecycle impact of materials.
Don't Foul Their Nests	Health & Happiness: Healthy Interior Environment	Both emphasize creating healthy and nontoxic environments.	Biomimicry focuses on emulating nature's ability to create clean and healthy ecosystems, while the Living Building Challenge emphasizes minimizing exposure to harmful pollutants and optimizing indoor air quality.	Prioritize natural materials with low off-gassing potential, ensure adequate ventilation, and implement air filtration systems to maintain a healthy indoor environment.
Don't Draw Down Resources	Place: Limits to Growth Place	Both emphasize living within ecological limits and respecting natural resources.	Biomimicry focuses on emulating nature's ability to operate within resource constraints, while the Living Building Challenge emphasizes minimizing the lab's impact on local ecosystems and resources.	Employ water-efficient fixtures, implement rainwater harvesting systems, and reduce reliance on nonrenewable resources.
Remain in Balance with the Biosphere	Place: Habitat Exchange	Both emphasize maintaining a balance with the natural world.	Biomimicry focuses on emulating nature's ability to maintain equilibrium and resilience, while the Living Building Challenge emphasizes restoring and enhancing biodiversity.	Integrate biophilic design elements, create habitats for wildlife, and promote sustainable practices that minimize the lab's impact on the surrounding environment.

Table 5. *Cont.*

Biomimicry Principle	Living Building Challenge Petal: Imperative	Similarities	Difference	Application to Laboratories
Run on Information	Beauty: Education + Inspiration	Both emphasize the importance of knowledge and learning.	Biomimicry focuses on emulating nature's ability to gather and process information, while the Living Building Challenge emphasizes creating a culture of learning and innovation.	Foster a culture of knowledge sharing, encourage interdisciplinary collaboration, and incorporate educational elements into the lab's design and operation.
Shop Locally	Materials: Living Economy Sourcing	Both emphasize supporting local communities and economies.	Biomimicry focuses on emulating nature's interconnectedness and reliance on local resources, while the Living Building Challenge emphasizes sourcing materials and equipment from local businesses.	Prioritize locally sourced materials and equipment, support local businesses, and engage with the local community throughout the lab's development and operation.

4.6. *Nature-Inspired and Living Laboratory (NILL 1.0)™ Building Assessment Index*

The authors propose a novel index system for the assessment of laboratory buildings based on nature-inspired strategies and the living building concept. The index is based on a novel approach for enhancing laboratory buildings by leveraging the combined potential of biomimicry, biophilic design, and Living Building Challenge (LBC) principles, all evaluated through the NILL assessment index. The NILL index introduces the concept of a Living Laboratory—a platform that fosters curiosity, sparks innovation, and actively contributes to the advancement of nature-inspired solutions. By integrating biomimicry and biophilic design as core elements, the NILL index would create a framework for sustainable design strategies within the Living Laboratory. This dynamic environment would engage users, researchers, and the public through multisensory experiences, nurturing a collaborative learning culture. It is important to note that the current index system is a qualitative guideline, which will be further developed and quantitatively validated in future work.

The core concept of the NILL index lies in the synergy between biomimicry, biophilic design, and the interconnectedness of the human body's systems. The LBC framework, which emphasizes buildings functioning in harmony with nature, aligns beautifully with this biological principle. The NILL index takes this analogy beyond function. The human body thrives due to a network of integrated systems—respiratory, circulatory, nervous, etc.—working together seamlessly. Similarly, the NILL index encourages laboratories to operate as a cohesive unit, with each design element (ventilation, lighting, and water management) contributing to a holistic and sustainable environment that reflects biomimicry and biophilic design principles. By integrating these core elements with LBC principles, the NILL index and the living laboratory concept can transform laboratory buildings into models of sustainable design, user-centric innovation, and a newfound understanding of the built environment's role in supporting human health and well-being. Table 6 shows the overlay between the main themes of the NILL index and the connection with the LBC petals and the human body systems.

Table 6. Overview of the Nature-inspired & Living Laboratory (NILL 1.0)™ Building Assessment Index.

Index Theme	LBC Petal	Analogy	Connection
Foundation and Structure	Place	Skeletal and Muscular Systems	A strong foundation and musculoskeletal system provide support and stability, just as a well-designed site and building envelope are crucial for a building.
Hydration and Flow	Water	Circulatory System	The circulatory system efficiently transports water and nutrients throughout the body, similar to how a water-positive building manages and utilizes water resources.
Energy and Power	Energy	Metabolic System	The metabolic system converts food into energy, mirroring a building's ability to generate and utilize renewable energy.

Table 6. *Cont.*

Index Theme	LBC Petal	Analogy	Connection
Wellbeing and Resilience	Health and Happiness	Nervous and Immune Systems	The nervous and immune systems ensure overall health and adaptation, just as a healthy building environment promotes occupant well-being and resilience.
Skin and Breath	Materials	Integumentary System (Skin) and Respiratory System	The skin protects us from the environment, and the respiratory system facilitates healthy air exchange, analogous to how a building's materials and ventilation systems manage internal and external interactions.
Balance and Harmony	Equity	Endocrine System	The endocrine system regulates various bodily functions. Similarly, the LBC's equity petal promotes social justice and creates a balanced and equitable environment for all.
Senses and Inspiration	Beauty and Spirit	Sensory System and Central Nervous System	The sensory system and central nervous system allow us to perceive and experience the world, similar to how a building's aesthetics and functionality can inspire and uplift its occupants.
Operation Center	Integration of all petals	Human Brain	The human body integrates all its different systems that are controlled by the brain, while the LBC relies on the collective results of all its petals to control its overall performance and become a regenerative self-sustained building.

Furthermore, the proposed NILL index provides a connection between each of its themes, the parallel LBC petal and the relevant Biomimicry Life Principles (Table 7) to showcase the essence of the index and its main purpose of mimicking nature's best and implementing nature's best practices and unique strategies in achieving the highest level of performance for a laboratory building.

Table 7. Alignment between the NILL index main themes, the relevant LBC petal, and the Biomimicry Life Principles.

Index Theme	LBC Petal	Aligned Biomimicry Principles
Foundation and Structure	Focuses on minimizing site disturbance, restoring ecological functions, and integrating biomimicry strategies for sustainable site management.	Reshuffle Information Build from the Bottom Up Combine Modular and Nested Components
Hydration and Flow	Emphasizes achieving net positive water use through biomimicry-inspired water management strategies.	Leverage Cyclic Processes Use Feedback Loops
Energy and Power	Focuses on achieving net positive energy through biomimicry-inspired renewable energy generation.	Use Low Energy Processes Leverage Cyclic Processes Integrate the Unexpected
Wellbeing and Resilience	Integrates biophilic design principles inspired by nature to enhance occupant well-being and connection with the environment, while also fostering a beautiful and inspiring workspace.	Embody Resilience (variation and decentralization) Maintain Integrity through Self-Renewal Cultivate Co-operative Relationships
Skin and Breath	Prioritizes minimizing environmental impact through sustainable material selection, biomimicry-inspired design features, and ensuring an accessible and inclusive environment for all users.	Recycle All Materials Use Multi-Functional Design Fit Form to Function Break Down Products into Benign Constituents Build Selectively with a Small Subset of Elements Do Chemistry in Water
Balance and Harmony	Ensures the laboratory fosters a sense of community, collaboration, and a safe and healthy work environment.	Incorporate Diversity Cultivate Co-operative Relationships
Senses and Inspiration	Combines biophilic design elements that appeal to the senses with biomimicry-inspired solutions to foster creativity, innovation, and a sense of connection with nature, and integrating the Living Lab concept to design the laboratory as a research platform for biomimicry and sustainable technologies, actively collecting and sharing data to advance the field.	Reshuffle Information Use Feedback Loops Self-Organize Build from the Bottom Up Combine Modular and Nested Components

Table 7. *Cont.*

Index Theme	LBC Petal	Aligned Biomimicry Principles
Operation Center	Brains and building control centers are both information hubs. While the brain processes senses and commands to control the body, the laboratory operation center analyzes sensor data (ex. temperature) to adjust lab systems for optimal performance.	Replicate Strategies that Work Adapt To Changing Conditions Use Feedback Loops Be Locally Attuned and Responsive Use Readily Available Materials and Energy Be Resource Efficient (Material and Energy)

To further explain the respective themes within the NILL index, each theme encompasses multiple categories or constructs that form the main criteria of assessment within the general theme, supported by a selection of features/indicators describing the applicable features or characteristics that inform the design, construction, and operation of a nature-inspired and living laboratory building, as shown in Table 8.

In Theme 1 (Foundation and Structure), brownfield redevelopment and minimal site disturbance are given priority to guarantee that the laboratory is environmentally conscious. By using biomimicry principles to create habitat for native species, we can promote biodiversity and deepen our relationship with the natural world. Take bioswales or rain gardens, which mimic natural filtration for stormwater management, as an illustration of how the laboratory can be inspired by and integrate natural processes into its design.

In Theme 2 (Hydration and Flow), water conservation and management are emphasized. By utilizing rainwater collection, greywater reuse, and wastewater treatment to achieve net positive water use, the laboratory raises the bar for water sustainability. Biomimicry-inspired water purification systems demonstrate how the lab can benefit from understanding the efficient natural processes. Bigger water features in a range of patterns and styles help manage water and create a connection with an important element.

In Theme 3 (Energy and Power), the transition to clean energy is promoted. The laboratory can run with the least possible environmental impact thanks to net positive energy generation, which uses on-site renewable energy sources like solar or wind turbines. Biomimicry-inspired energy production technologies demonstrate how the inventiveness of nature can be used to the laboratory's advantage. Energy-efficient laboratory equipment and HVAC systems minimize energy consumption, and evaluations of energy demand and equipment load optimize operations for long-term sustainability.

In Theme 4 (Wellbeing and Resilience), the priority is given to the happiness and wellbeing of the occupants. Skylights, well-placed windows, and indoor plants are examples of natural elements that can improve air quality and create a peaceful atmosphere. With less reliance on mechanical systems, natural ventilation techniques inspired by biomimicry offer a comfortable working environment. Designing with natural forms and patterns creates a space that is both harmonious and aesthetically pleasing. Furthermore, integrating the previous strategies not only satisfies buildings users and enhances their wellbeing but also ensures that laboratory buildings achieve resilience through promoting durability, flexibility, and ongoing functionality in the face of expected and unexpected challenges. As a result, laboratory buildings are capable of enduring and adjusting to diverse environmental and operational obstacles.

In Theme 5 (Skin and Breath), a strong emphasis is placed on material selection that is responsible and inclusive. By choosing low-impact, recycled, or bio-based materials over those on the LBC Red List, the laboratory reduces its environmental impact. Naturalistic design features, such as self-cleaning surfaces, can lessen the need for abrasive chemicals. Universal access features ensure that the laboratory is accessible to all, fostering equity and a sense of community. The adverse effects of waste management practices on the environment are mitigated by sustainable methods.

Table 8. The main themes, dimension, and indicators of the Nature-inspired & Living Laboratory (NILL 1.0)^{TM} Building Assessment Index.

Theme	Dimension	Indicators	Description	Reference
Foundation & Structure (FD)	Site & Location (SL)	Sustainable Site Selection & Development	Minimal site disturbance with priority to refurbishment of old sites, brownfield redevelopment prioritization, or location near ancient site.	[31,32]
		Building Orientation	Well-oriented building and an envelope that is tightly sealed prevent unwanted heat transfer	[33]
		Habitat Creation	Rooftops, pollinator gardens, or green walls as habitats for native species mimicking natural ecosystems.	[31,32]
		Proximity to Public Transit	Availability of nearby bus station; main road access.	[34,35]
		Pedestrian Area	Designated walking area with minimal vehicular access.	[31]
	Nature-inspired Structures (NIS)	Nature-inspired Facades	Reactive facades resembling the mechanism of pinecone in responding to environmental stimuli (open and close); integrated solar panels into exterior façade resembling the lotus flower's ability to absorb sunlight efficiently or integrated photobioreactors in building façade with microalgae (to reduce thermal loads with the absorption of radiation)	[36–38]
		Nature-inspired Surrounding	Use of artwork showcasing biomimicry examples.	[39]
		Green Space Coverage	Maximize green space coverage ratio and promote plant canopies for shading and sheltering. Prioritize the use of native species to enhance biodiversity or use plants that provide shading.	[4,40]
		Access to Nature	Provide balconies, terraces, courtyards, or roof slopes for direct connection to nature.	[31,41]
Hydration & Flow (HF)	Water Sourcing (WS)	Waste Water Treatment System	Greywater treatment housed on the grounds or local system affiliated with central treatment plant; can be inspired by mussel filtration.	[42,43]
		Seawater Desalination	Seawater desalination system housed on the grounds or local system affiliated with central desalination plant.	[31,44]
		Rain Water Capturing	Rainwater collection system.	[31,42, 43]
		Dew & Condensate Capturing	Dew harvesting or condensation capture.	
		Nature-based Water Systems	Biofiltration systems based on plants' inherent ability to filter air or bioswales; wetlands or rain gardens mimicking natural filtration for stormwater management.	[45]
	Water Management (WM)	Water Use Optimization	Closed-loop systems in nature.	[46,47]
		Water Efficient Practices	Low-flow fixtures and water-saving technologies (foot pedals) in laboratory sinks and equipment.	[48]
		Treated Wastewater Repurposing	Using treated wastewater for toilet flushing.	[42,43]
		Treated Grey Water Repurposing	Greywater reuse for irrigation.	

Table 8. *Cont.*

Theme	Dimension	Indicators	Description	Reference
Energy & Power (EP)	Energy Generation (EG)	Renewable Energy	Solar or PV panels (mimicking photosynthesis) or wind turbines (inspired by bird wings) to generate net positive energy.	[49,50]
		Excess Energy Sharing	Excess renewable energy shared with the local grid.	[51,52]
		Passive Heating/Cooling	Natural ventilation or solar chimneys	[4,36]
	Energy Management (EM)	Equipment Load	Energy-efficient equipment selection and placement for minimized energy use throughout the lab's lifespan	[48,53]
		Energy Demand Optimization	Demand-controlled ventilation (DCV) systems, LED lights, and automatic light control.	[36,54–56]
		Energy-efficient HVAC System	Heat recovery ventilation (HRV) or energy recovery ventilation (ERV) systems mimicking natural ventilation processes, or integrating PV panels with HVAC Systems.	[36,57]
		Energy-efficient Cold Storage	Optimization of cold storage units, including insulation levels and practices for reducing unnecessary cooling cycles.	[58,59]
		Energy-efficient Laboratory Equipment	Low-power microscopes, fume hoods with occupancy sensors, and autoclaves with heat recovery systems with certifications like ENERGY STAR or equivalent.	[60–62]
		Equipment Utilization	Practices like scheduling, shared use protocols, and right-sizing equipment for specific needs; equipment auction or swapping.	[48]
		Equipment Management	Smart Energy Management Systems or software and practices for regular maintenance and calibration of equipment to ensure optimal performance and efficiency.	[36,63,64]
		Energy Conservation	Energy conservation measures like charged batteries for equipment.	[43,63]
Wellbeing & Resilience (WR)	Biophilic Design & Sensory Engagement (BDSE)	Visualizing Nature & Water	Maximize windows (floor-to-ceiling) with views of nature and calming water features.	[19,41,65–67]
		Natural light	Natural light patterns through strategically placed skylights and operable windows.	[31,68]
		Natural Sounds	Introduce natural sounds (e.g., water fountains or waterfalls and birdsong recordings).	[69–71]
		Quiet zones	Create designated quiet zones for focused work.	
		Indoor planting	Utilize indoor plants with air-purifying properties and aromatic benefits (e.g., lavender and rosemary).	
		Sustainable Cleaning	Avoid harsh chemical odors through sustainable cleaning practices and material selection.	[72–74]
		Natural Materials	Incorporate natural materials like wood and stone into the design for a connection with the natural world.	[75,76]
		Heating System	Floor heating system to maintain a constant laboratory temperature.	[49]
	Restorative Environment (RE)	Nature-inspired Ventilation	Ventilation inspiration by termite mounds or equivalent/alternative innovative approach.	[77,78]
		Natural Forms & Patterns	Incorporation of natural forms and patterns for a harmonious space and features promoting human amusement and the rooting of culture, spirit, and place	[31]
		Shading Solution	Maximize the direct exposure of the solar rays to the glazing facades in the cooling season and minimize the same gains in the heating season.	[36]
		Sustainable Materials & Finishes	Low-VOC paints, sealants, and adhesives to minimize indoor air pollution.	[79,80]

Table 8. *Cont.*

Theme	Dimension	Indicators	Description	Reference
Skin & Breath (SB)	Materials & Assets (MA)	Building Material	Use of thermal mass materials that passively absorb and radiate heat	[81]
		Building Envelope	Lightweight building envelope structures with good insulation, high light penetration, and diffusion.	[36,82]
		Prohibited Materials	Avoidance of materials on the LBC Red List.	[31]
		Low-impact Material	Durable and comfortable furniture (e.g., bamboo), mushroom-based packaging for lab supplies, and materials with little maintenance.	[31]
		Recycled Material	Durable and reusable labware made from recycled materials whenever possible.	[83,84]
		Biobased Material	Self-healing material that heals and patch up small cracks or composite binders (including waterproof and frost-resistant gypsum binders).	[85,86]
		Locally-sourced material	Use locally sourced wood, stone, or other natural building materials	[31]
		Nature-inspired Asset Design	Design of self-cleaning surfaces inspired by lotus leaves to minimize harsh chemicals.	[87]
	Waste Management (WM)	Internal Waste Segregation	Segregation of plastics, glass, paper, biohazard waste, chemical waste, laboratory wastewater, and grey water.	[48,88,89]
		External Waste Segregation	Segregation to landfills or treatment facilities.	
		Waste Recycling	Clean, repackage, and reuse non-sharp equipment. Recycling/composting bins or membership in recycling programs.	[48]
		Waste Minimization	Replace single-use plastics with sustainable alternatives like metal loops and reusable wooden sticks or use of bio-based consumables where possible.	[90]
		Waste Bioremediation	Utilize plants or microbes to break down pollutants in soil or water.	[91,92]
		Hazard Reduction	Use of less hazardous solvents whenever possible, prioritizing options with lower toxicity and environmental impact and solvent recycling.	[93,94]
Balance & Harmony (BH)	Occupant & Public Engagement (OPE)	Universal Access Features	Accessible doors and smooth ramps for elderly and special needs.	[31,36]
		Interactive Laboratory Layout	Open floor plan with designated collaborative zones.	[31]
		Culturally inclusive Design Elements and Amenities	Prayer and Meditation Spaces, Lactation Rooms and Nurseries, General Food Court, Washrooms for users with special needs and elderly people, and Multilingual Signage/systems.	[31,68,95]
		Public Education & Awareness	Develop educational programs and tours for occupants and the public to showcase the lab's sustainable features and biomimicry inspiration and integrate biomimicry and sustainability education into the research priorities.	[31,96,97]
	Safety & Security (SS)	Occupant Safety Programs	Implementation of a comprehensive laboratory safety program that prioritizes the well-being of lab users and minimizes environmental impact.	[98,99]
		Building Safety Features	Proper ventilation systems for fume hoods, emergency eyewash stations, and clear signage for hazardous materials in the building. Gas and fire detectors.	[31,100–102]
		Occupant Safety Measures	Personal Protective Equipment (PPE)	[103,104]
		Building Security Measures	Presence of surveillance camera and controlled door access; emergency exits.	[31]
		Continuous Building Operation	Presence of back-up energy generators.	[105]

Table 8. *Cont.*

Theme	Dimension	Indicators	Description	Reference
Senses & Inspiration (SI)	Living Laboratory (LL)	Real-Time Dashboards	Integrate real-time dashboards showcasing the lab's environmental performance, including energy use, water consumption, waste generation through a design inspired by biomimicry's focus on feedback loops and biophilic design's connection with nature.	[106–108]
		Living Wall Research Platform	Design a living wall or vertical garden that serves as a research platform for studying plant-based air purification technologies and their effectiveness in laboratory settings.	[36]
		Data Collection & Sharing	Comprehensive data collection system to monitor energy use (motion sensors), water consumption, waste generation, and indoor air quality.	[106, 109]
	Adaptive & Evolving Design	Growth & Expansion	Modular Construction or equivalent/alternative innovative approach.	[110–112]
		Reconfiguration & Enhancement	Universal Lab Shell with adaptable Mechanical, Electrical, and Plumbing (MEP) or equivalent/alternative innovative approach.	[113, 114]
		Building Innovation	Emphasis on innovative design, construction, and operational practices that push the boundaries of sustainability in living laboratories, considering potential for advancements in energy-efficient laboratory equipment or closed-loop waste management systems within research activities (open category)	[115]
Operation Center (OC)	Feedback Loop (FL)	Occupant Perception	Regular surveys to measure occupant awareness of building features and their significance.	[116, 117]
		User Satisfaction	Regular surveys to measure level of satisfaction	[118]
	Logical Reasoning (LR)	Material Selection Criterion	Life Cycle Assessment (LCA): to evaluate the environmental impact of the bio-inspired materials or processes throughout their life cycle, considering factors like energy consumption, material extraction, and end-of-life disposal.	[119, 120]
		Asset Selection Criterion	Feasibility Study (FS) and/or Cost–Benefit Analysis (CBA)	[121, 122]
		Awareness & Preparedness	Risk Assessment (RA) and Contingency plans	[123, 124]
	Knowledge & Intellect Treasury (KIT)	Conceptual Record Preservation	Preservation of design plans, blueprints, and log records of construction process, including origins of nature-inspired ideas.	[125]
		Data Preservation	Back-up servers for databases and archive of printed records	[126, 127]

In Theme 6 (Balance and Harmony), a safe and co-operative atmosphere is created in the lab. Universal access features promote inclusivity, and the lab's layout fosters collaboration and communication among researchers. Culturally sensitive design cues ensure that each researcher is treated with dignity and respect. A comprehensive safety program for laboratories prioritizes the well-being of researchers while minimizing any negative effects on the environment. Integrated safety measures, such as sufficient ventilation and emergency eyewash stations, protect researchers from potential hazards. Scientists can operate safely and effectively by funding training courses and personal safety gear.

In Theme 7 (Senses and Inspiration), the laboratory is transformed into a dynamic learning environment that stimulates the senses, spurs creativity, and advances sustainable design and biomimicry concepts. It turns into evidence of the ability of nature to serve as both an inspiration and a model for a more sustainable future.

In Theme 8 (Operation Center), the assessment index is transformed from a static building evaluation into a dynamic, living laboratory that draws inspiration from nature. This modification emphasizes the importance of continuous learning and growth. Occupant survey responses are an essential source of data for data-driven optimization of the building's architectural features. Furthermore, life cycle assessments use rational thinking to ensure cost-effective and environmentally conscious design choices. Ultimately, the category archives design documents and guiding principles to highlight intellectual stewardship. With the help of this "living laboratory" strategy, the building itself can become a knowledge base for upcoming design projects, resulting in an ongoing cycle of learning and development.

5. Conclusions

This article investigated the growing prominence of nature-inspired design philosophies within the built environment, emphasizing their potential to promote sustainability and human well-being in laboratory settings. Focusing on biomimicry and biophilic design as key approaches, the paper explored how mimicking nature's innovations and fostering connections to natural elements could benefit laboratory buildings. Additionally, the article delved into emerging methods like living building, which prioritizes user experience through co-creation and integrates sustainable and regenerative structures. By integrating these nature-inspired and living building concepts, the paper argued for the creation of dynamic and adaptable laboratory spaces that prioritized both environmental responsibility and a positive human experience. As a novel contribution to the field, this work proposed the Nature-inspired & Living Laboratory (NILL 1.0)™ Building Assessment index. This index served as a guideline for the design and construction of laboratory buildings, drawing inspiration from nature and the analogy of human body systems. By embracing these nature-inspired design principles, laboratory buildings could be transformed into models for future built environments that promote a healthier and more sustainable future.

In summary, the novel assessment index that has been suggested provides a comprehensive system for evaluating the degree to which a nature-inspired and living laboratory building integrates sustainable practices, biomimicry principles, biophilic design elements, and living building features to stimulate scientific innovation and establish a comprehensive environment that encourages occupant well-being, environmental responsibility, and scientific innovation. It synchronizes the seven LBC petals with the human body system, going beyond energy and water usage, resource utilization, or material selection. The end goal is to design structures that are not only useful but also resilient, environmentally friendly, self-sustaining, regenerative, and adaptable to the surrounding environment.

Moreover, the limitations of the current work shall be highlighted. First, the current research is limited to examining human-centric approaches, such as biomimicry and biophilia, due to the increased number of laboratory facilities post COVID-19 pandemic and the developments in R&D industries, leading to an increase in the amount of time spent by laboratory users in the buildings and the need to prioritize their wellbeing and provide productive and comfortable working environments that are inspired by nature and its processes to further enhance such experiences. Consideration of nature-based and nature-positive design can definitely broaden the scope of the topic and could serve as a future expansion for the current work. To further illustrate, the Commandments of Biomimicry, similar to principles found in Cradle-to-Cradle design and Positive Development, focus on mimicking nature's solutions to improve sustainability in human systems. Unlike net positive design frameworks, which prioritize actively increasing natural capital to counter biodiversity loss and environmental degradation, biomimicry primarily aims to optimize resource use and efficiency through bio-inspired innovations. Enhancing landscaping around buildings can improve local biodiversity and aesthetics, yet it may not adequately offset the broader impacts of biodiversity loss over the building's lifecycle. Therefore, while biomimicry offers valuable strategies for sustainable design, integrating net positive principles would more comprehensively address the critical need to restore and regenerate natural ecosystems globally.

Furthermore, the current scope of the research mentions net positive energy, water, and waste as a critical pillar within the proposed NILL building assessment scheme. Achieving net positive energy, water, and waste reduction at the site level marks significant progress in sustainable building design. However, integrating a broader systemic perspective is essential. Addressing environmental impacts beyond immediate site boundaries is crucial for ensuring that sustainability efforts contribute effectively to overarching goals such as climate resilience, resource conservation, and ecosystem health. Therefore, expanding the scope of the current work addressing nature-inspired and living laboratory buildings to encompass systemic impacts is necessary for initiatives like the Living Building Challenge to achieve the sustainable development objectives successfully. As such, there is potential for future expansion of the current work to overcome this limitation by exploring the systems impact of designing and constructing NILL laboratory buildings.

To further solidify the NILL index's position as a valuable tool for the design, construction, and operation of laboratory facilities, several future refinements can be explored. First, implementing a weighted scoring system would enable a more nuanced evaluation by assigning different weights to various criteria within biomimicry, biophilic design, and living building principles to allow for a more comprehensive assessment, reflecting the relative importance of each aspect in achieving the desired outcomes. Second, integrating a life cycle assessment framework could provide a more holistic evaluation by encompassing the environmental impact of materials, construction processes, and operational energy use throughout the building's lifespan, aligning with the core principles of sustainable design. Third, the NILL index could be enhanced by allowing for regional and project-specific customization through incorporating factors like local climate, available resources, and specific laboratory functions, which could lead to more contextually relevant and achievable design goals. Finally, developing a database of NILL-rated laboratories, along with detailed case studies, would be a valuable resource for designers and building owners, as this would showcase successful implementations and best practices, inspiring future projects and demonstrating the real-world impact of the NILL index. By addressing these potential areas for improvement, the NILL index can evolve into an even more robust and practical tool, guiding the design and construction of future laboratory buildings that embrace a sustainable and human-centered approach.

Author Contributions: Conceptualization, M.A. and S.B.; Methodology, M.A. and S.B.; Writing—original draft preparation, M.A.; Writing—review and editing, M.A., S.B. and S.A.; Supervision and Project Administration, S.B. and S.A. All authors have read and agreed to the published version of the manuscript.

Funding: This research received no external funding.

Institutional Review Board Statement: Not applicable.

Data Availability Statement: Data are contained within the article.

Acknowledgments: The authors would like to thank the College of Engineering at the American University of Sharjah for supporting the publication fee. This paper represents the opinions of the authors and does not mean to represent the position or opinions of the American University of Sharjah.

Conflicts of Interest: The authors declare no conflicts of interest.

References

1. Bajçinovci, B. Biomimicry and Biophilic Design: Multiple Architectural Precepts. *J. Sci. Humanit. Arts* **2019**, *6*, 1–10. [CrossRef]
2. Aanuoluwapo, O.O.; Ohis, A.C. Biomimetic Strategies for Climate Change Mitigation in the Built Environment. *Energy Procedia* **2017**, *105*, 3868–3875. [CrossRef]
3. Oguntona, O.; Aigbavboa, C. Nature inspiration, imitation, and emulation: Biomimicry thinking path to sustainability in the construction industry. *Front. Built Environ.* **2023**, *9*, 1085979. [CrossRef]
4. Zhong, W.; Schröder, T.; Bekkering, J. Biophilic design in architecture and its contributions to health, well-being, and sustainability: A critical review. *Front. Archit. Res.* **2022**, *11*, 114–141. [CrossRef]
5. Wijesooriya, N.; Brambilla, A. Bridging biophilic design and environmentally sustainable design: A critical review. *J. Clean. Prod.* **2021**, *283*, 124591. [CrossRef]
6. Kellert, S.R.; Wilson, E.O. *The Biophilia Hypothesis*; Island Press: Washington, DC, USA, 1995; Volume 4, pp. 187–188.
7. Kellert, S.R.; Heerwagen, J.; Mador, M. *Biophilic Design: The Theory, Science and Practice of Bringing Buildings to Life*; John Wiley & Sons: Hoboken, NJ, USA, 2011.
8. Benyus, J.M. *Biomimicry: Innovation Inspired by Nature*; William Morrow: New York, NY, USA, 1997.
9. Gruber, P.; Imhof, B. Patterns of Growth—Biomimetics and Architectural Design. *Buildings* **2017**, *7*, 32. [CrossRef]
10. Knippers, J.; Speck, T.; Nickel, K.G. Biomimetic research: A dialogue between the disciplines. In *Biomimetic Research for Architecture and Building Construction: Biological Design and Integrative Structures*; Springer: Cham, Switzerland, 2016; pp. 1–5.
11. Hakio, K.; Mattelmäki, T. Future Skills of Design for Sustainability: An Awareness-Based Co-Creation Approach. *Sustainability* **2019**, *11*, 5247. [CrossRef]
12. Cole, R.J. Transitioning from green to regenerative design. *Build. Res. Inf.* **2012**, *40*, 39–53. [CrossRef]
13. du Plessis, C. Towards a regenerative paradigm for the built environment. *Build. Res. Inf.* **2012**, *40*, 7–22. [CrossRef]
14. Assem, A.; Hassan, D.K. Biophilia in the workplace: Apilot project for a living wall us ng an interactive parametric design approach. *Archit. Eng.* **2024**, *9*, 3–16. [CrossRef]
15. Yeter, I.H.; Tan, V.S.; Le Ferrand, H. Conceptualization of Biomimicry in Engineering Context among Undergraduate and High School Students: An International Interdisciplinary Exploration. *Biomimetics* **2023**, *8*, 125. [CrossRef]
16. Jiang, Y.; Li, N.; Yongga, A.; Yan, W. Short-term effects of natural view and daylight from windows on thermal perception, health, and energy-saving potential. *Build. Environ.* **2022**, *208*, 108575. [CrossRef]
17. Crawford, A.; In-na, P.; Caldwell, G.; Armstrong, R.; Bridgens, B. Clay 3D printing as a bio-design research tool: Development of photosynthetic living building components. *Archit. Sci. Rev.* **2022**, *65*, 185–195. [CrossRef]
18. Stefanova, A.; In-na, P.; Caldwell, G.S.; Bridgens, B.; Armstrong, R. Photosynthetic textile biocomposites: Using laboratory testing and digital fabrication to develop flexible living building materials. *Sci. Eng. Compos. Mater.* **2021**, *28*, 223–236. [CrossRef]
19. Ko, W.H.; Schiavon, S.; Zhang, H.; Graham, L.T.; Brager, G.; Mauss, I.; Lin, Y. W. The impact of a view from a window on thermal comfort, emotion, and cognitive performance. *Build. Environ.* **2020**, *175*, 106779. [CrossRef]
20. Hoffman, A. Building on the inherent strengths of green space environments: Promoting trust, democracy, and resilience among ethnically diverse groups. *J. Prev. Interv. Community* **2020**, *48*, 210–224. [CrossRef]
21. Baja, F.D.F.; Bajracharya, S.; Freeman, M.A.; Gray, A.J.; Haglund, B.T.; Kuipers, H.R.; Opatola, O.R. Leed gold but not equal: Two case study buildings. *Int. J. Des. Nat. Ecodyn.* **2019**, *14*, 52–62. [CrossRef]
22. Cianfrani, C.M.; Hews, S.; Ior, J.; Jewhurst, J.J.; Shillington, C.; Raymond, M. The R.W. Kern Center as a Living Laboratory: Connecting Campus Sustainability Goals with the Educational Mission at Hampshire College, Amherst, MA. *J. Green Build.* **2018**, *13*, 123–145. [CrossRef]
23. May, E. Modelling to drive design: Honing the SU RE house through performance simulations. *Archit. Des.* **2018**, *88*, 72–81.
24. Pritchard, R., Kelly, S. Realising Operational Energy Performance in Non-Domestic Buildings: Lessons Learnt from Initiatives Applied in Cambridge. *Sustainability* **2017**, *9*, 1345. [CrossRef]
25. Azari, R. Chapter 5—Life Cycle Energy Consumption of Buildings; Embodied + Operational. In *Sustainable Construction Technologies*; Tam, V.W.Y., Le, K.N., Eds.; Butterworth-Heinemann: Oxford, UK, 2019; pp. 123–144. [CrossRef]
26. Cheshire, D.; Menezes, A.C. *Evaluating Operational Energy Performance of Buildings at the Design Stage*; Chartered Institution of Building Services Engineers: London, UK, 2013.
27. MacKinno, R. *An Introduction to Life's Principles*; The Biomimicry Institute: Missoula, MT, USA, 2021; Volume 2024.
28. DesignLens: Life's Principles. Available online: https://biomimicry.net/the-buzz/resources/designlens-lifes-principles/ (accessed on 21 March 2024).
29. Living Building Challenge. Available online: https://living-future.org/lbc/ (accessed on 1 June 2024).
30. Living Building Challenge Basics. Available online: https://living-future.org/lbc/basics4-0/ (accessed on 1 June 2024).

31. Hegazy, I.; Seddik, W.; Ibrahim, H. The living building: Integrating the built environment with nature evaluating the Bibliotheca of Alexandria according to the challenge imperatives. *Int. J. Low Carbon Technol.* **2017**, *12*, 244–255. [CrossRef]
32. Petrovski, A.; Pauwels, E.; González, A.G. Implementing Regenerative Design Principles: A Refurbishment Case Study of the First Regenerative Building in Spain. *Sustainability* **2021**, *13*, 2411. [CrossRef]
33. Baird, G. Sustainable Buildings–best and worst performers in terms of Comfort, Health and Productivity. In *CIB World Building Congress Proceedings*; CIB World: West Lafayette, IN, USA, 2013; p. 359.
34. Halil, F.M.; Nasir, N.M.; Hassan, A.A.; Shukur, A.S. Feasibility Study and Economic Assessment in Green Building Projects. *Procedia—Soc. Behav. Sci.* **2016**, *222*, 56–64. [CrossRef]
35. Rostami, R.; Khoshnava, S.M.; Ahankoob, A.; Rostami, R. Sustainable Site planning and Design. In Proceedings of the Management in Construction Research Association (MiCRA) Post Graduate Conference, Universiti Teknologi Malaysia, Johor Bahru, Malaysia, 5 December 2012.
36. Caetano, N.S.; Carvalho, R.R.; Franco, F.R.; Afonso, C.A.R.; Felgueiras, C. Sustainable engineering labs—A Portuguese perspective. *Energy Procedia* **2018**, *153*, 455–460. [CrossRef]
37. Sommese, F.; Badarnah, L.; Ausiello, G. Smart materials for biomimetic building envelopes: Current trends and potential applications. *Renew. Sustain. Energy Rev.* **2023**, *188*, 113847. [CrossRef]
38. Kamińska, P.; Michalak, H. Innovative, Modular Building Facades-as a Tool to Counteract The Effects of and to Prevent Climate Change. *Civ. Environ. Eng. Rep.* **2022**, *32*, 184–209. [CrossRef]
39. Hawsawi, H. Nature Inspired Interior Design Principles in the Hot Arid Climate of Saudi Arabia. Master's Thesis, Arizona State University, Tempa, AZ, USA, 2016.
40. Winkler, J.; Jeznach, J.; Koda, E.; Sas, W.; Mazur, Ł.; Vaverková, M.D. Promoting Biodiversity: Vegetation in a Model Small Park Located in the Research and Educational Centre. *J. Ecol. Eng.* **2021**, *23*, 146–157. [CrossRef]
41. Kuwahara, R.; Kim, H. Studying the Indoor Environment and Comfort of a University Laboratory: Air-Conditioning Operation and Natural Ventilation Used as a Countermeasure against COVID-19. *Buildings* **2022**, *12*, 953. [CrossRef]
42. Aulia, A.N.; Dewi, O.C. Wastewater management optimization in the integrated teaching laboratory building. *AIP Conf. Proc.* **2022**, *2534*, 030005. [CrossRef]
43. Helling, K.; Bölsche, D. Innovative Water Management at the Environmental Campus Birkenfeld. *J. Sustain. Perspect.* **2021**, *1*, 21–31. [CrossRef]
44. Hidayat, F.; Dewi, O.C. Strengthening Sustainable Context of Water Management in Educational Building Universitas Indonesia. *IOP Conf. Ser. Earth Environ. Sci.* **2022**, *1058*, 012009. [CrossRef]
45. Kraakman, N.J.R.; González-Martín, J.; Pérez, C.; Lebrero, R.; Muñoz, R. Recent advances in biological systems for improving indoor air quality. *Rev. Environ. Sci. Bio/Technol.* **2021**, *20*, 363–387. [CrossRef]
46. Roy, M.; Linnanen, L.; Chakrabortty, S.; Pal, P. Developing a Closed-Loop Water Conservation System at Micro Level Through Circular Economy Approach. *Water Resour. Manag.* **2019**, *33*, 4157–4170. [CrossRef]
47. Wanjiru, E.; Xia, X. Optimal energy-water management in urban residential buildings through grey water recycling. *Sustain. Cities Soc.* **2017**, *32*, 654–668. [CrossRef]
48. Leak, L.B.; Tamborski, J.; Commissaris, A.; Brophy, J.A.N. Forging a path toward a more sustainable laboratory. *Trends Biochem. Sci.* **2023**, *48*, 5–8. [CrossRef]
49. Noor, L.H.W.; van der Kroef, D.A.; Wattam, D.; Pinnock, M.; van Rossum, R.; Smit, M.G.; Brussaard, C.P.D. Innovative transportable laboratories for polar science. *Polar Rec.* **2018**, *54*, 18–28. [CrossRef]
50. Papanicolas, C.N.; Lange, M.; Fylaktos, N.; Montenon, A.; Kalouris, G.; Fintikakis, N.; Fintikaki, M.; Kolokotsa, D.; Tsirbas, K.; Pavlou, C.; et al. Design, construction and monitoring of a near-zero energy laboratory building in Cyprus. *Adv. Build. Energy Res.* **2015**, *9*, 140–150. [CrossRef]
51. Bartolini, A.; Carducci, F.; Muñoz, C.B.; Comodi, G. Energy storage and multi energy systems in local energy communities with high renewable energy penetration. *Renew. Energy* **2020**, *159*, 595–609. [CrossRef]
52. Yang, Y.; Jia, Q.S.; Deconinck, G.; Guan, X.; Qiu, Z.; Hu, Z. Distributed Coordination of EV Charging With Renewable Energy in a Microgrid of Buildings. *IEEE Trans. Smart Grid* **2018**, *9*, 6253–6264. [CrossRef]
53. Felgueiras, C.; Santos, R.; Fonseca, L.; Caetano, N. Buildings Sustainability: The HVAC Contribution. *J. Clean Energy Technol.* **2015**, *4*, 375–379. [CrossRef]
54. Shehabi, A.; Ganeshalingam, M.; DeMates, L.; Mathew, P.A.; Carter, D. *Characterizing the Laboratory Market*; Lawrence Berkeley National Laboratory: Berkeley, CA, USA, 2017.
55. Chinery, M.B. Using Sensor-based Demand Controlled Ventilation to Realize Energy Savings in Laboratories. Master's Thesis, Air Force Institute of Texhnology, Wright-Patterson Air Force Base, OH, USA, 2014.
56. Raminhos, F.M.M.; Valdez, M.M.T.; Ferreira, C.M. Study of a clinical analysis laboratory's lighting system design. In Proceedings of the 2013 48th International Universities' Power Engineering Conference (UPEC), Dublin, Ireland, 2–5 September 2013; pp. 1–5.
57. Fiorentini, M.; Cooper, P.; Ma, Z.; Wall, J.L. Experimental investigation of an innovative HVAC system with integrated PVT and PCM thermal storage for a net-zero energy retrofitted house. In Proceedings of the ASHRAE Winter Conference 2015, Chicago, IL, USA, 24–28 January 2015.
58. Sevault, A.; Bøhmer, F.; Næss, E.; Wang, L. Latent heat storage for centralized heating system in a ZEB living laboratory: Integration and design. *IOP Conf. Ser. Earth Environ. Sci.* **2019**, *352*, 012042. [CrossRef]

59. Ghanbari, L. Applying Building Energy Models for Operation Scheduling Optimization. Master's Thesis, Master of Science Electrical Engineering. The University of New Mexico, Albuquerque, NM, USA, 2014.
60. Montenon, A.C.; Papanicolas, C. Economic Assessment of a PV Hybridized Linear Fresnel Collector Supplying Air Conditioning and Electricity for Buildings. *Energies* **2021**, *14*, 131. [CrossRef]
61. Im, H.; Srinivasan, R.S.; Maxwell, D.; Steiner, R.L.; Karmakar, S. The Impact of Climate Change on a University Campus' Energy Use: Use of Machine Learning and Building Characteristics. *Buildings* **2022**, *12*, 108. [CrossRef]
62. Kitzberger, T.; Kotik, J.; Pröll, T. Energy savings potential of occupancy-based HVAC control in laboratory buildings. *Energy Build.* **2022**, *263*, 112031. [CrossRef]
63. Moya, F.D.; Torres-Moreno, J.L.; Álvarez, J.D. Optimal Model for Energy Management Strategy in Smart Building with Energy Storage Systems and Electric Vehicles. *Energies* **2020**, *13*, 3605. [CrossRef]
64. Elston, H.J. Preventative maintenance systems and the laboratory safety professional: You get what you inspect, not what you expect. *J. Chem. Health Saf.* **2007**, *14*, 5–9. [CrossRef]
65. Boubekri, M.; Cheung Ivy, N.; Reid Kathryn, J.; Wang, C.-H.; Zee Phyllis, C. Impact of Windows and Daylight Exposure on Overall Health and Sleep Quality of Office Workers: A Case-Control Pilot Study. *J. Clin. Sleep Med.* **2014**, *10*, 603–611. [CrossRef] [PubMed]
66. Edwards, L.; Torcellini, P. *Literature Review of the Effects of Natural Light on Building Occupants*; U.S. Department of Energy: Washingto, DC, USA, 2002.
67. Yu, G.; Yang, H.; Luo, D.; Cheng, X.; Ansah, M.K. A review on developments and researches of building integrated photovoltaic (BIPV) windows and shading blinds. *Renew. Sustain. Energy Rev.* **2021**, *149*, 111355. [CrossRef]
68. Tilley, A.R.; Rider, A.H.; Maze, C.; Cannon, D.; McGee, C.; Steinfeld, E.; White, J.; Perry, L.G.; Lehner, R.; Mohnke, D.; et al. Functional Planning. In *Architectural Graphic Standards Reference*; John Wiley & Sons, Inc.: Hoboken, NJ, USA, 2019; pp. 3–72. [CrossRef]
69. Sheikha, R.H.E.; Abdel-Salam, H.M.; Bakr, A.F. Optimizing the Performance of Health Facilities in Egypt: Using a Hybrid Approach Combining Biophilic Design and BIM. *Int. J. Struct. Civ. Eng. Res.* **2023**, *12*, 99–105. [CrossRef]
70. Scott, M.; Lennon, M.; Haase, D.; Kazmierczak, A.; Clabby, G.; Beatley, T. Nature-based solutions for the contemporary city/Re-naturing the city/Reflections on urban landscapes, ecosystems services and nature-based solutions in cities/Multifunctional green infrastructure and climate change adaptation: Brownfield greening as an adaptation strategy for vulnerable communities?/Delivering green infrastructure through planning: Insights from practice in Fingal, Ireland/Planning for biophilic cities: From theory to practice. *Plan. Theory Pract.* **2016**, *17*, 267–300. [CrossRef]
71. Noble, L.; Devlin, A.S. Perceptions of Psychotherapy Waiting Rooms: Design Recommendations. *HERD Health Environ. Res. Des. J.* **2021**, *14*, 140–154. [CrossRef] [PubMed]
72. Bartczak, C.; Dunbar, B.; Bohren, L. Incorporating Biophilic Design through Living Walls: The Decision-Making Process. In *Constructing Green: The Social Structures of Sustainability*; The MIT Press: Cambridge, MA, USA, 2013. [CrossRef]
73. Aydogan, A.; Cerone, R. Review of the effects of plants on indoor environments. *Indoor Built Environ.* **2020**, *30*, 442–460. [CrossRef]
74. Ceballos, D.M.; Burr, G.A. Evaluating a Persistent Nuisance Odor in an Office Building. *J. Occup. Environ. Hyg.* **2012**, *9*, D1–D6. [CrossRef] [PubMed]
75. Dalay, L. The Impact of Biophilic Design Elements on the Atmospheric Perception of the Interior Space. *Uluslararası Peyzaj Mimar. Araştırmaları Derg. IJLAR* **2020**, *4*, 4–20.
76. Lei, Q.; Lau, S.S.; Yuan, C.; Qi, Y. Post-Occupancy Evaluation of the Biophilic Design in the Workplace for Health and Wellbeing. *Buildings* **2022**, *12*, 417. [CrossRef]
77. Chen, X.; Chen, X.; Su, R.; Cao, B. Optimization Analysis of Natural Ventilation in University Laboratories Based on CFD Simulation. *Buildings* **2023**, *13*, 1770. [CrossRef]
78. Wei, Y.; Lin, Z.; Wang, Y.; Wang, X. Simulation and Optimization Study on the Ventilation Performance of High-Rise Buildings Inspired by the White Termite Mound Chamber Structure. *Biomimetics* **2023**, *8*, 607. [CrossRef]
79. Thibaudeau, P. Integrated Design is Green. *J. Green Build.* **2008**, *3*, 78–94. [CrossRef]
80. Smith, J.; Earl, G.G. Australia's First 6-Star Green Education Building: Design and Performance. In Proceedings of the SASBE 2009 CIB International Conference: Smart and Sustainable Built Environments, Delft, The Netherlands, 1 June–3 October 2009.
81. Amirifard, F.; Sharif, S.A.; Nasiri, F. Application of passive measures for energy conservation in buildings—A review. *Adv. Build. Energy Res.* **2019**, *13*, 282–315. [CrossRef]
82. Pohl, G.; Nachtigall, W. Biomimetics for Buildings. In *Biomimetics for Architecture & Design: Nature—Analogies—Technology*; Pohl, G., Nachtigall, W., Eds.; Springer International Publishing: Cham, Switzerland, 2015; pp. 25–52. [CrossRef]
83. Capet, X.; Aumont, O. Decarbonization of academic laboratories: On the trade-offs between CO_2 emissions, spending, and research output. *Clean. Environ. Syst.* **2024**, *12*, 100168. [CrossRef]
84. Seydel, C. Greening the lab. *Nat. Methods* **2023**, *20*, 1449–1453. [CrossRef] [PubMed]
85. Amran, M.; Onaizi, A.M.; Fediuk, R.; Vatin, N.I.; Muhammad Rashid, R.S.; Abdelgader, H.; Ozbakkaloglu, T. Self-Healing Concrete as a Prospective Construction Material: A Review. *Materials* **2022**, *15*, 3214. [CrossRef] [PubMed]
86. Lesovik, V.; Fediuk, R.; Amran, M.; Vatin, N.; Timokhin, R. Self-Healing Construction Materials: The Geomimetic Approach. *Sustainability* **2021**, *13*, 9033. [CrossRef]

87. Rabajczyk, A.; Zielecka, M.; Klapsa, W.; Dziechciarz, A. Self-Cleaning Coatings and Surfaces of Modern Building Materials for the Removal of Some Air Pollutants. *Materials* **2021**, *14*, 2161. [CrossRef] [PubMed]
88. Alyeddin, W.; Peters, S.; Zembrzycka, A.A.; Hudson, L.; Tun, S. Bleeding green: Sustainability in practice in a clinical skills teaching laboratory. *J. Clim. Change Health* **2022**, *8*, 100149. [CrossRef]
89. Lehmann, S. Resource recovery and materials flow in the city: Zero waste and sustainable consumption as paradigms in urban development. *Sustain. Dev. Law Policy* **2011**, *11*, 13. [CrossRef]
90. Alves, J.; Sargison, F.A.; Stawarz, H.; Fox, W.B.; Huete, S.G.; Hassan, A.; McTeir, B.; Pickering, A.C. A case report: Insights into reducing plastic waste in a microbiology laboratory. *Access Microbiol.* **2021**, *3*, 1–7. [CrossRef] [PubMed]
91. Mankiewicz, P.; Borsuk, A.; Ciardullo, C.; Hénaff, E.; Dyson, A. Developing design criteria for active green wall bioremediation performance: Growth media selection shapes plant physiology, water and air flow patterns. *Energy Build.* **2022**, *260*, 111913. [CrossRef]
92. Gunasinghe, Y.H.K.I.S.; Rathnayake, I.V.N.; Deeyamulla, M.P. Plant and Plant Associated Microflora: Potential Bioremediation Option of Indoor Air Pollutants. *Nepal J. Biotechnol.* **2021**, *9*, 63–74. [CrossRef]
93. Winterton, N. The green solvent: A critical perspective. *Clean Technol. Environ. Policy* **2021**, *23*, 2499–2522. [CrossRef]
94. Sangle, M.; Shirsath, A.; Gondkar, S.; Kulkarni, A. Fundamentals and recent progress on green high performance liquid chromatography. *Int. J. Res. Anal. Rev.* **2022**, *9*, 69.
95. Al Rashdi, S.A.; Sudhir, C.V.; Basha, J.S.; Saleel, C.A.; Soudagar, M.E.M.; Yusuf, A.A.; El-Shafay, A.S.; Afzal, A. A case study on the electrical energy auditing and saving techniques in an educational institution (IMCO, Sohar, Oman). *Case Stud. Therm. Eng.* **2022**, *31*, 101820. [CrossRef]
96. Schiller, C. Buildings as Teaching Tools: A Case Study Analysis to Determine Best Practices that Teach Environmental Sustainability. Ph.D. Thesis, Carnegie Mellon University, Pittsburgh, PA, USA, 2012.
97. Newman, J.; Fernandez, L. Strategies for institutionalizing sustainability in higher education. In *Yale School of the Environment Publications Series*; Yale University: New Haven, CT, USA, 2007.
98. McCarthy, J.F.; Fragala, M.A.; Baker, B.J. Analyzing the Risk: Balancing Safety and Efficiency in Laboratory Ventilation. *ACS Chem. Health Saf.* **2022**, *29*, 434–440. [CrossRef]
99. Aliyo, A.; Edin, A. Assessment of Safety Requirements and Their Practices Among Teaching Laboratories of Health Institutes. *Microbiol. Insights* **2023**, *16*, 11786361231174414. [CrossRef]
100. Fajri, H.; Alamsyah, W.; Basrin, D.; Pratama, M.R.; Yunita, I. Evaluation of Fire Protection System on Educational Laboratory Building of Universitas Samudra. *J. Serambi Eng.* **2023**, *8*, 6307–6315. [CrossRef]
101. Schaufele, M.E. Toxic and flammable gases in research laboratories: Considerations for controls and continuous leak detection. *J. Chem. Health Saf.* **2013**, *20*, 8–14. [CrossRef]
102. Nikolaou, C.; Surti, N. Laboratories. In *Metric Handbook*; Routledge: Oxfordshire, UK, 2015; pp. 441–470.
103. Nath, N.D.; Behzadan, A.H.; Paal, S.G. Deep learning for site safety: Real-time detection of personal protective equipment. *Autom. Constr.* **2020**, *112*, 103085. [CrossRef]
104. Wagner, H.; Kim Angella, J.; Gordon, L. Relationship between Personal Protective Equipment, Self-Efficacy, and Job Satisfaction of Women in the Building Trades. *J. Constr. Eng. Manag.* **2013**, *139*, 04013005. [CrossRef]
105. Salameh, T.; Olabi, A.G.; Abdelkareem, M.A.; Masdar, M.S.; Kamarudin, S.K.; Sayed, E.T. Integrated Energy System Powered a Building in Sharjah Emirates in the United Arab Emirates. *Energies* **2023**, *16*, 769. [CrossRef]
106. Hua, Y.; Oswald, A.; Yang, X. Effectiveness of daylighting design and occupant visual satisfaction in a LEED Gold laboratory building. *Build. Environ.* **2011**, *46*, 54–64. [CrossRef]
107. Ballon, P.; Schuurman, D. Living labs: Concepts, tools and cases. *Info* **2015**, *17*. [CrossRef]
108. O'Brien, W.; Doré, N.; Campbell-Templeman, K.; Lowcay, D.; Derakhti, M. Living labs as an opportunity for experiential learning in building engineering education. *Adv. Eng. Inform.* **2021**, *50*, 101440. [CrossRef]
109. Hebert, P.R.; Kang, M.; Thompsen, R.J. Energy saving via lighting study at US National Laboratories. *Facilities* **2014**, *32*, 396–410. [CrossRef]
110. Stocchero, A.; Seadon, J.K.; Falshaw, R.; Edwards, M. Urban Equilibrium for sustainable cities and the contribution of timber buildings to balance urban carbon emissions: A New Zealand case study. *J. Clean. Prod.* **2017**, *143*, 1001–1010. [CrossRef]
111. Munaro, M.R.; Tavares, S.F. Design for adaptability and disassembly: Guidelines for building deconstruction. *Constr. Innov.* 2023; *ahead-of-print*. [CrossRef]
112. Kuiri, L.; Leardini, P. Design for adaptability and disassembly: Towards zero construction waste in the Australian housing sector. In Proceedings of the 1st International e-Conference on Green and Safe Cities 2022 (IeGRESAFE), Online, 20–21 September 2022; pp. 744–762.
113. Stark, S.; Gaughan, R. Costs for renovations and non-domestic new construction: Renovation costs are on the upswing; international costs vary widely. *Rev. Optom.* **2004**, *141*, 14–16.
114. Marinelli, M. Emergency Healthcare Facilities: Managing Design in a Post Covid-19 World. *IEEE Eng. Manag. Rev.* **2020**, *48*, 65–71. [CrossRef]
115. Save, P.; Terim Cavka, B.; Froese, T. Evaluation and Lessons Learned from a Campus as a Living Lab Program to Promote Sustainable Practices. *Sustainability* **2021**, *13*, 1739. [CrossRef]

116. Francisco, A.; Truong, H.; Khosrowpour, A.; Taylor, J.E.; Mohammadi, N. Occupant perceptions of building information model-based energy visualizations in eco-feedback systems. *Appl. Energy* **2018**, *221*, 220–228. [CrossRef]
117. Curry, E.; Fabritius, W.; Hasan, S.; Kouroupetroglou, C.; ul Hassan, U.; Derguech, W. A Model for Internet of Things Enhanced User Experience in Smart Environments. In *Real-Time Linked Dataspaces: Enabling Data Ecosystems for Intelligent Systems*; Curry, E., Ed.; Springer International Publishing: Cham, Switzerland, 2020; pp. 271–294. [CrossRef]
118. Quality, B.B.; Huizenga, C.; Laeser, K.L.; Arens, E.A. *A Web-Based Occupant Satisfaction Survey for Benchmarking Building Quality*; UC Berkeley: Berkeley, CA, USA, 2002.
119. Figueiredo, K.; Pierott, R.; Hammad, A.W.A.; Haddad, A. Sustainable material choice for construction projects: A Life Cycle Sustainability Assessment framework based on BIM and Fuzzy-AHP. *Build. Environ.* **2021**, *196*, 107805. [CrossRef]
120. Soust-Verdaguer, B.; Gutiérrez Moreno, J.A.; Llatas, C. Utilization of an Automatic Tool for Building Material Selection by Integrating Life Cycle Sustainability Assessment in the Early Design Stages in BIM. *Sustainability* **2023**, *15*, 2274. [CrossRef]
121. Hubbard, B.; Wang, H.; Leasure, M.; Ropp, T.; Lofton, T.; Hubbard, S.; Lin, S. Feasibility study of UAV use for RFID material tracking on construction sites. In Proceedings of the 51st ASC Annual International Conference Proceedings, TX, USA, 22–25 April 2015; pp. 669–676.
122. Berlec, T.; Tanšek, B.; Kušar, J. Selection of the most suitable material handling system in production. *Int. J. Simul. Model.* **2021**, *20*, 64–75. [CrossRef]
123. Plan, C. *Newborn Screening Contingency Plan (CONPLAN)*; Department of Health and Human Services: Washington, DC, USA, 2010.
124. Manouchehr, O.; Mansouri, N. Fire and Spillage Risk Assessment Pattern in Scientific Laboratories. *Int. J. Occup. Hyg.* **2015**, *6*, 68–74.
125. Tunstall, G. *Managing the Building Design Process*; Routledge: Oxfordshire, UK, 2006.
126. Pan, J.; Liu, X. A Central Control System for the Environmental Laboratory. In Proceedings of the 2011 IEEE Ninth International Symposium on Parallel and Distributed Processing with Applications Workshops, Busan, Republic of Korea, 26–28 May 2011; pp. 189–192.
127. Rehiara, A.; Musa, A.Y.; Stepanus, J.B. Energy auditing and electricity saving opportunities in BPOM laboratory of manokwari. *Soc. Ecol. Econ. Sustain. Dev. Goals J.* **2023**, *1*, 1–16. [CrossRef]

biomimetics

MDPI

Article

Ecuadorian Woods: Building Material Selection Using an Entropy-COPRAS Comparative Analysis Based on the Characterization of Ecuadorian Oak and Guayacan Timber

Juan Francisco Nicolalde [1,2,3], Javier Martínez-Gómez [2,*], Paúl Dávila [1], Johanna Medrano-Barboza [3] and Jaime Vinicio Molina-Osejos [3]

1 Facultad de Ciencias Técnicas, Universidad Internacional Del Ecuador UIDE, Quito 170411, Ecuador; junicolaldego@uide.edu.ec (J.F.N.); padavilaal@uide.edu.ec (P.D.)

2 Universidad de Alcalá, Departamento de Teoría de la Señal y Comunicación, Área de Ingeniería Mecánica, Escuela Politécnica, 28805 Alcalá de Henares, Madrid, Spain

3 Facultad de Arquitectura e Ingenierías, Universidad Internacional SEK, Albert Einstein s/n and 5th, Quito 170302, Ecuador; johanna.medrano@uisek.edu.ec (J.M.-B.); jaime.molina@uisek.edu.ec (J.V.M.-O.)

* Correspondence: javier.martinezgomez@uah.es

Abstract: Considering that global awareness for sustainable development has risen to face environmental damages, different building materials have been considered from a mechanical perspective. In this sense, considering the richness of South America regarding its woods, the Guayacan and the Ecuadorian oak timbers have not been previously characterized. The present research has performed mechanical, thermal, and moisture content characterizations to acknowledge the benefits of considering these materials for the building industries. In this sense, Guayacan has been shown to have lower thermal conductivity, making it ideal for thermal insulation; the oak from Manabi showed the best compressive strength; while the oak from El Oro stands with the best tensile strength; and the oak from Loja showed the best modulus of elasticity. On the other hand, all the materials were compared by multicriteria decision methods to select the best, by using the COPRAS method driven by the objective entropy-weighted method, showing that the oak from Loja is the best choice considering the advantage that presents with the modulus of elasticity. In this sense, it is concluded that regarding the mechanical properties, there is not much difference for the compression, bending, and tensile strength; nevertheless, for the modulus of elasticity the oak from Loja stands out, making it a factor to be considered in the selection of a wood for building applications that is corroborated through multicriteria decision methods.

Keywords: multicriteria decision methods; material characterization; mechanical resistance; thermal conductivity

1. Introduction

The growing awareness of climate change has led to research on sustainable materials, where wood, as a renewable resource, can be considered as an excellent material with excellent qualities that can become a key element of the future bioeconomy [1]. On the other hand, wood is considered a renewable resource with mitigation impacts, and waste is reused for other wood products and energy production [2]. Consequently, a progressive societal demand for sustainable, effective, and durable materials must be considered.

In this sense, in the construction sector, wood has been considered a useful material for structural beams, frames, and flooring. This status was achieved considering its excellent strength-to-weight ratios, thermal and acoustic insulation properties, and especially, its high sustainability rates and recyclability characteristics [3]. However, traditional construction materials such as timber in Ibero-American countries have not been considered as a primary material [4]. Instead, brick buildings have been used as the most common construction practice. However, in the past, wood was used for buildings but resulted in poor-performance housing due to construction techniques with energy-inefficient results [5], which leaves a door open for sustainable development research in the sector [4]. Based on the energy efficiency of the materials for building purposes, it is important to consider the design of buildings with different materials and with different architectural forms, to reduce a large part of the energy losses, given that the transfer of energy between the environment and the building generates an energy demand [6]. On the other hand, it is important to state that the use of wood as a sustainable material for buildings must come with a responsible practice that in South America has been resolved by each country's government that has promoted the maintenance of an economically productive and sustainable forest industry [7]. In this sense, energy efficiency in buildings is an important topic to be disclosed; even more, it becomes more relevant considering that energy efficiency is considered a reliable strategy to achieve sustainable development, where developing countries are still in an improvement process compared to industrializations [8]. Taking in mind these considerations, in Ecuador there are many varieties of trees, the most commonly mobilized species are Laurel (*Laurus nobilis*), Doncel (*Zanthoxylum rhoifolium*), Zapote (*Pouteria sapota*), Balsa (*Ochroma pyramidale*), Ceibo (*Ceiba pentandra*) Chalviande (*Virola sebifera*), Arenillo (*Andira inermis*), Tamburo (*Vochysia leguina*), Colorado (*Cissus striata*) and Copal (*Bursera glabrifolia*) [9] while timbers such as pigue (*Piptocoma discolor*), jacaranda (*Jacaranda mimosifolia*), huayacan (*Guaiacum officinale*), laurel mascarey (Hieronyma Alchorneoides), pachaco (*Schizolobium parahyba*), cutanga (*Parkia multijuga*), batea caspi (*Cabralea canjerana*), cedro (*Cedrus*), Ferman Sanchez (*Triplaris cumingiana*), capirona de monte (*Calycophyllum spruceanum*), manglillo (*Hedyosmum mexicanum*), tachuelo (*Zanthoxylum precerum*), peine de mono (*Apeiba membranacea*), guayacan (*Tebebuia chrysantha*), and pechiche (*Vitex cymosa*) are commercial species that in early stages are identified by farmers and promoted in on-farm agroforestry [10].

Considering Ecuador's sustainable woods, the oak (*Nectandra montana*) is a traditional material that can be named. The term oak is used to refer to many species of trees of the genus *Quercus*, native to the Northern Hemisphere, and occasionally also to species of other genera of the same family (*Fagaceae*) or even of other families, as in the case of some South American species of Nothofagus (*fam. Nothofagaceae*) [11]. Oak wood has a straight grain, a coarse grain, and you can clearly distinguish sapwood from heartwood. Sawing is difficult, but wear and tear on saws is normal. Drying should be slow and gentle, as there is a risk of superficial flaws appearing. This timber is durable against the action of fungi, but sensitive to anobiums, lyctids, and cerambycids, and moderately durable to termites [12]. On the other hand, oak has good aptitude for obtaining flat panels if it is previously dried and does not present problems for gluing. The wood of this species of oak is used in interior carpentry, furniture, sleepers, and waterworks, and was formerly widely used in Gothic cathedrals and shipbuilding. It is also a wood that is used to make quality wine, being comparable to oaks of French or American origin [13]. On the other hand, local research has found that Guayacan wood (*Caesalpinia paraguariensis*) can be used as a material for buildings. Guayacan has been used traditionally as a source of fuelwood and charcoal, it has been used for furniture and indigenous people used it for tools. Even more, the tree can reach its maturity between 10 and 20 years, despite different climate conditions such as rainfalls or extreme temperatures, making it a solid candidate for studying its thermal properties in building applications [14].

The characterization of building materials is ruled by different standards that allow for analysis of the heat exchange of building materials. There are different methods such as the ISO 5151: 2010 [15] that reports total cooling capacity, the AISI/ASHRAE Standard 16–2016 for cooling and dehumidification [16], and the AHRI Standard 340/360:2015 for cooling and dehumidification [17], among others [18], such as the heat balance proposed in the regulations of air conditioning, refrigeration, and heating.

In this way, Guayacan is a potential material to be exploited as a construction timber, while oak has proven to be useful in traditional applications, and considering the region's common practices brick and cement are used for buildings. However, there are different characteristics to be considered, keeping in mind that the building components must be chosen to fulfil the requirements of serviceability and safety limits [19]. In this sense, while the thermal properties of materials allow them to determine the thermal behavior regarding different temperature variations [20], the susceptibility to humidity that might arise in the structure can mean a risk of biodegradability and infestation of insects and fungi, which should be considered over the reduction in mechanical properties [21]. Lastly, mechanical properties such as flexural strength, tensile strength, and compressive strength are commonly compared, considering that materials with flexural strength and low density are widely used as beams and girders in structural applications [22]. Considering these three alternatives, it is proposed to compare and assess these materials based on multicriteria decision methods (MCDM). In this sense, the research of Tupėnaitė et al. (2020) [23] used the simple additive weighting method to define the best utilization in high-rise buildings considering the criteria of wood utilization, height of the building, number of floors, building cost, length of implementation, reduction in CO_2 emissions, and use of energy. The MCDM are driven by a weighted matrix that uses the weight of every criterion to evaluate the candidates. In this sense, the entropy method as a weighting method allows calculation of the degree of difference between criteria and gives a more important value, allowing an objective evaluation [24]. In this way, the research of Reddy et al. (2019) [25] used the entropy method to make a material selection considering sustainable materials for building applications, where 10 sustainable criteria were considered. For the decision maker, this meant a challenge, considering the vagueness and impreciseness addressed by the fuzzy subjective weighting methodology. However, by using the objective criteria, these problems were solved by considering the utilization of the entropy concept. On the other hand, Movaffaghi and Yitmen (2021) [26], used the Complex Proportional Assessment (COPRAS) as the MCDM to study the best timber–concrete composite alternative for buildings. In the same way, Motuzienė et al., (2016) [27] used the COPRAS method to determine the best envelope material for a single-family house, choosing between masonry, log, and timber frames, considering the criteria of reduction in expenses, greenhouse gases, non-renewable primary energy, and ozone layer depletion.

In this way, the purpose of the following research is to characterize the Guayacan timber obtained in the province of Pichincha of Ecuador, manufactured by artisanal production companies, carrying out tests for compressive strength perpendicular to the wood fiber, compressive strength parallel to the wood fiber, flexural strength of wood, tensile strength of wood, moisture content of the wood, and thermal conductivity of the material.

In the same way, a parallel characterization will be performed on oak timber from Loja, El Oro, and Manabi for the same mechanical and thermal tests.

Lastly, the characterized timbers will be compared by multicriteria decision means, using the entropy method as an objective weighting tool and the COPRAS method as the MCDM.

2. Materials and Methods

In this section, the testing methodology for thermal conductivity, moisture content, and mechanical characterization will be presented, which includes compressive strength parallel to the grain of the wood, compression perpendicular to the grain of the wood, and resistance flexural and tensile strength. It is important to mention that the tested timbers were obtained from artisanal production, testing the samples in the state that is commonly used. In this sense, even though the density of the wood at a specific moisture content has an important role on the physical properties of the wood, it has not been altered towards the characterization.

2.1. Characterization of Ecuadorian Oak and Guayacan Timber

Different samples from different locations are studied considering that small and clear specimens of the same species can vary their mechanical properties, such as in the case of the modulus of elasticity, which could change in a range of approximately 7–20% [28]. In this sense, thermal and mechanical tests were carried out with hard oak wood specimens and Guayacan timber. The oak hardwood was obtained from three different provinces of Ecuador in the Coastal region: Manabí, El Oro, and Loja. In the case of the Guayacan timber, it was obtained from artisanal producers in the province of Pichincha in the highland region, as shown in Figure 1.

Figure 1. Artisanal production of timber.

2.2. Thermal Tests

For the thermal tests, 12 specimens were used: 3 specimens of oak from El Oro, 3 specimens from Manabí, 3 specimens from Loja, and 3 specimens of Guayacan from Pichincha. The tests for the determination of thermal conductivity are based on the ISO 8302:1991 standard [29] and ASTM-C177 [30] by the sheltered hot plate method, where the samples must have a homogeneous composition with an area of 150 × 150 mm² and thicknesses from 10 mm to 200 mm. The tests are carried out at temperatures between −10 °C and 50 °C. This procedure allows measuring conductivities ranging from 0.002 to 2.5 $W \times m^{-1} \times K^{-1}$. The specimens must previously be kept in a room with humidity and a controlled temperature of 23 °C for 24 h. After the necessary conditioning time has elapsed, measurements are taken with precision scales of ±0.05 g. It is verified with precision instruments that the sample complies with the ISO 8302 standard; then, 5 measurements of the specimen are taken to have an average or arithmetic mean. Then, the samples are covered with polypropylene plastic and placed on top of the ultrasonic gel sample, avoiding vacuum spaces.

Furthermore, to measure the thermal conductivity of the test samples, the hot plate thermal machine, Lambda-Messtechnik λ-Meter EP500e Version C from the thermal testing and energy efficiency laboratory of the Escuela Politécnica del Litoral (ESPOL), was used and displayed in Figure 2, along with Equation (1):

$$k = \frac{q \times L}{\Delta T \times A} \tag{1}$$

where q (W) is the heat flow, k ($W \times m^{-1} \times K^{-1}$) is the thermal conductivity (m^2), A is the area of the section, ΔT is the difference of temperatures through the sample, and L (m) is the thickness of the sample.

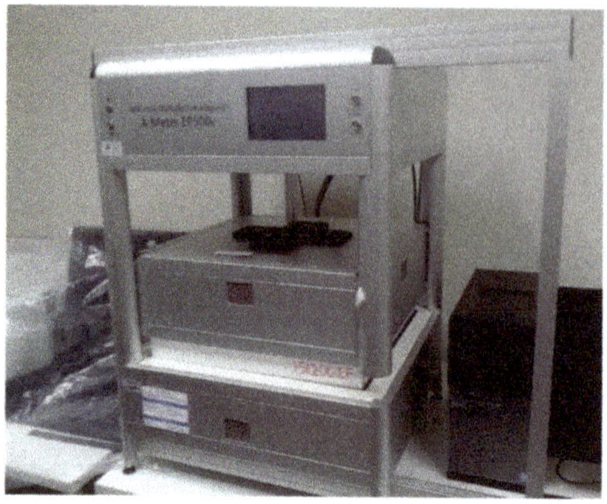

Figure 2. Thermal conductivity test machine.

2.3. Mechanical Test

For the mechanical tests, 80 specimens were needed, which were divided for the different tests of parallel compression, perpendicular compression, flexion, and tensile, in such a way that 5 samples from every location were used, taking into consideration the limitations for using the measurement equipment. Every sample was tested with the standard ASTM D143-14 [31] under the methodologies for parallel compression, perpendicular compression, flexion, and tension parallel. The tests were carried out with the WAW- 6OOC Computer Control Electra-hydraulic Servo-controlled Universal Testing Machine from the Particular International University (SEK), as shown in Figure 3.

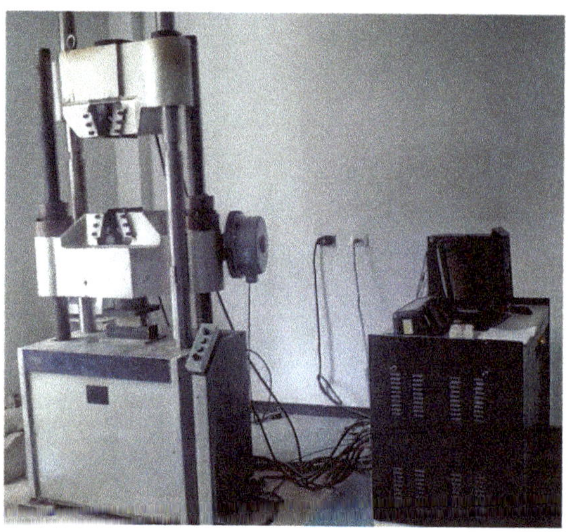

Figure 3. Universal testing machine.

2.3.1. Parallel Compression

The dimensions of the specimens for this test are 20 cm long with a rectangular section of 5 $\times$ 5 cm, considering that a prism must be perpendicular on its faces. A tolerance of $\pm$1 cm is considered, even if that is not part of the recommendations of the standard. The speed of application of load is 0.6 mm/min until the break occurs, and the failure stress to compression is obtained with Equation (2) [31]:

$$\sigma_c = \frac{F_c}{A} \tag{2}$$

where σ_c is the failure compressive stress parallel (MPa), F_c is the failure load (N) in compression, and A is the cross-sectional area of the sample.

The calculation of the modulus of elasticity uses Hooke's law considering σ_c as the parallel compressive stress (MPa), E is the modulus of elasticity or Young's modulus, (MPa) and ε is the unitary deformation, according to Equation (3), while Figure 4 shows how the test was conducted.

$$E = \frac{\sigma_{max1}}{\varepsilon} \tag{3}$$

Figure 4. Parallel compression test.

2.3.2. Perpendicular Compression

The test specimen is a prism with a rectangular section of 5 × 5 cm and a length of 15 cm that must be cut with these exact measurements. A metallic plate with dimensions of 5.11 × 11.6 cm with a thickness of approximately 0.5 is used to distribute the load to the center of the sample, in such a way that the failure occurs by crushing. The continuous load speed applied is 0.305 mm/min until it reaches failure or a maximum penetration of 2.5 mm. The resistance to the limit of proportionality–RLP (MPa) is calculated with Equation (4) [31]:

$$RLP = \frac{F_{lim}}{A_c} \tag{4}$$

where F_{lim} is the load at the proportional limit (N) and A_c is the contact area by the steel plate on the surface of the sample. On the other hand, the maximum unit resistance is calculated with Equation (5):

$$\sigma_c = \frac{F_p}{A_p} \tag{5}$$

where σ_c is the maximum unitary resistance to compression (MPa), F_P is the necessary load up to penetration of 2.5 mm (N) and A_p is the contact area by the plate steel on the surface of the sample. The sample tested for perpendicular compression is displayed in Figure 5.

Figure 5. Perpendicular compression test.

2.3.3. Bending Test

The test sample is a rectangular section prism 5 × 5 cm and 76 cm long, with ±3% tolerance. Five measurements are taken from the sample on the central part to obtain an average. The distance between supports must be 71 cm and a support must allow rotation in the axis of the test sample. The speed for this test is 2.5 mm/min until the test piece breaks with a limit load F_{lim} on the proportional limit, while the unit stress in the proportionality limit–σ_{UL} (MPa) is calculated with the following Equation (6). The test is displayed in Figure 6.

$$\sigma_{UL} = \frac{3F_{lim} \times L}{2b \times h^2} \quad (6)$$

where L is the length between supports (mm), b is the base of the specimen (mm), h is the height of the test sample (mm) and F_m is the maximum load. The maximum unit stress to flexion–σ_b (MPa) is calculated with Equation (7):

$$\sigma_b = \frac{3F_m \times L}{2b \times h^2} \quad (7)$$

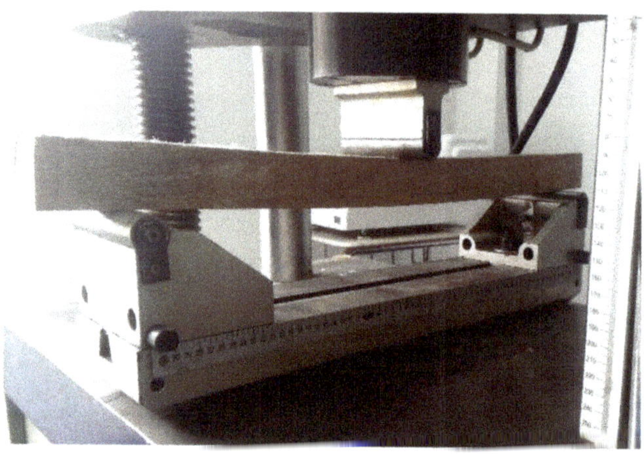

Figure 6. Bending test.

The modulus of elasticity on the unit proportional limit–E_{lim} (MPa) is calculated with Equation (8) and the modulus of elasticity–E (MPa) with Equation (9), where d_1 (mm) is the displacement of the test sample on the proportionality limit, σ_b (MPa) is the stress on the proportionality limit, and ε is the percentual unitary deformation.

$$E_{lim} = \frac{F_{lim} \times L^3}{2d_1 \times b \times h^3} \quad (8)$$

$$E = \frac{\sigma_b}{\varepsilon} \quad (9)$$

2.3.4. Tension Test

The test is performed by the application of an axial tensile force until failure. The dimensions of the specimens had a length of 46 cm, and the central section was 0.98 × 0.45 cm for 6.35 cm. The end supports had an extension of 10.1 cm and will have a square section measuring 2.54 × 2.54 cm. The speed for this test is 1 mm/min continuously until failure of the specimen. The maximum unit stress to tension–σ_t (MPa) is calculated with the area of the minimum section A_{min} (m^2) in the sample following Equation (10) and the unitary stress on the limit of proportionality–σ_{lim} is calculated using the tensile load F_t with Equation (11) and Figure 7 shows a sample being tested:

$$\sigma_t = \frac{F_{max}}{A_{min}} \quad (10)$$

$$\sigma_{lim} = \frac{F_t}{A_{min}} \quad (11)$$

Figure 7. Tension tests.

2.4. Moisture Content (MC) Tests

Water is present in the wood in three forms: hygroscopic or fixed, free water that fills the cell cavities within the wood, and bound water that is held by cellulose. In this sense, the moisture content test was carried out under the standard ASTM D4442 [32] considering that to determine the content of humidity, a relationship is established between the mass of free water and the mass of hygroscopic water contained in a sample, concerning the mass of the anhydrous sample (oven dry weight) [33]. The preparation of the specimens is later for each mechanical test; hence the number of test pieces was 80. The specimens must be cut and acquired from the nearby area to failure and stored under conditions that do not alter the content of humidity. The specimens were labeled for recognition and subsequently weighed before putting them in the oven. Drying time is 24 h at a temperature of 103 °C ± 2 °C. After this time, proceed to remove the specimens from the oven for cooling to room temperature of 20 °C ± 2 °C, before being weighed for the last time. The determination of the moisture content of wood is achieved by following Equation (12):

$$MC = \frac{w_w \times w_d}{w_d} \times 100\% \quad (12)$$

where MC is the moisture content (%), w_w is the weight of the wood in its initial wet state (g), and w_d is the weight of wood in a dry state (g). Lastly, Figure 8 shows all the tested samples submitted to the MC analysis.

Figure 8. MC tested samples.

2.5. Multicriteria Decision Methods

Previously, we had studied the utilization of alternative materials for building applications and the selection of the best material by MCDM, where it was found that the bamboo of Ecuador, also known as Guadua, presented the best properties regarding resistance as beams for social-interest housing [34]. In this sense, the same material was compared to Guayacan and the oak from different zones of origin in Ecuador. However, in this case, the thermal conductivity is considered as a factor for temperature insulation. In this sense, the following Table 1 displays the properties of the 5 materials to be assessed by MCDM, which is considered a matrix of i rows and j columns. Information was selected by CES-Edupack; Granta Design Limited [35].

Table 1. MCDM materials properties.

Material	Index	Thermal Conductivity ($W \times m^{-1} \times K^{-1}$)	Flexural Modulus (GPa)	Compressive Strength (MPa)	Bending Strength (MPa)	Tensile Strength (MPa)
Bamboo *	M1	0.18	17	60	35.9	160
Guayacan	M2	0.105	36.2	6.81	54.94	64.81
Oak El Oro	M3	0.162	72.4	19.23	82.56	173.54
Oak Manabi	M4	0.176	91.36	20.12	77.11	158.13
Oak Loja	M5	0.172	151.36	19.08	77.62	118.75

* Properties of the material extracted from software [35].

In this way, considering that the materials present different capabilities for the different samples, the present selection took the worst scenarios, showing that the material with the lowest thermal conductivity is Guayacan, making it the best for insulating the environmental temperature. The highest value is the worst, considering that the material will conduct external heat into the building; hence, the worst insulating material is the oak of Manabi. The flexural modulus has an important advantage for the oak of Loja and the worst performance for bamboo. The compressive strength is the best in the case of bamboo and the worst for Guayacan. In terms of bending strength, the oak from El Oro shows the best properties in contrast to the bamboo which is the worst, while in terms of the tensile strength, the oak of El Oro again shows the best performance and Guayacan, importantly, falls behind. In this sense, there is no optimal material, and the performance changes depending on the material's characteristics, which makes it suitable to use the MCDM to assess the best material.

Entropy-Weighted COPRAS Method

The research of Shi and Sun (2023) [36] considered that the entropy method delivers an objective evaluation of the weights of each criterion reflecting the importance of the information delivered. In this sense, the entropy method follows the subsequent steps.

The decision matrix is generated based on the values of every material characterized, via Equation (13), where the X_{ij} values represent every position of Table 1; for example, X_{11} represents the thermal conductivity value of the material M1.

$$P_{ij} = \frac{x_{ij}}{\sum_{j=1}^{n} x_{ij}} \tag{13}$$

1. These criteria are normalized and then calculated in entropy Equation (14).

$$E_i = \frac{\sum_{j=1}^{n} P_{ij} \times \ln P_{ij}}{\ln n} \tag{14}$$

2. The last step weighs each criterion with Equation (15) [37].

$$w_i = \frac{1 - E_i}{\sum_{i=1}^{m} (1 - E_i)} \tag{15}$$

On the other hand, the ranking of the COPRAS method is developed based on the decision or objective matrix (DOM) and evaluates the alternatives considering their relative importance [38]. The present method was guided by the following steps:

The normalization matrix is reached with the original DOM using the summation for normalization, as shown in Equation (16).

$$r_{ij} = \frac{X_{ij}}{\sum_{i=1}^{m} X_{ij}} \tag{16}$$

1. The weighted matrix is found by multiplying each objective value by the weighted value found in the entropy method, as shown in Equation (17).

$$y_{ij} = r_{ij} \times W_j \tag{17}$$

2. The weighted normalized scores are calculated considering the maximization of the objectives with Equation (18). This means that for all the criteria a higher score is better, except for the thermal conductivity where a lower score is better. Equation (19) refers to minimizations.

$$S_{+i} = \sum_{j=1}^{n} y_{+ij} \tag{18}$$

$$S_{-i} = \sum_{j=1}^{n} y_{-ij} \tag{19}$$

3. The calculation of comparative significance Qi determinates the relative importance of each alternative, as shown in Equation (20).

$$Q_i = S_{+i} + \frac{\sum_{i=1}^{m} S_{-i}}{S_{-i} \sum_{i=1}^{m} (S_{-i})^{-1}} \tag{20}$$

4. The level of choice utility is found at last (in Equation 21), which makes it possible to rank the alternatives [38,39].

$$U_i = \left(\frac{Q_i}{Q_{max}} \right) \times 100\% \tag{21}$$

3. Results and Discussion

3.1. Thermal Characterization of Guayacan and Ecuadorian Oak

The thermal conductivity results of the Ecuadorian oak (Figure 9) showed that in eight of the nine specimens, the thermal conductivity has a value between the range of 0.16 and 0.19 W $\times$ m^{-1} $\times$ K^{-1}. This range is defined for the test conditions according to the ASTM-C177 standard. In this way, test 2 does not comply with the standard of thermal conductivity since it showed a conductivity of 0.142 W $\times$ m^{-1} $\times$ K^{-1} established in the values of the ASTM-C177 standard [30]. In the same way, for Guayacan timber, the thermal conductivity average was found to be 0.105 W $\times$ m^{-1} $\times$ K^{-1}. Figure 9 compares all samples while the numerical results are displayed in the appendix. The thermal conductivity is greater for the oak rather than the Guayacan, while in the differentiation of the different zones from its origin, there is a slight difference that gives an advantage to the oak from Manabi.

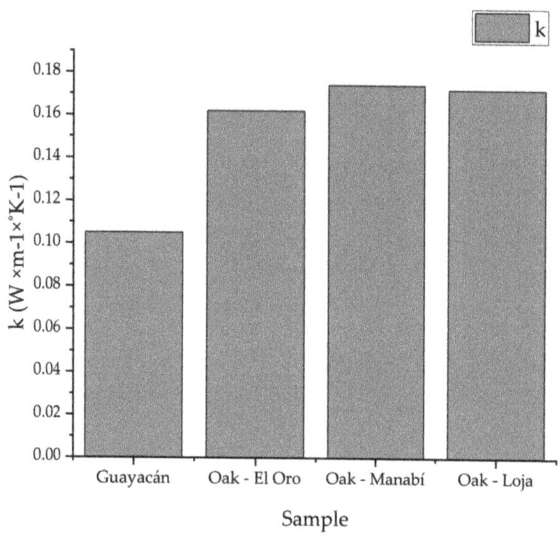

Figure 9. Thermal conductivity comparison.

3.2. Mechanical Tests of Ecuadorian Oaks and Guayacan Timber

The results of the mechanical test have been unified in Figure 10 where a comparison of the Guayacan timber vs. the Ecuadorian oaks from three different zones of the country can be seen, showing the strength against perpendicular compression, bending, and tension following the left x-axis and the modulus of elasticity from parallel compression in the right x-axis. It is important to mention that the modulus of elasticity has been calculated for both parallel compression and bending; however, the bending test was performed following the three bending configurations, which is not purely axial, and compared to the four points test it underestimated about the 19% of the modulus of elasticity [40]. For this reason, the modulus of elasticity from the parallel compression test is displayed. In the same way, the data obtained are displayed in Appendix A.

The five test samples of guayacan wood were subjected to compression with loads exceeding 45 kN. The samples G2PC and G5PC have a lower failure stress with a difference of 1.2835 MPa based on the standard deviation. On the other hand, for the five test specimens subjected to perpendicular compression by crushing with loads from 6.81 kN to 7.35 kN, an average of 7.19 (MPa) is recorded for the maximum compression stress resistance by crushing, with a standard deviation of 0.32 MPa which indicates that the variation is low, showing that all the test samples have the correct dimensions based on the standard. The flexural resistance test shows that the sample G2F has the greater resistance based on the applied load of 8.24 kN, while in the case of the tensile strength test, there is a variation of the load between the specimens due to dimensions that are not exact for this type of tests. However, the results show an average for maximum stress of 79.03 MPa, and the standard deviation indicates a variation of 14.62 MPa due to the variation of loads, possibly resulting from the moisture content that reduces the resistance of the samples.

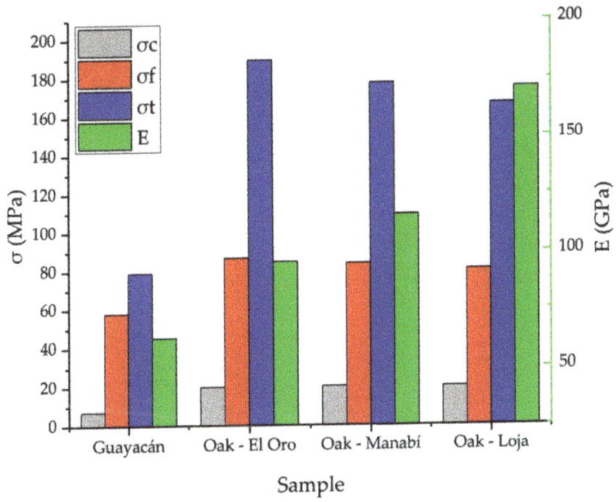

Figure 10. Mechanical properties.

On the other hand, the mechanical properties of the same oak species obtained from different parts of Ecuador were characterized, showing that the elastic modulus values are higher in the case of the oak that comes from the province of Loja and in the case of El Oro province, 40% of the test pieces did not meet the results according to the ASTM D143 standard [31]. Considering the results of the average maximum resistance stress to perpendicular compression σ_c, the tests comply with the ASTM D143-14 standard [31]. In the 15 specimens tested from the three regions, the maximum strength (MPa) varies between 19.08 and 20.69 MPa, with better performance for the oaks from Manabí, but without an important difference. On the other hand, the bending tests of the oak hardwood show a result ranging from 77.11 to 89.80 MPa, complying with the ASTM D143-14 standard [31]. The results obtained show how the oak from El Oro province is higher than those of other provinces, but considering the standard deviation, the oak from the coast region (El Oro and Manabi) could behave in the same way, while there is a slight difference for Loja. The results of the tensile tests showed that samples O3T-Manabi, O3T-Loja, and O5T-Loja had values below the reference values of the ASTM D143 14 standard, while all the samples coming from the province of El Oro complied with the standard [31]. In the same way, the oak from El Oro showed the best performance with an average tensile strength of 189.41 MPa but again considering the standard deviation, it is no different from the Manabi samples.

Comparatively, traditional Chinese structural timber has served as the main material for the ancient buildings of the region, making them one of the most famous building systems in the world for its configuration in joints. This traditional building has been characterized and was found to have a compressive strength of 30 MPa, a bending strength of 50 MPa, and a tensile strength of 91.4 MPa [41] which is comparable with the Guayacan timber, except for the compressive strength where it shows a deficiency. On the other hand, different research has studied alternative materials for building, intending to face environmental challenges. Within these materials, the development of compressed wood (CW) can be found, which is manufactured by thermos–mechanical compression, allows a substantial increase in the mechanical properties, making them competitive against other structural materials [42]. In this sense, the research of Haller and Wehsener (2004) [43] showed a bending and tensile strength of 169 Mpa and 185 MPa, respectively, which is comparable in the case of the tensile strength of the Ecuadorian oak, but not for the bending strength.

3.3. Moisture Content of Ecuadorian Oaks and Guayacan Timber

The weight of the specimens in the initial wet state and the anhydrous state (final weight), extracted from the samples of parallel compression, along with the average moisture content, are displayed in the appendix while a comparative graphic compares its behavior in Figure 11. It is observed that specimen G5MC has a significant reduction in moisture content compared to the other samples, while specimen G3MC has a higher moisture content for Guayacan, which is the material with the lowest moisture content compared to the oaks. On the other hand, the moisture content has a higher moisture content, but the sample O5MC-ElOro is shown to be higher. On average, the oak from Manabi shows a lower MC% among the oaks, a fact that may explain the best thermal conductivity. In the case of moisture content, none of the specimens of oak hardwood exceed 30%, as established in the ASTM D4442 standard [32].

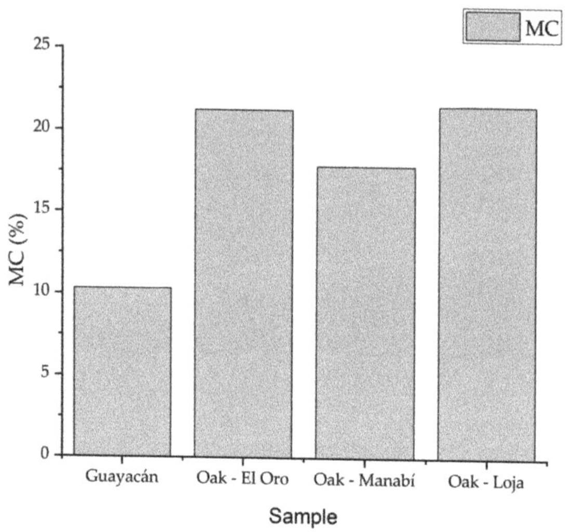

Figure 11. Moisture content comparison.

In normal environmental conditions, the moisture content can be a dangerous factor for building materials, considering that excessive moisture can lead to deteriorated habitation quality, and a reduction in the resistance to high temperatures. There are more possibilities to increase the mechanical stress of the material, salt transportation, and decay [44]. These considerations make Guayacan a material to be considered, regarding its low moisture content in comparison to the oak of all other regions. Furthermore, this abnormal moisture variation shows that the artisanal method is not the best and should be improved.

3.4. Comparison by Entropy-COPRAS MCDM

The comparison regarding MCDM began with the weighting of all criteria. As an objective method, entropy shows that the thermal conductivity is the most important characteristic, with a 0.99 entropy and a weight of 0.212, which is congruent with our study considering that the efficiency of thermal insulation depends on the thermal conductivity since this property allows the material or the combination of materials to slow down the heat flow from conduction, convection, or radiation origins [44]. On the other hand, regarding the mechanical properties, the bending strength is placed second but slightly over the tensile strength and with a more important weight compared to the modulus of elasticity. The comparison shows that the difference in thermal conductivity between candidates was considered as a more important factor [25,36], ensuring that the weight of the criteria has the expected objective component. The results of the entropy weighting method are presented in Table 2, obtained by the utilization of Equations (12)–(14).

Table 2. Entropy method result.

Criteria	Entropy	Weight	Rank
Thermal conductivity	0.99	0.212	1
Modulus of elasticity	0.87	0.187	4
Compressive strength	0.86	0.184	5
Bending strength	0.98	0.209	2
Tensile strength	0.97	0.208	3

Since the criteria weights have been delivered by the objective entropy method which gives a weight by mathematical means without the subjective intervention of the selector, the MCDM COPRAS allows to rank the materials, a step towards finding the best material. In this sense, even though thermal conductivity is the most important criterion and Guayacan presents the best property for insulation systems, its deficiency in terms of mechanical properties puts it at last with a significant difference, with a utility of 61%. On the other hand, even when compared to other cement-reinforced materials, bamboo proved to be the best [34]; compared to the characterized materials, it showed a utility of 86% which is not far from the oak of Manabi, which is the next best, with a utility of 88%. However, the oak from Loja is considered the best and shows a significant advantage over the oak from El Oro which is the second best. These results can be seen in Table 3.

Table 3. COPRAS method.

Material	Qi	Ui	Rank
Bamboo	0.203	86%	4
Guayacan	0.144	61%	5
Oak El Oro	0.209	89%	2
Oak Manabi	0.208	88%	3
Oak Loja	0.236	100%	1

Even when the oak from Loja is not the best in every criterion, it stands as the best for its modulus of elasticity, since in all the criteria there is not much differentiation but in the modulus of elasticity the oak of Loja shows an outstanding superiority. Objectively, all the materials have a building application that can be comparable to traditional ancient buildings [41] but oak can be considered the best among them. With an intrinsic characteristic of modulus of elasticity, it would be preferable to use the oak from Loja.

4. Conclusions

The research on climate change needs to explore all the alternatives to prevent greater damage to the planet. In this sense, the present research has characterized two kinds of woods: Guayacan and oak from Ecuador, to give an alternative to the building sector.

In this way, the results show that Guayacan has lower mechanical resistance and thermal conductivity. The material itself has these disadvantages that could be related to a significantly lower moisture content.

On the other hand, the oaks have shown good properties that can change depending on the origin zone of the wood. It was determined that the oak from Manabi has a slight advantage in thermal conductivity, which at the same time could be related to a lower moisture content. However, regarding the mechanical properties, there is not much difference for the compression, bending, and tensile strength, nevertheless, for the modulus of elasticity, the oak from Loja stands out, making this a factor to be considered in the selection of a wood for building applications.

Considering that different materials show different properties, in the case of the same species but from other regions there are similitudes. This makes it important to study new species with the appropriate standard.

Moreover, timbers can be used for different applications that require different properties, making the entropy method a good technique for assessing the weight of the criteria objectively and for general purposes. In this sense, the multicriteria decision methods driven by weighted objective techniques show that an assessment that can lead to better selection by mathematical means can be performed. This is the case when using an entropy-COPRAS method that has not been thoroughly explored for building material selection. In this sense, this research allows us to analyze a different approach for material selection for buildings.

On the other hand, the materials were studied considering their artisanal production. It is recommended to continue the research with a more technical treatment of the material, taking into account the variation of moisture presented.

Author Contributions: Conceptualization, J.F.N. and J.M.-G.; methodology, J.F.N. and J.M.-G.; software, J.F.N. and J.M.-G.; validation, J.F.N., J.M.-G., P.D., J.M.-B. and J.V.M.-O.; formal analysis, J.F.N. and J.M.-G.; investigation, J.F.N. and J.M.-G.; resources, J.F.N., J.M.-G.; data curation, J.F.N. and J.M.-G.; writing—original draft preparation, J.F.N. and J.M.-G.; writing—review and editing, P.D., J.M.-B. and J.V.M.-O.; visualization, J.M.-G.; supervision, J.M.-G.; project administration, J.M.-G.; funding acquisition, J.M.-G. All authors have read and agreed to the published version of the manuscript.

Funding: This research takes part in the project selection, characterization, and simulation of phase change materials for thermal comfort, cooling, and energy storage. This project is part of the INEDITA call for R&D research projects in the field of energy and materials. This research takes part in project P121819, Parque de Energías Renovables, founded by Universidad Particular International SEK.

Institutional Review Board Statement: Not applicable.

Data Availability Statement: Will be provided on demand.

Acknowledgments: The authors also thank the support provided by Mario Bustamente Crespo in the development of this research.

Conflicts of Interest: The authors declare no conflicts of interest.

Appendix A

Appendix A.1. Thermal Conductivity Results

Table A1. Thermal conductivity for Guayacan.

Sample	k ($W \times m^{-1} \times K^{-1}$)
G1TC	0.092
G2TC	0.115
G3TC	0.109
Average	0.105
Std. Deviation	0.009

Table A2. Thermal conductivity for Ecuadorian oak from El Oro.

Sample	k ($W \times m^{-1} \times K^{-1}$)
O1TC-ElOro	0.176
O2TC-ElOro	0.142
O3TC-ElOro	0.169
Average	0.162
Std. Deviation	0.014

Table A3. Thermal conductivity for Ecuadorian oak from Manabí.

Sample	k ($W \times m^{-1} \times K^{-1}$)
O1TC-Manabi	0.174
O2TC-Manabi	0.181
O3TC-Manabi	0.168
Average	0.174
Std. Deviation	0.005

Table A4. Thermal conductivity for Ecuadorian oak from Loja.

Sample	k ($W \times m^{-1} \times K^{-1}$)
O1TC-Loja	0.168
O2TC-Loja	0.174
O3TC-Loja	0.176
Average	0.172
Std. Deviation	0.003

Appendix A.2. Mechanical Results

Table A5. Mechanical resistance for Guayacan.

Sample	E (GPa)	Sample	σ_c (MPa)	Sample	σ_b (MPa)	Sample	σ_t (MPa)
G1PC	46.43	G1PrC	7.35	G1B	56.62	G1T	70.79
G2PC	56.15	G2PrC	6.93	G2B	65.13	G2T	91.38
G3PC	41.20	G3PrC	7.57	G3B	54.94	G3T	74.03
G4PC	36.20	G4PrC	6.81	G4B	57.27	G4T	99.34
G5PC	49.21	G5PrC	7.34	G5B	56.48	G5T	64.81
Average	45.33		7.19		57.98		79.03
Std. Deviation	7.621		0.32		4.03		14.62

Table A6. Mechanical resistance for Ecuadorian oak from El Oro.

Sample	E (GPa)	Sample	σ_c (MPa)	Sample	σ_b (MPa)	Sample	σ_t (MPa)
O1PC-ElOro	92.84	O1PrC-ElOro	20.49	O1B-ElOro	86.05	O1T-ElOro	191.25
O2PC-ElOro	99.75	O2PrC-ElOro	20.46	O2B-ElOro	88.18	O2T-ElOro	173.54
O3PC-ElOro	72.48	O3PrC-ElOro	19.34	O3B-ElOro	89.80	O3T-ElOro	177.50
O4PC-ElOro	72.4	O4PrC-ElOro	19.23	O4B-ElOro	86.73	O4T-ElOro	201.88
O5PC-ElOro	86.22	O5PrC-ElOro	20.62	O5B-ElOro	82.56	O5T-ElOro	202.92
Average	84.73		20.02		86.66		189.41
Std. Deviation	10.91		0.61		2.42		12.12

Table A7. Mechanical resistance for Ecuadorian oak from Manabi.

Sample	E (GPa)	Sample	σ_c (MPa)	Sample	σ_b (MPa)	Sample	σ_t (MPa)
O1PC-Manabi	92.55	O1PrC-Manabi	20.44	O1B-Manabi	86.31	O1T-Manabi	200.63
O2PC-Manabi	113.69	O2PrC-Manabi	20.39	O2B-Manabi	88.35	O2T-Manabi	192.71
O3PC-Manabi	118.41	O3PrC-Manabi	20.12	O3B-Manabi	77.11	O3T-Manabi	158.13
O4PC-Manabi	91.36	O4PrC-Manabi	20.20	O4B-Manabi	85.63	O4T-Manabi	167.08
O5PC-Manabi	128.86	O5PrC-Manabi	20.69	O5B-Manabi	82.90	O5T-Manabi	169.38
Average	108.9		20.36		84.06		177.58
Std. Deviation	14.74		0.19		3.88		16.22

Table A8. Mechanical resistance for Ecuadorian oak from Loja.

Sample	E (GPa)	Sample	σ_c (MPa)	Sample	σ_b (MPa)	Sample	σ_t (MPa)
O1PC-Loja	171.95	O1PrC-Loja	20.20	O1B-Loja	79.07	O1T-Loja	171.04
O2PC-Loja	193.66	O2PrC-Loja	20.42	O2B-Loja	81.88	O2T-Loja	182.08
O3PC-Loja	151.36	O3PrC-Loja	19.08	O3B-Loja	80.43	O3T-Loja	148.75
O4PC-Loja	167.60	O4PrC-Loja	19.49	O4B-Loja	77.62	O4T-Loja	173.96
O5PC-Loja	191.72	O5PrC-Loja	20.52	O5B-Loja	84.35	O5T-Loja	157.92
Average	175.25		19.94		80.67		166.75
Std. Deviation	15.81		0.56		2.32		11.89

Appendix A.3. Moisture Content Results

Table A9. Moisture content for Guayacan.

Sample	w_w (gr)	w_d (gr)	MC (%)
G1MC	30.9	28	10.36
G2MC	32.5	29.3	10.92
G3MC	29.3	26.4	10.98
G4MC	29.4	26.6	10.53
G5MC	28	25.7	8.95
	Average		10.31
	Std. Deviation		0.82

Table A10. Moisture content for Ecuadorian oak from El Oro.

Sample	w_w (gr)	w_d (gr)	MC (%)
O1MC-ElOro	64.08	52.33	22.45
O2MC-ElOro	65.5	53.01	23.56
O3MC-ElOro	64.76	55	17.74
O4MC-ElOro	60.24	48.24	24.87
O5MC-ElOro	66.56	53.08	25.39
	Average		21.25
	Std. Deviation		2.52

Table A11. Moisture content for Ecuadorian oak from Manabí.

Sample	w_w (gr)	w_d (gr)	MC (%)
O1MC-Manabi	64.33	53.98	19.17
O2MC- Manabi	61.14	54.50	17.68
O3MC- Manabi	63.70	54.59	16.68
O4MC- Manabi	61.15	52.15	17.25
O5MC- Manabi	65.65	53.91	21.77
	Average		17.84
	Std. Deviation		1.02

Table A12. Moisture content for Guayacan.

Sample	w_w (gr)	w_d (gr)	MC (%)
O1MC-Loja	66.25	54.44	21.69
O2MC- Loja	65.33	53.52	22.06
O3MC- Loja	67.63	55.08	22.78
O4MC- Loja	67.06	56.09	19.55
O5MC- Loja	67.05	55.17	21.53
	Average		22.17
	Std. Deviation		0.45

References

1. Schubert, M.; Panzarasa, G.; Burgert, I. Sustainability in Wood Products: A New Perspective for Handling Natural Diversity. *Chem. Rev.* **2023**, *123*, 1889–1924. [CrossRef] [PubMed]
2. Hossain, M.U.; Poon, C.S. Comparative LCA of wood waste management strategies generated from building construction activities. *J. Clean. Prod.* **2018**, *177*, 387–397. [CrossRef]
3. Asdrubali, F.; Ferracuti, B.; Lombardi, L.; Guattari, C.; Evangelisti, L.; Grazieschi, G. A review of structural, thermo-physical, acoustical, and environmental properties of wooden materials for building applications. *Build. Environ.* **2017**, *114*, 307–332. [CrossRef]
4. Araya, R.; Guillaumet, A.; Valle, D.; Duque, M.d.P.; Gonzalez, G.; Cabrero, J.M.; De León, E.; Castro, F.; Gutierrez, C.; Negrão, J.; et al. Development of Sustainable Timber Construction in Ibero-America: State of the Art in the Region and Identification of Current International Gaps in the Construction Industry. *Sustainability* **2022**, *14*, 1170. [CrossRef]
5. Nunes, G.; de Melo Moura, J.D.; Güths, S.; Atem, C.; Giglio, T. Thermo-energetic performance of wooden dwellings: Benefits of cross-laminated timber in Brazilian climates. *J. Build. Eng.* **2020**, *32*, 101468. [CrossRef]
6. Wang, Z.; Xue, Q.; Ji, Y.; Yu, Z. Indoor environment quality in a low-energy residential building in winter in Harbin. *Build. Environ.* **2018**, *135*, 194–201. [CrossRef]
7. Keller, M.; Asner, G.P.; Blate, G.; McGlocklin, J.; Merry, F.; Peña-Claros, M.; Zweede, J. Timber production in selectively logged tropical forests in South America. *Front. Ecol. Environ.* **2007**, *5*, 213–216. [CrossRef]
8. Silvero, F.; Rodrigues, F.; Montelpare, S.; Spacone, E.; Varum, H. The path towards buildings energy efficiency in South American countries. *Sustain. Cities Soc.* **2019**, *44*, 646–665. [CrossRef]
9. Carrasco, A.; Cristian, T.; Crespo, E.; Mejía, E. Domestic timber market. *For. Use Timber Mark. Ecuadorian Amaz.* **2014**, *111*, 28–44.
10. Peck, R.B.; Bishop, J.P. Management of secondary tree species in agroforestry systems to improve production sustainability in Amazonian Ecuador. *Agrofor. Syst.* **1992**, *17*, 53–63. [CrossRef]
11. Vázquez, F.J.A.; Guitián, M.A.R. Bioclimatic and phytosociological diagnosis of the species of the *Nothofagus* genus (Nothofagaceae) in South America. *Int. J. Geobot. Res.* **2011**, *1*, 1–20. [CrossRef]
12. Sánchez-Pinillos, M.; Coll, L.; De Cáceres, M.; Ameztegui, A. Assessing the persistence capacity of communities facing natural disturbances on the basis of species response traits. *Ecol. Indic.* **2016**, *66*, 76–85. [CrossRef]
13. Muñoz, G.R.; Gete, A.R.; Saavedra, F.P. Implications in the design of a method for visual grading and mechanical testing of hardwood structural timber for designation within the European strength classes. *For. Syst.* **2011**, *20*, 235. [CrossRef]
14. Aronson, J.; Toledo, C.S. Caesalpinia paraguariensis (Fabaceae): Forage tree for all seasons. *Econ. Bot.* **1992**, *46*, 121–132. [CrossRef]
15. *ISO 5151:2017*; Non-Ducted Air Conditioners and Heat Pumps—Testing and Rating for Performance. International Standard Organization: Geneva, Switzerland, 2017; p. 71. Available online: https://www.iso.org/standard/63409.html (accessed on 15 May 2024).
16. *ANSI/ASHRAE Standard 16-2016*; Method of Testing For Rating Room Air Conditioners, Packaged Terminal Air Conditioners, and Packaged Terminal Heat Pumps for Cooling and Heating Capacity. ANSI/ASHRAE: Peachtree Corners, GA, USA, 2016; pp. 1–44.
17. *AHRI Standard 340/360*; Standard for Performance Rating of Commercial and Industrial Unitary Air-conditioning and Heat Pump Equipment. AHRI: Arlington, VA, USA, 2015; pp. 1–62. Available online: https://www.ahrinet.org/system/files/2023-06/AHRI_Standard_340-360_2015.pdf (accessed on 9 June 2024).
18. Ghani, S.; Gamaledin, S.M.A.; Rashwan, M.M.; Atieh, M.A. Experimental investigation of double-pipe heat exchangers in air conditioning applications. *Energy Build.* **2018**, *158*, 801–811. [CrossRef]
19. Kodur, V.K.R.; Harmathy, T.Z. Properties of Building Materials. In *SFPE Handbook of Fire Protection Engineering*; Springer: New York, NY, USA, 2016; pp. 277–324. [CrossRef]
20. Damfeu, J.C.; Meukam, P.; Jannot, Y.; Wati, E. Modelling and experimental determination of thermal properties of local wet building materials. *Energy Build.* **2017**, *135*, 109–118. [CrossRef]
21. Kouroussis, G.; Fekih, L.B.; Descamps, T. Assessment of timber element mechanical properties using experimental modal analysis. *Constr. Build. Mater.* **2017**, *134*, 254–261. [CrossRef]

22. Darmono; Ma'arif, F.; Widodo, S.; Pamungkas, S. Analysis of Mechanical Properties of Timber Materials. *J. Phys. Conf. Ser.* **2019**, *1387*, 012093. [CrossRef]
23. Tupėnaitė, L.; Žilėnaitė, V.; Kanapeckienė, L.; Sajjadian, S.M.; Gečys, T.; Sakalauskienė, L.; Naimavičienė, J. Multiple criteria assessment of high-rise timber buildings. *Eng. Struct. Technol.* **2020**, *11*, 87–94. [CrossRef]
24. Chen, C.-H. A Novel Multi-Criteria Decision-Making Model for Building Material Supplier Selection Based on Entropy-AHP Weighted TOPSIS. *Entropy* **2020**, *22*, 259. [CrossRef]
25. Reddy, A.S.; Kumar, P.R.; Raj, P.A. Entropy-based fuzzy TOPSIS framework for selection of a sustainable building material. *Int. J. Constr. Manag.* **2019**, *22*, 1194–1205. [CrossRef]
26. Movaffaghi, H.; Yitmen, I. Multi-criteria decision analysis of timber–concrete composite floor systems in multi-storey wooden buildings. *Civ. Eng. Environ. Syst.* **2021**, *38*, 161–175. [CrossRef]
27. Motuzienė, V.; Rogoža, A.; Lapinskienė, V.; Vilutienė, T. Construction solutions for energy efficient single-family house based on its life cycle multi-criteria analysis: A case study. *J. Clean. Prod.* **2016**, *112*, 532–541. [CrossRef]
28. Cavalli, A.; Cibecchini, D.; Togni, M.; Sousa, H.S. A review on the mechanical properties of aged wood and salvaged timber. *Constr. Build. Mater.* **2016**, *114*, 681–687. [CrossRef]
29. *ISO 8302:1991*; Thermal Insulation Determination of Steady-State Thermal Resistance and Related Properties. Guarded Hot Plate Apparatus. International Organization for Standardization: Geneva, Switzerland, 1991; p. 47.
30. *ASTM C177-19*; Standard Test Method for Steady-State Heat Flux Measurements and Thermal Transmission Properties by Means of the Guarded-Hot-Plate Apparatus. ASTM International: West Conshohocken, PA, USA, 2023; p. 23. [CrossRef]
31. *ASTM D143-14*; Standard Test Methods for Small Clear Specimens of Timber. ASTM International: West Conshohocken, PA, USA, 2021; p. 31. [CrossRef]
32. *ASTM D4442-20*; Standard Test Methods for Direct Moisture Content Measurement of Wood and Wood-Based Materials. ASTM International: West Conshohocken, PA, USA, 2020; p. 5.
33. Otten, K.A.; Brischke, C.; Meyer, C. Material moisture content of wood and cement mortars—Electrical resistance-based measurements in the high ohmic range. *Constr. Build. Mater.* **2017**, *153*, 640–646. [CrossRef]
34. Nicolalde, J.F.; Yaselga, J.; Martínez-Gómez, J. Selection of a Sustainable Structural Beam Material for Rural Housing in Latin América by Multicriteria Decision Methods Means. *Appl. Sci.* **2022**, *12*, 1393. [CrossRef]
35. Granta-Design, L. *CES-Edupack*; Granta Design Limited: Cambridge, UK, 2019.
36. Shi, J.; Sun, J. Prefabrication Implementation Potential Evaluation in Rural Housing Based on Entropy Weighted TOPSIS Model: A Case Study of Counties in Chongqing, China. *Sustainability* **2023**, *15*, 4906. [CrossRef]
37. Zhu, Y.; Tian, D.; Yan, F. Effectiveness of Entropy Weight Method in Decision-Making. *Math. Probl. Eng.* **2020**, *2020*, 3564835. [CrossRef]
38. Nabavi, S.R.; Wang, Z.; Rangaiah, G.P. Sensitivity Analysis of Multi-Criteria Decision-Making Methods for Engineering Applications. *Ind. Eng. Chem. Res.* **2023**, *62*, 6707–6722. [CrossRef]
39. Mousavi-Nasab, S.H.; Sotoudeh-Anvari, A. A new multi-criteria decision making approach for sustainable material selection problem: A critical study on rank reversal problem. *J. Clean. Prod.* **2018**, *182*, 466–484. [CrossRef]
40. Brancheriau, L.; Bailleres, H.; Guitard, D. Comparison between modulus of elasticity values calculated using 3 and 4 point bending tests on wooden samples. *Wood Sci. Technol.* **2002**, *36*, 367–383. [CrossRef]
41. Chun, Q.; Meng, Z.; Han, Y. Research on Mechanical Properties of Main Joints of Chinese Traditional Timber Buildings with the Type of Post-and-Lintel Construction. *Int. J. Archit. Herit.* **2017**, *11*, 247–260. [CrossRef]
42. Namari, S.; Drosky, L.; Pudlitz, B.; Haller, P.; Sotayo, A.; Bradley, D.; Mehra, S.; O'Ceallaigh, C.; Harte, A.M.; El-Houjeyri, I. Mechanical properties of compressed wood. *Constr. Build. Mater.* **2021**, *301*, 124269. [CrossRef]
43. Haller, P.; Wehsener, J. Festigkeitsuntersuchungen an Fichtenpressholz (FPH). *Holz als Roh und Werkst.* **2004**, *62*, 452–454. [CrossRef]
44. Anh, L.D.H.; Pásztory, Z. An overview of factors influencing thermal conductivity of building insulation materials. *J. Build. Eng.* **2021**, *44*, 102604. [CrossRef]

biomimetics

MDPI

Article

Insight into the Use of Brewers' Spent Grain as a Low-Carbon Aggregate in Building Materials

Badreddine El Haddaji [1,2], Mohammed-Hichem Benzaama [1], Marc Quiertant [1,*] and Yassine El Mendili [1]

1 Institut de Recherche ESTP, 94230 Cachan, France; belhaddaji@estp.fr (B.E.H.); hbenzaama@estp.fr (M.-H.B.); yelmendili@estp.fr (Y.E.M.)

2 ESITC Paris, 94110 Arcueil, France

* Correspondence: mquiertant@estp.fr

Abstract: This study investigates the use of Brewers' Spent Grains (BSGs) as a sustainable biocomposite building materials, using cornstarch as a biopolymer binder. BSG aggregates are compared with hemp shives, a conventional aggregate known for its thermal properties. Starch is employed as a natural binder in three different formulations to further reduce the carbon footprint of the building material. Considering aggregates, the first formulation contains only BSGs, the second consists of half BSGs and half hemp shives, and the third uses only hemp shives. In addition, morphological analysis using Scanning Electron Microscopy (SEM) is conducted to examine the microstructure and porosity of the raw BSG and hemp shives. Hygrothermal properties are measured using Heat Flow Meter (HFM) and Dynamic Vapor Sorption (DVS) techniques, while mechanical properties are also assessed. Results indicate that the thermal conductivity of the BSG formulation (0.131 W/(m·K)) is double that of the hemp shives formulation (0.067 W/(m·K)), whereas the mixed BSG/hemp shives formulation exhibits a thermal conductivity of 0.106 W/(m·K). However, DVS measurements reveal better hygrothermal properties for the BSG formulation compared to the hemp shives formulation. Lastly, mechanical properties are found to be nearly equivalent across the three formulations. These findings suggest that BSG waste has potential as a viable material for use in construction. Further work on formulation optimization and durability is necessary to fully realize the potential of this waste in promoting a circular economy within the building materials industry.

Keywords: brewers' spent grains; gelatinization of starch; bio-organic binders; thermal conductivity; sustainable materials; thermal insulation; hygrothermal properties; mechanical characterization

1. Introduction

As global environmental concerns intensify, the search for sustainable materials has become a key priority across industries. This is particularly the case in the construction sector, one of the largest contributors to carbon emissions [1–3]. Studies have shown that material emissions contribute a substantial portion of total emissions in buildings, ranging from 32 to 48% [4]. Embodied carbon emissions from construction materials have been quantified in various studies, with steel and concrete identified as the most carbon-emitting materials [5]. Strategies to reduce carbon emissions in buildings include using local and reused materials, which have been shown to have a significant impact on reducing embodied carbon emissions [5]. Among these strategies for reducing carbon emissions is the reuse and transformation of local waste from across the spectrum of human activities into unexpected sources of low-carbon alternatives for building materials. Multiple studies in the literature use this strategy with wastes from different origins. Example of such wastes includes waste sludges, demolition waste, plastic waste, and seashells wastes, which show the potential of some waste in building materials [6–10]. Properties of building materials containing such waste remain as good as those of conventional material for multiple applications [11,12].

Brewers' Spent Grain (BSG) is a by-product of the brewing process that has the potential for valorization in various industries, including construction [13]. Most BSGs end up in animal feed, in landfills, and as biogas, with lesser volumes being used for polymers, soil fertilizer, and paper, as shown in Figure 1. These applications are possible thanks to the fractions that can be found in BSGs, such as lignin, cellulose, hemicellulose, proteins, and lipids [14]. In the EU alone, there is an estimated 10.02 million tons of BSGs generated each year (2020) [15]. The current fetch price of this waste is EUR 40 a wet ton, and it can reach, for example, a market price of around EUR 180 a ton after being processed for bio-refinery applications [16]. However, storage conditions are important due to the microbiological activity of BSGs, as it can lead to fungal development [17]. This can be detrimental for the final application of this product and should be taken into consideration.

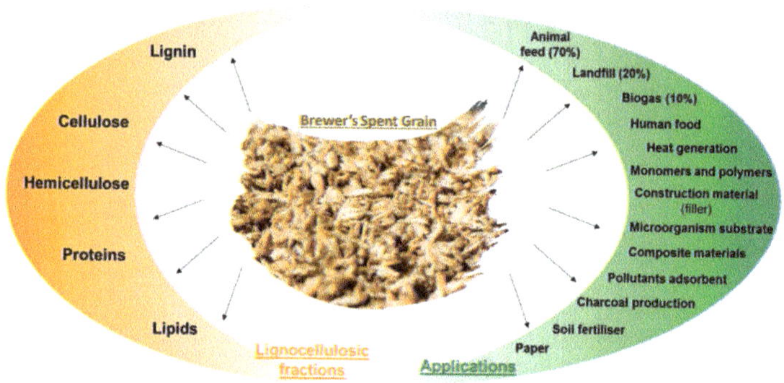

Figure 1. Fractions present in BSGs' lignocellulosic biomass and their potential applications [14].

The use of spent grains in building materials is a topic that has gained attention in recent years. Brewers' Spent Grains have been used in the production of lignin, a structural component with efficient thermal properties [18]. One study found that spent grains could be used to increase porosity in bricks, making them suitable for building materials [19]. Additionally, spent grains have been investigated for their potential in thermally insulating cellulose nanofiber aerogels, highlighting their versatility in different applications [20]. Furthermore, the valorization of Brewer's Spent Grain has been explored to produce biopolymer composite materials and bio-based building blocks [14]. BSG has emerged through the literature as a promising alternative to traditional building material. Indeed, this by-product of the brewing industry holds potential to contribute to sustainable construction practices.

The components of BSG (Figure 1) are used in building materials. These components are commonly known as lignocellulosic biomass, derived from agricultural waste, and are composed of various constituents, such as cellulose, hemicellulose, lignin, pectin, and protein [21]. These components have been explored for their potential use in building materials for insulation purposes. Cellulose, hemicellulose, and lignin have been identified as key components in this regard [22]. Furthermore, the antimicrobial properties of these natural biopolymers have been investigated, showcasing their potential as effective agents in building materials [23]. Additionally, the production of algal cellulose has been explored, emphasizing its potential use in building materials due to its composition of lipids, carbohydrates, and proteins [24]. In search of sustainable building materials, the use of biodegradable polymers has gained traction to reduce non-biodegradable plastic waste [25]. Bioplastics derived from cellulose and hemicellulose have also been studied for their potential applications in building materials [26]. Hence, common sense indicates that introducing BSGs into building materials is not a far-fetched idea and can bring benefits to the building materials industry with an additional purpose of reusing widely available waste in some regions.

In this work, we investigate the potential of using BSG as an aggregate in building materials. First, preconditioning of the raw BSG is studied and an adequate protocol is used to hamper the fungal development and ensure accurate dosing when formulating. Then, we choose to use raw BSG materials as a material for isolation that could increase energy efficiency of buildings. To achieve this goal, we have compared several formulations using cornstarch as a low carbon binder. We have chosen hemp shives as a commonly studied reference aggregate for comparison. Therefore, three formulations are studied, the first one with only BSG as aggregates, the second formulation with half hemp shives and half BSG and the third formulation with only hemp shives. We then have characterized the physical, mechanical and thermal properties of the studied formulations for potential use in insulations panels.

2. Materials and Methods

2.1. Materials

2.1.1. Brewers' Spent Grain

Brewers' Spent Grain (BSG) is a by-product of the beer-brewing process, consisting primarily of leftover husks, seeds, and other solid materials from malted barley. This material is typically rich in fibers, proteins, and other organic components. Sourced from local breweries, BSG is collected after the brewing process, where it would otherwise go to waste or be used for limited purposes such as animal feed.

In this study, we focus on repurposing BSG to produce sustainable insulation panels. The fibrous nature of BSG makes it a promising candidate for use as a bio-based aggregate in building materials, particularly in panels designed for thermal insulation. By using BSG, we aim to reduce the reliance on conventional, carbon-intensive materials, while supporting circular economic principles and promoting the local sourcing of raw materials from breweries.

The composition of BSG from literature ranges from 19.2% to 40% for hemicellulose, 12% to 33% for cellulose, 1% to 2.7% for starch, 14.2% to 26.7% for protein, 11.5% to 22% for lignin, 3% to 13% for lipids, 1.1% to 4.6% for Ash and 0.7% to 2% for phenolics [13].

The SEM image of BSG in Figure 2a shows a relatively dense surface with some visible micropores scattered across the structure. These pores are primarily formed by the fibrous remnants of the grain husks, which contribute to its moderate porosity. However, the compactness of the BSG structure limits the overall pore volume.

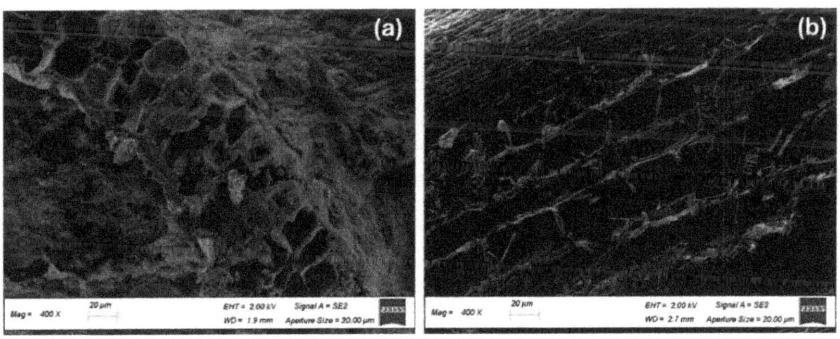

Figure 2. SEM images of (**a**) Brewers' Spent Grain and (**b**) hemp shives.

2.1.2. Hemp Shives

Hemp shives, the woody core of the hemp plant, are sourced from local farmers in Bar-sur-Aube, situated in the Champagne-Ardenne region of France. These shives are a by-product of hemp cultivation, primarily used for fiber production, and consist of lightweight, fibrous material with excellent thermal and acoustic properties. The thermal conductivity of hemp shives is 0.05 W/(m·K).

In this project, hemp shives are repurposed for the production of sustainable insulation panels. Their natural insulating capacity, coupled with their local availability, makes them an ideal bio-based aggregate for environmentally friendly building materials. By using hemp shives, we aim to reduce the environmental impact of insulation materials, support local agriculture, and promote the use of renewable, low-carbon resources in construction.

The composition of hemp shives from the literature ranges from 44% to 51.6% for cellulose, from 6.4% to 27% for hemicellulose, from 8.1% to 28% for lignin, from 1.0% to 29.4% for soluble compounds, from 0.6% to 8.8% for proteins, and from 1.0% to 6.6% for ashes [27]. The used hemp shives were provided from ISOCANNA® Saint-Astier in Saint-Astier, France.

In contrast to what was observed for BSGs, the SEM image of hemp shives in Figure 2b reveals a significantly more porous structure, with large, well-defined macropores. The fibrous and woody nature of hemp shives creates a highly open and irregular texture, allowing for greater pore connectivity and a higher overall porosity compared to BSGs. This increased porosity is advantageous for applications requiring enhanced insulation and moisture regulation.

Table 1 provides a summary of the physical properties of Brewers' Spent Grains and hemp shives.

Table 1. Physical properties of BSGs and hemp shives.

	Diameter (mm)	Density (kg·m^{-3})	Initial Water Content (%)
Hemp shives	1–5	105 ± 10	7.6
Brewers' Spent Grains	1–3	640 ± 50	10.3

2.1.3. Binder

Table 2 presents the chemical composition of cornstarch. The analysis of its composition shows that cornstarch is distinguished by a relatively high concentration of polysaccharides, specifically amylopectin and amylose, while the levels of extractives, including crude fats, proteins, and ash, are notably low.

Table 2. Chemical composition of the cornstarch.

Component	Content
Amylose (%)	26.55
Amylopectin (%)	73.67
Crude fats (%)	5.43
Crude proteins (%)	9.45
Ash (%)	0.87
Phosphor (%)	0.07
Moisture content (%)	10.8
Density (kg/m^3)	1416

The amylose content, which is the primary component of native starches, was measured at 24.6 g per 100 g of starch. This concentration falls within the typical range of amylose content for native plant starches, which is between 14% and 29%.

Cornstarch, primarily composed of amylose and amylopectin, undergoes gelatinization when heated in water. At 80 °C, the starch granules absorb water, causing them to swell and break the hydrogen bonds that hold the glucose molecules together. This swelling disrupts the crystalline structure of amylopectin, leading to a loss of birefringence. As the temperature rises, some of the amylose is released into the surrounding solution, while amylopectin remains in the swollen granules, increasing the viscosity of the mixture. The interaction of amylose and amylopectin with water forms a viscoelastic network, which traps water and gives the material its thickening and stabilizing properties. Upon cooling, amylose and amylopectin chains may partially recrystallize, a process known as retrogradation, which can strengthen the structure of the gel but may also cause syneresis, or water release.

In this study, we used gelatinized cornstarch as a bio-organic binder to formulate the targeted sustainable building materials. Corn-derived gelatinized starch, processed at 80 °C, forms a natural adhesive that promotes particle cohesion in plant fiber-based composites or biosourced aggregates. Acting as a natural polymer, it creates a viscoelastic network that improves the mechanical stability and moisture retention of the material. The use of cornstarch as a binder contributes to reducing the carbon footprint of building materials, aligning with circular economic principles, and providing an eco-friendly alternative to synthetic binders while maintaining comparable mechanical performance.

2.1.4. Water

In this study, gelatinized starch was prepared using local tap water. The elemental composition of the tap water used for mixing is presented in Table 3. The pH level of the tap water was recorded at 7.73. Understanding the composition of the water, particularly its pH, is essential, as it can influence the characteristics of the material. Moreover, this information is crucial for replicating and comprehending the experimental conditions.

Table 3. Chemical composition of tap water used for the preparation of gelatinized starch.

Element	Ca	Mg	Na	Cl	K	P	F	Zn	SO_4^-	NO_3^-	Fe	Cu	Mn
Concentration (mg/L)	95.3	6.8	12.5	18.1	3	0.7	0.2	0.2	37.6	43.4	<0.1	<0.1	<0.1

2.2. Methods

2.2.1. Sample Preparation

As already mentioned, we have developed three distinct formulations for biocomposite building materials using BSGs and hemp shives as aggregates, with gelatinized starch serving as the binder:

- Formulation 1 (BSG100): This formulation contains only BSGs.
- Formulation 2 (BSG50): This formulation consists of equal parts (50% each) of BSGs and hemp shives.
- Formulation 3 (BSG0): This formulation uses only hemp shives.

To cast the biocomposite materials, we prepared two types of samples: cubic samples for hygroscopic and thermal testing, and cylindrical samples for compressive strength testing. The dimensions and specifications for the samples are as follows:

- Cylindrical samples: These were produced in disposable molds with dimensions of 11 cm in diameter and 22 cm in height. The cylindrical geometry was chosen to facilitate mechanical testing, particularly compressive strength, which benefits from consistent load distribution across a circular cross-section.
- Cubic samples: These were produced using reusable wooden molds with dimensions of 25 cm in height, 25 cm in length, and 7 cm in thickness. The cubic shape was selected to optimize thermal property measurements, such as thermal conductivity, as the larger flat surface area ensures accurate placement of sensors and uniform heat flow during testing.

In typical applications for starch-based insulating panels, the starch content in the mixture is set at 12.5% of the total mass of the panel.

To cast the biocomposite materials (Figure 3), we began by heating water to 80 °C for the gelatinization of starch. Once the desired temperature was reached, we gradually added starch to the water while stirring continuously until a homogeneous paste was obtained;

this paste served as the binder for our material. While the starch was gelatinized, we prepared the BSGs in a mixer. Once the starch paste was ready, we incorporated it into the prepared BSGs in the mixer. We then mixed the combined ingredients for approximately 10 min, pausing for 5 min during this period to homogenize the mixture, ensuring that any non-impregnated fibers located on the periphery of the mixer were fully integrated. This step is crucial for achieving a homogeneous blend that optimally uses the properties of the individual components. After obtaining a consistent mixture, we poured the material into the prepared cylindrical and cubic molds. Finally, once the samples were molded, we stored them in an oven at 23 °C and 50% humidity for 28 days. After this curing period, the characterization test could begin.

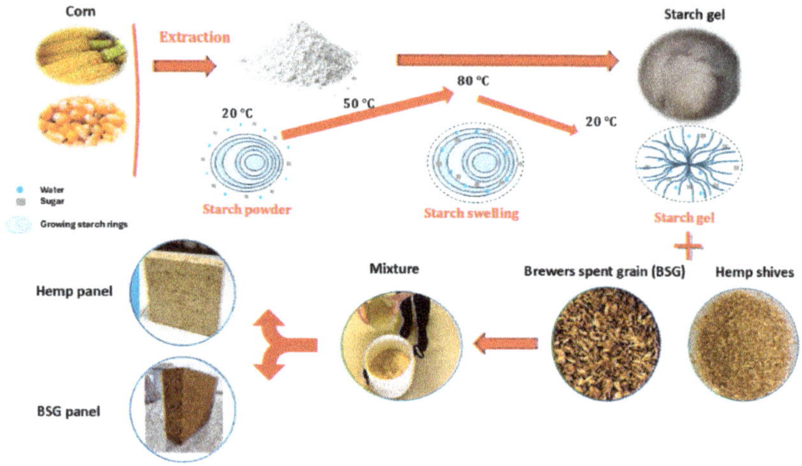

Figure 3. Overview of the manufacturing process for biocomposite building materials using Brewers' Spent Grains and hemp shives, with gelatinized starch as a binder.

By following this procedure, we can create effective biocomposite materials with the potential for use in sustainable building applications, leveraging the natural properties of BSG and hemp shives, along with the binding capacity of gelatinized starch.

2.2.2. Characterization Techniques

The compressive tests were performed in accordance with EN-12390-3 on 11–22 cm cylindrical specimens. After the specified curing period, each specimen was removed from the curing environment and measured to confirm that the diameter was 11 cm and the height was 22 cm. The compressive strength of samples was evaluated using an IGM testing equipment with a 50 kN capacity. The load was applied at a controlled rate of 1 mm/min until the specimen failed. Results were reported in megapascals (MPa).

Heat Flux Meter (HFM) measurements were conducted using a (Netzsch HFM 446, Selb, Germany) to evaluate the thermal transmittance of the building material. The panel samples, with dimensions of 25 × 25 × 7 cm^3, were placed in a testing chamber with controlled temperature conditions. The HFM sensors were affixed to the surface of the material to measure the rate of heat flow through it. Temperature sensors were installed on both sides of the panel to record the internal and external surface temperatures. The system was maintained at a steady-state temperature gradient, and data on heat flux and temperature differences were continuously recorded. The thermal transmittance (U-value) was computed based on the measured heat flux and temperature differential, providing an assessment of the material's insulation performance.

Dynamic Vapor Sorption (DVS) measurements were performed using a Vsorp Moisture Sorption Analyzer apparatus supplied from (Mercer Instruments, Passy, France) to assess the moisture sorption characteristics of the BSG samples. Small samples were placed in the chamber, where they were exposed to controlled humidity levels ranging from 0% to 95% relative humidity (RH) in incremental steps. The weight changes of the sample were continuously monitored as it absorbed or desorbed moisture. Equilibrium was assumed once the sample weight stabilized at each RH step. The data were used to generate sorption isotherms, providing insights into the material's moisture sensitivity and capacity.

The chemical composition of the tap water was analyzed using a (Thermo Scientific DIONEX ICS3000 CC, Villebon-sur-Yvette, France) ion chromatography system. This system is equipped with a dual-pump module, a chromatography detector module, an eluent generator module, and an autosampler. The column pressure and eluent flow rate were set to 1.0 mL/min. Ionic species were identified and quantified by interpolating against a suitable calibration curve, with all measurements performed at room temperature.

The fibers and Brewers' Spent Grains (BSGs) were analyzed to determine their specific surface and absolute density. The specific surface area was measured using the Brunauer–Emmett–Teller (BET) method, which relies on nitrogen gas adsorption [28]. Samples were dried at 50 °C for 24 h to eliminate moisture before undergoing adsorption in a BET apparatus. Absolute density was measured using a helium pycnometer (Accupyc II 1340, Microméritics, Mérignac, France), where dried samples were placed in a chamber with helium gas to determine their true volume by gas displacement, and density was calculated as the ratio of mass to volume.

Morphological characterization was conducted using a scanning electron microscope (SEM; SUPRA 55 SAPPHIRE; Carl Zeiss, Jena, Germany). BSG and hemp shives were cleaned with ethanol and oven-dried at 50 °C for 24 h. They were embedded in an epoxy resin mixed with hardener, poured into molds, and cured as per the resin's specifications. The cured sample was polished to expose the samples and create a smooth surface (<0.3 µm). To ensure conductivity, a thin silver coating was applied via sputtering. The prepared sample was mounted on a conductive SEM stub using carbon adhesive, ensuring proper charge dissipation for SEM imaging at 20 kV.

To assess the apparent porosity of the panel samples, the gravimetric technique outlined in ISO 5017 was employed. Initially, the samples were saturated with water and weighed while suspended on a scale. They were then submerged in water (M_w) before being weighed again in the air (M_o). After drying in an oven, the dry weight (M_d) was recorded. The apparent porosity was subsequently computed using the following Equation (1):

$$\rho_d = \rho_w * \frac{M_d}{M_o - M_w} \quad (1)$$

where ρ_d is the apparent density of the panel sample.

3. Results

3.1. Moisture Sorption Isotherm of Panel Specimens

Figure 4 illustrates the typical mass change during adsorption and desorption processes at various RH levels. In this figure, it can be seen that the moisture sorption curves for all samples exhibited a sigmoidal shape, indicating Type II isotherm behavior according to the IUPAC classification. This isotherm type is characteristic of porous materials, such as the formulations investigated in this study.

The mass differential values increased from BSG0 to BSG50, reflecting the influence of porosity on this phenomenon. Notably, BSG100, despite having the highest density and lowest porosity, exhibits a significant capacity for moisture absorption due to the hydrophilic nature of the BSG. In contrast, BSG0, which consists solely of hemp shives, demonstrates a different moisture absorption profile; while it has higher porosity, its ability to retain moisture is lower compared to BSG100. BSG50, with equal parts of both materials, shows intermediate absorption characteristics.

The comparison between hemp shives and BSGs reveals that while hemp shives contribute to the overall porosity and can enhance moisture retention, BSGs have a stronger capacity to absorb moisture due to their composition. This high absorption capacity of

BSG100 can inhibit surface condensation by promoting uniform moisture distribution throughout the formulation structure.

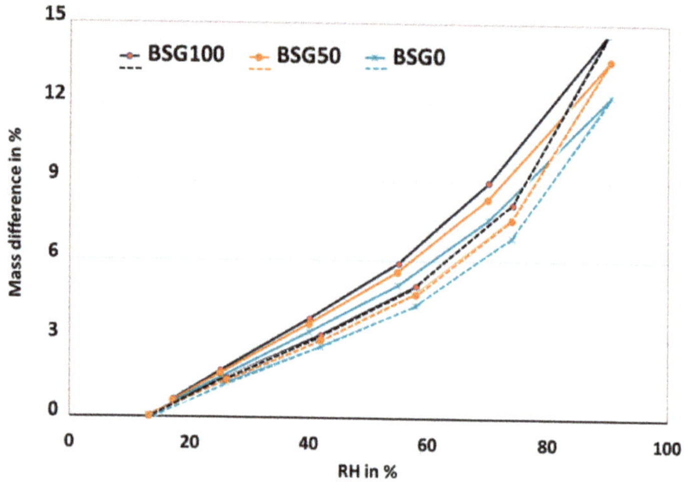

Figure 4. Moisture sorption isotherms for the studied formulations at room temperature. Continuous line: adsorption. Dashed line: desorption.

Figure 5a shows the decrease in density from approximately 350 kg/m^3 with 100% BSGs to 175 kg/m^3 with 100% hemp shives; this decrease can be explained by the inherent differences in the bulk densities of the raw materials, as detailed in Table 1. Specifically, raw BSGs have a bulk density nearly six times greater than that of hemp shives. This difference results in lower mass per unit volume for formulations containing hemp shives, leading to reduced overall density. Additionally, the observed increase in porosity, from 65% with 100% BSGs to 81% with 100% hemp shives, is influenced by the structural characteristics of the materials. Hemp shives, being less compact and more fibrous, naturally introduce larger void spaces within the mixture, enhancing porosity. The hydrophilic nature of both materials further contributes to this behavior by facilitating water absorption and swelling during sample preparation, thus increasing the volume without substantially affecting the mass. Furthermore, the compaction process during panel formation may align the particles in a manner that accentuates void creation, particularly in samples with a higher hemp shive content. This phenomenon is most pronounced in BSG0, where the combination of lower material density and increased porosity aligns with the properties of the constituent aggregates.

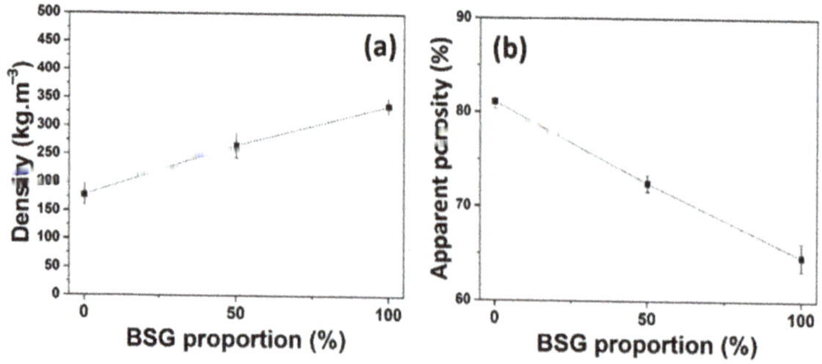

Figure 5. (**a**) Density of specimens with different proportions of BSG and (**b**) apparent porosity of panels with different proportions of BSG.

3.2. Thermal Conductivity of Panel Specimens

As shown in Figure 6, the thermal conductivity of the panels formulated solely with BSGs was measured to be 0.131 W/(m·K) at 20 °C, which is significantly higher than that of the formulation containing only hemp shives, as the latter exhibits a thermal conductivity of 0.067 W/(m·K). The mixed formulation (BSG50) shows an intermediate thermal conductivity of 0.106 W/(m·K).

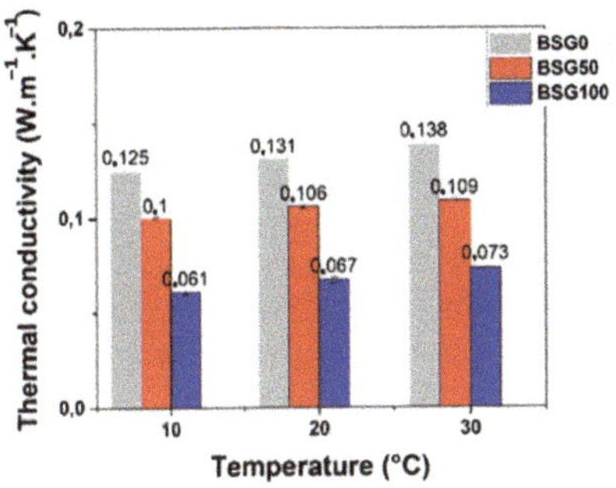

Figure 6. Thermal conductivity vs. temperature of specimens with different proportions of BSGs.

These results indicate that the incorporation of BSGs into the panel significantly influences thermal conductivity. The lower thermal conductivity of the hemp shive formulation suggests its superior insulating properties, while the mixed formulation reflects a balance between the two materials' thermal characteristics.

The insulating potential of the panels can be further enhanced through increased porosity within the material. The incremental substitution of hemp shives with BSGs shows that organic additives like BSGs contribute to pore development based on the physical properties of the organic additives, leading to improved insulation performance and mechanical properties. Additionally, the higher specific heat capacity of the BSG formulation plays a crucial role in reducing thermal gradients, which is essential for maintaining comfortable indoor temperatures.

Incorporating BSGs in panel construction improves thermal characteristics, minimizing heat transfer and enhancing thermal comfort for occupants. The thermal insulation capabilities of both BSGs and hemp shives are noteworthy. This relationship highlights the significance of energy efficiency and environmental considerations in the thermal performance of panels. Structures equipped with the formulated panels consume less energy for thermal comfort, ultimately reducing fuel and energy consumption.

3.3. Specific Heat Capacity of Panel Specimens

The specific heat capacity values for the formulated panels range from 1327 to 1682 $J \cdot kg^{-1} \cdot K^{-1}$ at 20 °C, as shown in Figure 7. Notably, at all measured temperatures, the specific heat capacity (Cp) of panels containing only Brewers' Spent Grains as granular (BSG100) is significantly higher than that of the formulation with hemp shives (BSG0) or the mixed formulation (BSG50), as illustrated in Figure 6. This increase in specific heat capacity is attributed to the unique properties of BSGs, which exhibit low thermal conductivity, alongside a high specific heat capacity.

In the construction industry, materials with elevated calorific capacity are essential for enhancing energy efficiency. The specific thermal capacity is crucial for maintaining indoor comfort for building occupants. Walls constructed with materials having high thermal capacity can effectively regulate temperatures, keeping spaces cooler in summer without the need for additional energy inputs. Conversely, in winter, such materials can help retain heat within buildings for extended periods.

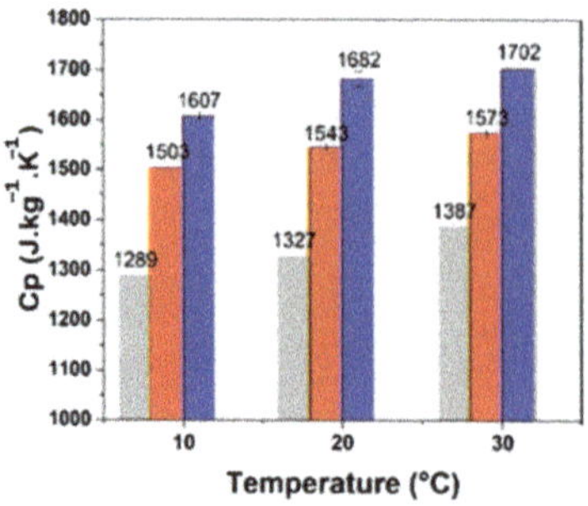

Figure 7. Specific heat capacity of specimens with different proportions of BSGs.

In conclusion, the results indicate that the formulated panels, particularly those incorporating BSGs, demonstrate improved thermal performance compared to formulations composed solely of hemp shives or a blend of the two materials.

3.4. Mechanical Strength of Panel Specimens

The compressive strength of building materials is a crucial factor that determines their suitability for construction applications. In this study, we evaluated the compressive strength of the three considered formulations: BSG0, BSG50, and BSG100. The results of the compressive strength tests are presented in Table 4.

Table 4. Mean compressive strength results.

Formulations	Compressive Strength (MPa)
BSG0	0.327 ± 0.03
BSG50	0.162 ± 0.02
BSG100	0.098 ± 0.02

As illustrated in Table 4, the formulation containing only hemp shives (BSG0) exhibits the highest compressive strength at 0.327 MPa. This superior strength can be attributed to the inherent properties of hemp shives, which are known for their fibrous structure that contributes to load-bearing capacity [29]. Conversely, the mixed formulation (BSG50) shows a reduced compressive strength of 0.162 MPa, while the formulation comprising entirely BSGs (BSG100) demonstrates the lowest compressive strength, at 0.098 MPa.

The decrease in compressive strength observed in the BSG50 and BSG100 formulations can be linked to the porous nature and lower density of BSGs, thus affecting the overall mechanical performance of the material. Studies indicate that materials with higher porosity typically exhibit lower compressive strength due to the increased voids that compromise the structural integrity [30].

Additionally, the use of organic materials, such as BSGs, can enhance the sustainability of building materials, although they may contribute to lower compressive strength compared to traditional materials [31]. It is crucial to balance the benefits of using recycled and biosourced materials with their mechanical performance to optimize their application in construction.

In conclusion, while the BSG0 formulation demonstrates the highest compressive strength, the BSG100 formulation highlights the trade-off between incorporating sustainable materials and achieving required mechanical properties. Further research is needed to explore the potential of enhancing the compressive strength of BSG-based formulations, possibly through the incorporation of additives or composite strategies.

4. Discussion

The results of this study demonstrate the hygrothermal and mechanical properties of building panels formulated with BSGs and hemp shives, providing insights into their moisture sorption behavior, specific heat capacity, and mechanical strength. The incorporation of BSG as an alternative material in building construction presents both advantages and trade-offs. In this section, the findings are discussed in comparison with the existing literature, focusing on the implications for building material performance and sustainability.

4.1. Moisture Sorption Behavior

The sorption isotherms for all formulations exhibited a Type II isotherm, characteristic of porous materials, and this finding is consistent with findings in the literature on bio-based materials such as hemp and recycled agricultural waste [32,33]. Notably, BSG100, despite its lower porosity and higher density, displayed a significant capacity for moisture absorption, which can be attributed to the hydrophilic nature of BSGs. This aligns with previous studies that reported high moisture affinity in materials with similar organic compositions [34]. The porosity of hemp shives in BSG0 contributed to higher mass differentials during moisture absorption, yet the ability to retain moisture was lower than that of BSG100. This trend of lower moisture retention with higher porosity has been documented in materials like hemp composites [35].

4.2. Thermal Conductivity

Thermal conductivity testing revealed that the panels made solely from hemp shives (BSG0) exhibited superior insulation properties, with a thermal conductivity of 0.067 W/(m·K), which is significantly lower than the 0.131 W/(m·K) of the BSG100 formulation. The observed thermal conductivity values for BSG0 are in line with previous studies on hemp-based materials, where hemp–lime and hemp–shive composites have shown thermal conductivity ranging from 0.05 to 0.08 W/(m·K), as shown in Table 5. Meanwhile, the higher thermal conductivity of BSG100 can be attributed to the denser structure of BSG, thus reducing its insulating potential. This result is expected, as an increased density tends to give a material with higher thermal conductivity. The chemical compositions of the BSG- and hemp shive-based composites play a significant role in determining their thermal and mechanical performance. Brewer's Spent Grains (BSGs), as seen in the Materials section, are composed of hemicellulose (22–40%), cellulose (12 33%), protein (14.2–26.7%), and lignin (11.5–27.8%), which contribute to the material's porosity and density, factors that are crucial in defining its thermal conductivity and mechanical strength. In contrast, hemp shives, which consist of cellulose (44–51.6%), hemicellulose (6.4–27%), and lignin (8.1–28%), provide excellent thermal insulation due to their fibrous structure and low density. The higher cellulose content in hemp shives, compared to BSGs, is likely a key contributor to the lower thermal conductivity of the BSG0 formulation (0.067 W/(m·K)), which falls within the range typically observed for hemp-based materials. Meanwhile, the BSG100 formulation, with a thermal conductivity of 0.131 W/(m·K), demonstrates higher conductivity due to the denser structure of BSG, which reduces its insulating potential. This aligns with the general principle that higher density materials tend to have higher thermal conductivity. The increased porosity of the BSG100 formulation, driven by the BSG content, also compromises its mechanical strength, as seen in the lower compressive strength values.

Thermal conductivity is a key criterion in evaluating insulation materials, and the values obtained for BSG formulations, especially BSG50 and BSG0, align with the ranges reported for commonly used materials. This demonstrates that BSG-based composites can deliver competitive performance while leveraging an abundant industrial by-product. Additionally, the comparison of thermal conductivity with other studies highlights consistency with renewable aggregates like hemp shives and flax, reinforcing the potential of BSG as an efficient, eco-friendly insulation material.

4.3. Specific Heat Capacity

The specific heat capacity of the panels varied from 1327 $J \cdot kg^{-1} \cdot K^{-1}$ to 1682 $J \cdot kg^{-1} \cdot K^{-1}$, with BSG100 displaying the highest values across all temperatures tested. This observation corresponds with findings from studies on bio-based materials, where a high calorific capacity was noted for materials such as wood waste [36]. The high specific heat capacity of BSG formulations allows for better thermal regulation, a desirable property for energy-efficient buildings.

Table 5. Comparison of mechanical and thermal properties within the literature.

Property	Formulation	Used Materials	Measured Value	Literature Range	Literature Materials	References
Thermal Conductivity (W/(m·K))	BSG0	Hemp shives	0.067	0.05–0.08	Wheat/hemp shives+starch/coconut husk/rice husk	[37–40]
	BSG50	50% BSG + 50% hemp	0.106	0.1–0.15	Hemp shives+flax/wood fibers/corn cob	[38,41,42]
	BSG100	Brewers' Spent Grains	0.131			
Compressive Strength (MPa)	BSG0	Hemp shives	0.327	0.2–0.4	Rye/hemp shives/reed/beet pulp+ starch	[38,43–45]
	BSG50	50% BSG + 50% hemp	0.162	0.06–0.2	Wheat/wheat starch+wheat straw/lavender/corn cob	[37,38,46,47]
	BSG100	Brewers' Spent Grains	0.098			

4.4. Mechanical Strength

The mechanical strength of the formulated panels, especially in terms of compressive strength, indicated that BSG0 exhibited the highest value (0.327 MPa), followed by BSG50 (0.162 MPa) and BSG100 (0.098 MPa). The decreased compressive strength observed with increasing BSG content can be attributed to the increased porosity of BSG, which compromises the structural integrity of the material. Similar trends have been reported in the literature for bio-based and porous materials, for which high porosity results in reduced mechanical performance, as shown in Table 5. However, the lower strength of BSG100 may be counterbalanced by its environmental benefits and thermal advantages.

Table 5 gives a comparison of the results presented in this paper with some other building materials used for insulation (values issued from the literature). Both thermal properties and mechanical properties are equivalent when considering the materials of this study or other existing materials. The materials chosen as examples and found in the literature (Table 5) are not necessarily sourced from wastes, and, in some cases, they are produced specifically to be incorporated in building materials. With our approach focused on sustainability and circular economy, the BSG can come not as a replacement but as an alternative in localities where this waste is abundant. Then, using BSG would reduce the stress on conventional resources destined for building materials. Although fungal development was not observed on the samples for the period studied, we recommend additional studies toward understanding the durability of materials containing BSGs.

4.5. Recommendations

Based on the findings of this study, several recommendations can be made for the application of BSG- and hemp-based materials in construction:

Optimizing the blend of materials. The use of a mixed formulation (BSG50) provides a balance between moisture sorption, thermal conductivity, and mechanical strength. Further optimization of the blend ratio, possibly combined with additives, could enhance the overall performance of the material.

Focus on thermal performance for energy efficiency: The high specific heat capacity of BSG100 makes it a suitable material for passive thermal regulation in buildings. Incorporating BSG into panel systems could reduce energy demands for heating and cooling, contributing to sustainability goals in the construction industry.

Addressing fungal development possibilities: The high moisture absorption capacity of BSGs can be beneficial for regulating humidity in indoor environments. However, strategies to prevent excessive moisture retention, such as surface treatments or incorporating moisture barriers [48], should be considered to avoid potential issues with mold growth or material degradation.

5. Conclusions

This study demonstrates the potential of BSGs and hemp shives, along with cornstarch, as sustainable building materials. The main conclusions are as follows:

- The moisture sorption analysis reveals that all formulations exhibit Type II isotherm behavior, with BSG displaying superior moisture absorption due to its hydrophilic nature. Thermal conductivity tests show that while the BSG formulation has higher thermal conductivity than hemp shives alone, it still contributes to energy efficiency through enhanced thermal regulation.
- Mechanical strength assessment indicates that the formulation with only hemp shives (BSG0) exhibits the highest compressive strength, while BSG formulations (BSG50 and BSG100) demonstrate lower strength due to the increased porosity of BSG aggregates. However, the environmental benefits of using BSGs, including waste reduction and improved thermal properties, could outweigh these limitations.

These findings underscore the importance of balancing sustainability with mechanical performance in building materials. By integrating BSGs into construction practices, the industry can advance toward a more circular economy. Future research should focus on optimizing material blends, enhancing mechanical properties, and exploring fungal development to maximize the performance of BSG-based building materials. This study lays the foundation for further exploration into sustainable construction practices that leverage abundant by-products, contributing to both environmental goals and resource efficiency in the building sector.

Author Contributions: Conceptualization, B.E.H., M.-H.B. and Y.E.M.; methodology, B.E.H., M.-H.B. and Y.E.M.; investigation, M.-H.B. and Y.E.M.; writing—original draft preparation, B.E.H., M.-H.B. and Y.E.M.; writing—review and editing, B.E.H., M.-H.B., M.Q. and Y.E.M. All authors have read and agreed to the published version of the manuscript.

Funding: This research received no external funding.

Institutional Review Board Statement: Not applicable.

Data Availability Statement: The experimental and computational data presented in this paper are available from the corresponding author upon request.

Acknowledgments: We would like to express our sincere gratitude to M. Thierry Masurier for their generous support in supplying the Brewers' Spent Grains (BSGs) used in this study.

Conflicts of Interest: The authors declare no conflicts of interest.

References

1. Whang, S.-W.; Kim, S. Determining sustainable design management using passive design elements for a zero emission house during the schematic design. *Energy Build.* **2014**, *77*, 304–312. [CrossRef]
2. Hildebrandt, J.; Hagemann, N.; Thrän, D. The contribution of wood-based construction materials for leveraging a low carbon building sector in Europe. *Sustain Cities Soc.* **2017**, *34*, 405–418. [CrossRef]
3. Zhang, X.; Wang, F. Analysis of embodied carbon in the building life cycle considering the temporal perspectives of emissions: A case study in China. *Energy Build.* **2017**, *155*, 404–413. [CrossRef]
4. Ahmed, K.; Hajian, H.; Hasu, T.; Kurnitski, J. Kouvola Housing fair NZEB houses energy, cost and carbon analyses. *E3S Web Conf.* **2020**, *172*, 13001. [CrossRef]
5. Liu, K.; Leng, J. Quantitative research on embodied carbon emissions in the design stage: A case study from an educational building in China. *J. Asian Archit. Build. Eng.* **2022**, *21*, 1182–1192. [CrossRef]
6. Johnson, O.A.; Napiah, M.; Kamaruddin, I. Potential uses of Waste Sludge in Construction Industry: A Review. *Res. J. Appl. Sci. Eng. Technol.* **2014**, *8*, 565–570. [CrossRef]

7. Lockrey, S.; Nguyen, H.; Crossin, E.; Verghese, K. Recycling the construction and demolition waste in Vietnam: Opportunities and challenges in practice. *J. Clean. Prod.* **2016**, *133*, 757–766. [CrossRef]
8. Hussan, A.; El Haddaji, B.; Zelloufi, M.; Sebaibi, N. Innovative Cutting and Valorization of Waste Fishing Trawl and Waste Fishing Rope Fibers in Cementitious Materials. *Appl. Sci.* **2024**, *14*, 3985. [CrossRef]
9. El Mendili, Y.; Benzaama, M.-H. Investigation of Mechanical and Thermal Performance of Concrete with Scallop Shells as Partial Cement Replacement: Alternative Binder and Life Cycle Assessment. *CivilEng* **2022**, *3*, 760–778. [CrossRef]
10. Hussan, A.; Sebaibi, N.; El Haddaji, B.; Zelloufi, M. Comparative Effects of Polyethylene and Polypropylene-Based Waste Fishing Gear Fibers on Mechanical Parameters and Porosity of Cementitious Materials. In *RILEM Spring Convention and Conference*; Springer: Cham, Switzerland, 2025; pp. 132–140. [CrossRef]
11. Affan, H.; El Haddaji, B.; Ajouguim, S.; Khadraoui, F. A Review—Durability, Mechanical and Hygrothermal Behavior of Building Materials Incorporating Biomass. *Eng* **2024**, *5*, 992–1027. [CrossRef]
12. Benzaama, M.-H.; Touati, K.; El Mendili, Y.; Le Guern, M.; Streiff, F.; Goodhew, S. Machine Learning-Based Indoor Relative Humidity and CO_2 Identification Using a Piecewise Autoregressive Exogenous Model: A Cob Prototype Study. *Energies* **2024**, *17*, 243. [CrossRef]
13. Lynch, K.M.; Steffen, E.J.; Arendt, E.K. Brewers' spent grain: A review with an emphasis on food and health. *J. Inst. Brew.* **2016**, *122*, 553–568. [CrossRef]
14. Pérocheau Arnaud, S. Valorisation of Brewer's Spent Grain: Lignocellulosic Fractionation and its Potential for Polymer and Composite Material Applications. *Chem. Afr.* **2024**, *7*, 2989–3010. [CrossRef]
15. Zeko-Pivač, A.; Tišma, M.; Žnidaršič-Plazl, P.; Kulisic, B.; Sakellaris, G.; Hao, J.; Planinić, M. The Potential of Brewer's Spent Grain in the Circular Bioeconomy: State of the Art and Future Perspectives. *Front. Bioeng. Biotechnol.* **2022**, *10*, 870744. [CrossRef] [PubMed]
16. Buffington, J. The Economic Potential of Brewer's Spent Grain (BSG) as a Biomass Feedstock. *Adv. Chem. Eng. Sci.* **2014**, *4*, 308–318. [CrossRef]
17. Bianco, A.; Budroni, M.; Zara, S.; Mannazzu, I.; Fancello, F.; Zara, G. The role of microorganisms on biotransformation of brewers' spent grain. *Appl. Microbiol. Biotechnol.* **2020**, *104*, 8661–8678. [CrossRef] [PubMed]
18. Gbenebor, O.P.; Olanrewaju, O.A.; Usman, M.A.; Adeosun, S.O. Lignin from Brewers' Spent Grain: Structural and Thermal Evaluations. *Polymers* **2023**, *15*, 2346. [CrossRef]
19. Russ, W.; Mörtel, H.; Meyer-Pittroff, R. Application of spent grains to increase porosity in bricks. *Constr. Build. Mater.* **2005**, *19*, 117–126. [CrossRef]
20. Heidari, N.A.; Fathi, M.; Hamdami, N.; Taheri, H.; Siqueira, G.; Nyström, G. Thermally Insulating Cellulose Nanofiber Aerogels from Brewery Residues. *ACS Sustain. Chem. Eng.* **2023**, *11*, 10698–10708. [CrossRef]
21. Mujtaba, M.; Fraceto, L.F.; Fazeli, M.; Mukherjee, S.; Savassa, S.M.; Araujo de Medeiros, G.; do Espírito Santo Pereira, A.; Mancini, S.D.; Lipponen, J.; Vilaplana, F. Lignocellulosic biomass from agricultural waste to the circular economy: A review with focus on biofuels, biocomposites and bioplastics. *J. Clean. Prod.* **2023**, *402*, 136815. [CrossRef]
22. Ali, S.S.; Al-Tohamy, R.; Mohamed, T.M.; Mahmoud YA, G.; Ruiz, H.A.; Sun, L.; Sun, J. Could termites be hiding a goldmine of obscure yet promising yeasts for energy crisis solutions based on aromatic wastes? A critical state-of-the-art review. *Biotechnol. Biofuels Bioprod.* **2022**, *15*, 35. [CrossRef] [PubMed]
23. Lobo, F.C.M.; Franco, A.R.; Fernandes, E.M.; Reis, R.L. An Overview of the Antimicrobial Properties of Lignocellulosic Materials. *Molecules* **2021**, *26*, 1749. [CrossRef]
24. Zanchetta, E.; Damergi, E.; Patel, B.; Borgmeyer, T.; Pick, H.; Pulgarin, A.; Ludwig, C. Algal cellulose, production and potential use in plastics: Challenges and opportunities. *Algal Res.* **2021**, *56*, 102288. [CrossRef]
25. Samir, A.; Ashour, F.H.; Hakim, A.A.A.; Bassyouni, M. Recent advances in biodegradable polymers for sustainable applications. *NPJ Mater. Degrad.* **2022**, *6*, 68. [CrossRef]
26. Abe, M.M.; Martins, J.R.; Sanvezzo, P.B.; Macedo, J.V.; Branciforti, M.C.; Halley, P.; Botaro, V.R.; Brienzo, M. Advantages and Disadvantages of Bioplastics Production from Starch and Lignocellulosic Components. *Polymers* **2021**, *13*, 2484. [CrossRef]
27. Diakité, M.-S.; Lenormand, H.; Lequart, V.; Arufe, S.; Martin, P.; Leblanc, N. Cell Wall Composition of Hemp Shiv Determined by Physical and Chemical Approaches. *Molecules* **2021**, *26*, 6334. [CrossRef]
28. Brunauer, S.; Emmett, P.H.; Teller, E. Adsorption of Gases in Multimolecular Layers. *J. Am. Chem. Soc.* **1938**, *60*, 309–319. [CrossRef]
29. Barnat-Hunek, D.; Smarzewski, P.; Fic, S. Mechanical and thermal Properties of Hemp-Lime Composites. *Compos. Theory Pract.* **2015**, *15*, 21–27.
30. Shalwan, A.; Alajmi, A.; Yousif, B.F. Theoretical Study of the Effect of Fibre Porosity on the Heat Conductivity of Reinforced Gypsum Composite Material. *Polymers* **2022**, *14*, 3973. [CrossRef]
31. Boros, A.; Tőzsér, D. The Emerging Role of Plant-Based Building Materials in the Construction Industry—A Bibliometric Analysis. *Resources* **2023**, *12*, 124. [CrossRef]
32. de Bruijn, P.; Johansson, P. Moisture fixation and thermal properties of lime–hemp concrete. *Constr. Build. Mater.* **2013**, *47*, 1235–1242. [CrossRef]

33. Segerholm, B.K.; Walkenström, P.; Nyström, B.; Wålinder, M.E.P.; Larsson Brelid, P. Micromorphology, moisture sorption and mechanical properties of a biocomposite based on acetylated wood particles and cellulose ester. *Wood Mater. Sci. Eng.* **2007**, *2*, 106–117. [CrossRef]
34. Gurunathan, T.; Mohanty, S.; Nayak, S.K. A review of the recent developments in biocomposites based on natural fibres and their application perspectives. *Compos. Part A Appl. Sci. Manuf.* **2015**, *77*, 1–25. [CrossRef]
35. Florea, I.; Manea, D.L. Analysis of Thermal Insulation Building Materials Based on Natural Fibers. *Procedia Manuf.* **2019**, *32*, 230–235. [CrossRef]
36. Cetiner, I.; Shea, A.D. Wood waste as an alternative thermal insulation for buildings. *Energy Build.* **2018**, *168*, 374–384. [CrossRef]
37. Ismail, B.; Belayachi, N.; Hoxha, D. Optimizing performance of insulation materials based on wheat straw, lime and gypsum plaster composites using natural additives. *Constr. Build. Mater.* **2020**, *254*, 118959. [CrossRef]
38. Viel, M.; Collet, F.; Lanos, C. Development and characterization of thermal insulation materials from renewable resources. *Constr. Build. Mater.* **2019**, *214*, 685–697. [CrossRef]
39. Panyakaew, S.; Fotios, S. New thermal insulation boards made from coconut husk and bagasse. *Energy Build.* **2011**, *43*, 1732–1739. [CrossRef]
40. Marques, B.; Tadeu, A.; António, J.; Almeida, J.; de Brito, J. Mechanical, thermal and acoustic behaviour of polymer-based composite materials produced with rice husk and expanded cork by-products. *Constr. Build. Mater.* **2020**, *239*, 117851. [CrossRef]
41. Brzyski, P.; Barnat-Hunek, D.; Suchorab, Z.; Łagód, G. Composite Materials Based on Hemp and Flax for Low-Energy Buildings. *Materials* **2017**, *10*, 510. [CrossRef]
42. Muthuraj, R.; Lacoste, C.; Lacroix, P.; Bergeret, A. Sustainable thermal insulation biocomposites from rice husk, wheat husk, wood fibers and textile waste fibers: Elaboration and performances evaluation. *Ind. Crops Prod.* **2019**, *135*, 238–245. [CrossRef]
43. Bakatovich, A.; Davydenko, N.; Gaspar, F. Thermal insulating plates produced on the basis of vegetable agricultural waste. *Energy Build.* **2018**, *180*, 72–82. [CrossRef]
44. Bakatovich, A.; Gaspar, F. Composite material for thermal insulation based on moss raw material. *Constr. Build. Mater.* **2019**, *228*, 116699. [CrossRef]
45. Karpaky, H.; Maalouf, C.; Bliard, C.; Gacoin, A.; Lachi, M.; Polidori, G. Mechanical and thermal characterization of a beet pulp starch composite for building applications. *E3S Web Conf.* **2019**, *85*, 08005. [CrossRef]
46. Abbass, O.A.; Salih, A.I.; Al Hurmuzy, O.M. Study of the mechanical and physical properties of bio-composite material based on wheat starch and wheat straw fibers. *IOP Conf. Ser. Mater. Sci. Eng.* **2020**, *745*, 012075. [CrossRef]
47. Ratiarisoa, R.V.; Magniont, C.; Ginestet, S.; Oms, C.; Escadeillas, G. Assessment of distilled lavender stalks as bioaggregate for building materials: Hygrothermal properties, mechanical performance and chemical interactions with mineral pozzolanic binder. *Constr. Build. Mater.* **2016**, *124*, 801–815. [CrossRef]
48. El Haddaji, B.; Keita, E.; Fabien, A.; Aimedieu, P.; Sebaibi, N.; Collet, P.; Andres, V.; Porcheron, T.; Van Damme, H.; Audibert-Hayet, A.; et al. Nanosilica-based post-treatment of hardened cement-based materials: The underlying physics. *Cem. Concr. Res.* **2023**, *171*, 107211. [CrossRef]

biomimetics

MDPI

Review

Plants Inspired Biomimetics Architecture in Modern Buildings: A Review of Form, Function and Energy

Maryam Bijari [1], Ardalan Aflaki [1,*] and Masoud Esfandiari [2]

1 Faculty of Architecture and Art, University of Guilan, Rasht 41996-13776, Iran
2 SYSTRA S.p.A, 38123 Trento, Italy
* Correspondence: ar.aflaki@guilan.ac.ir

Abstract: Biomimicry architecture provides innovative solutions to contemporary environmental challenges by drawing inspiration from nature's strategies to enhance sustainability and energy efficiency in the built environment. Plants, with their remarkable ability to adapt to changes in light, temperature, and humidity, serve as a central model for biomimetic design due to their potential to optimize energy use and improve building performance. By leveraging these natural principles, biomimetic architecture can significantly reduce carbon emissions and create eco-friendly structures that respond dynamically to environmental conditions. This approach not only addresses the urgent need for sustainable development but also fosters harmony between human-made environments and the natural world. This study offers a comprehensive review of biomimetic technologies, focusing on their role in improving energy efficiency and building performance. Also, it examines a range of global case studies that have successfully implemented biomimicry, showcasing its versatility and effectiveness across diverse environmental and architectural contexts. Based on these insights, this research proposes a novel design inspired by the moonflower plant, which adapts to its environment by responding to external stimuli. The proposed design applies these adaptive strategies to architectural solutions, enabling buildings to optimize performance by dynamically interacting with environmental conditions such as light and temperature. By analyzing biomimetic principles and their applications, this study contributes to the growing body of knowledge on sustainable architecture. It highlights the potential of biomimicry to balance environmental sustainability with economic growth, offering valuable insights for architects, designers, and policymakers seeking to create greener, more efficient built environments.

Keywords: adoptable building façade; biomimicry architecture; daylighting; energy-efficient building; nature-based solution

1. Introduction

The increasing impact of global climate change has sparked worries about environmental harm, and the release of greenhouse gases from these activities leads to irreversible damage [1,2]. Consequently, cutting down on greenhouse gas emissions has been a key to this in recent years to curb the rise in global temperatures [3]. One important consideration for reducing energy consumption and minimizing the carbon footprint is to manage the use of electricity and artificial lighting in buildings. Artificial lighting typically requires significant energy, often sourced from non-renewable resources such as oil [4]. By controlling consumption, artificial lighting not only can improve the quality of the environment and reduce psychological and physiological issues in humans, but also helps to reduce

https://doi.org/10.3390/biomimetics10020124

the use of non-renewable energy sources and minimize the undesirable impacts on the environment [5].

Nature provides an abundant supply of solutions for numerous scientific and technical challenges. Over billions of years, biological systems have evolved to create advantageous relationships between structure, properties, and performance that are essential for survival [6]. By emulating these natural systems, buildings can be designed to maximize natural light exposure, reduce glare and heat gain, and utilize dynamic shading systems that adapt to changing conditions [7,8].

Biomimicry, as a scientific field, involves an interdisciplinary approach and has the capacity to offer sustainable solutions through the collaboration of biologists, physicists, chemists, engineers, and architects [9]. The importance of this approach in architecture has increased because biomimicry brings several inspirations from nature and introduces great potential to create a sustainable, energy-efficient built environment. This great opportunity is more tangible, particularly today, because new building materials and new construction techniques can be seen more than ever in the past [10,11]. On the other hand, plants utilize various mechanisms to harness sunlight efficiently, making them a prime model to emulate for sustainable structures that maximize natural light utilization [12].

Despite the studies and activities carried out in this field, there are still many unresolved problems and issues that require more attention. One of these issues is designing biomimicry architecture, which requires the development of cognitive systems that support the early stages of the innovation process, including problem-based information retrieval and formation and estimation. So, this article aims to examine the implementation of biomimicry with a focus on inspiration from plants in a built environment to reduce the use of non-renewable energy resources and fossil fuels for lighting. It assesses the effectiveness of this approach by looking at various aspects, such as maximizing natural light and decreasing the use of artificial lighting. Ultimately, it seeks to improve and address existing challenges in this area with creative and efficient solutions, taking a step forward toward employing biomimicry architecture to reduce environmental pollution and carbon footprint.

2. A Literature Review

2.1. Biomimicry

Contreras Lezcano [13] is known as the founder of the biomimicry movement. She set out all her theories in the book "*Biomimicry: Innovations Inspired by Nature*". A year later, together with Dayna Baumeister, they founded biomimicry, which puts into practice a deep understanding of the biological adaptation of organisms to help architects, engineers, and designers solve design and engineering problems in a sustainable way. Biomimicry is a relatively new scientific field that involves studying nature's designs and mimicking them to solve human problems. Essentially, designers look at nature as a "model, measure, and mentor", with one of its key goals being sustainability and environmental preservation [13].

The term biomimicry comes from the Greek word "bios", meaning life, and mimesis, meaning imitation. Biomimicry is defined as "the abstract design of good ideas from nature" or "emerging disciplines that mimic designs and processes of nature to create healthier and more sustainable planets" [14]. The idea behind this concept is to emphasize sustainability as a main goal of biomimicry, which, when applied to building design in order to increase the strength of materials through self-healing and self-assembly properties, can provide better solutions to increase building performance, save energy, and provide material cost savings by eliminating waste [15].

Biomimicry is all about reimagining humans and about what to do to make industrial life possible. This perspective introduces nature as a mentor to walk alongside and respect it [16]. Biomimicry is a brilliant approach to looking for sustainable solutions to human problems by mimicking and emulating nature in its analogies, phenomena, and patterns. Biomimicry's main aim is to make a great design by mimicking the different living organisms, which have been evolving for 3.8 billion years [17]. Biomimicry imitates living biological systems in non-living systems based on the characteristics of a living biological system [18]. In architecture, biomimicry manifests as both a tool and ideology, leading to holistic designs that reflect biological processes [19].

2.2. Biomimicry in Architecture

In recent years, a nature-inspired approach to architecture has emerged as a way to mitigate the negative impacts of the industrial age, enhancing urban living while creating visually appealing super-tall buildings [20,21]. In fact, biomimicry facilitates the expansion of sustainable thinking through principles such as connectivity and system integration. In architecture, it is an interdisciplinary design approach that has not yet been fully explored, but it can expand the designer's realm of ideas across various fields [22]. While nature is imitated, it is not merely reproduced; instead, it requires further abstraction to derive solutions inspired by natural models. These solutions align with life's principles, potentially reducing environmental burdens by being less reliant on fossil fuels and more self-organizing and multifunctional [23,24].

Depending on the usage, biomimicry in the literature has various definitions. In sustainability, it refers to mimicking biological forms, processes, and systems to develop sustainable solutions. Ecosystem design strategies involve applying lessons from the natural world to support sustainable architecture and urban planning. Furthermore, design takes an interdisciplinary approach, studying and adapting concepts from nature to address design challenges [25].

As shown in Figure 1, biomimicry approaches are usually divided into two categories: defining human design needs or problems and searching for solutions from other organisms or ecosystems to solve this problem or designing based on biology or identifying a specific feature, behavior, or function in an organism or ecosystem and translating it into human designs. This utilization of nature and living organisms can be used in three methods: copy, abstraction, and inspiration [26,27].

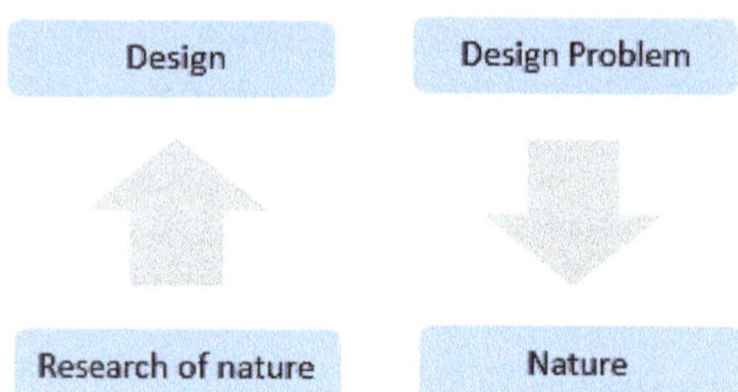

Figure 1. Biomimicry approaches.

The organism level requires the designer to become inspired by a specific organism, like a plant or an animal, while also analyzing how it functions, using the whole organism or a part of it [28]. The behavioral level involves mimicking the behavior of a living being in its environment to draw inspiration from a structure or a process in design [29]. The ecosystem level includes imitating how a living being interacts with its environment and the number of its constituent elements, which are examined on a broader scale [27].

Biomimicry architecture has three levels of imitation, including imitation of organisms, behavior, and ecosystems. On the organism level, buildings may imitate the characteristics of an individual organism. On the behavioral level, the design may be inspired by how the organism behaves or relates to its larger context [30]. On the ecosystem level, design may draw from the entire ecosystem of an organism and its surroundings. It emphasizes natural processes and cycles of the greater environment [31,32]. Three levels of biomimicry determine which aspect of 'bio' can be 'mimicked' and applied to a design problem: the organism, the behavior, or the ecosystem level repeated [30]. As shown in Figure 2, at each of these levels, there are five other possible dimensions for imitation, including form, material, structure, process, and function [33,34].

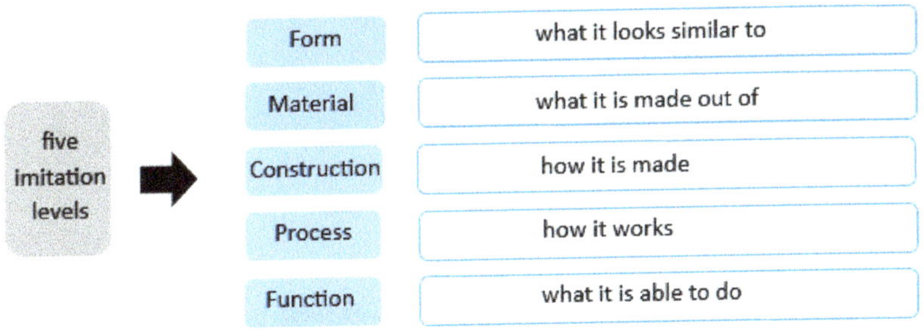

Figure 2. Level of biomimicry.

2.3. Plants in Biomimetic Architecture

Inspiration from plants is often seen in biomimicry architecture, which is motivated by nature [35]. Trees and plants are flexible structures that are sensitive to climatic conditions, and as a response, they have developed a number of techniques and features that help them overcome such conditions [36]. Throughout history, there has always been a tendency to mimic natural forms of plants and incorporate their beauty into the designs. This has laid the foundations of botanical design. With advancements in natural sciences, simple imitation of plants has gradually transformed into a more detailed examination of their structure and function [37].

There is a design approach aimed at enhancing the mechanical, adaptive, and environmental performance of buildings by mimicking the morphology and structure of plants [38]. Plants, being a rich and extensive source of inspiration, are utilized in various ways based on studies conducted on nature-inspired structures. Some plants are used for natural ventilation, others for efficient lighting solutions, and some can even contribute to acoustic considerations.

While some plants are only considered for aesthetics in architecture, others can be inspired or used to significantly improve building performance and efficiency [39,40]. For instance, the ability of plants to dispose of light as an energy source and to reuse it in case of need is unparalleled [41], which inspires controlling entry for sunlight in buildings. Also, regulating internal temperature is another significant characteristic of buildings inspired by plants or flowers [39].

Plant biomimetic design research primarily focuses on the plant's form, color, texture, function, and structure. This design philosophy emphasizes integrating knowledge from ergonomics, art, new materials, and energy efficiency into the design of buildings, products, and materials. This approach contributes to multiple goals, including enhancing design aesthetics and practicality, balancing technology and aesthetics, and harmonizing the delicate combination of tradition and modernity [42].

When drawing inspiration from the structure and function of plants, this process can be referred to as Phyto-mimetics. Plants and buildings exhibit similarities in how they interact with the environment. Due to their predominantly immobile nature, plants have to autonomously manage their temperature, exposure to light, and other essentials. Similarly, building architecture needs to integrate mechanisms for managing solar heat gain, air circulation, temperature regulation, and provision of natural light to cater to the occupants [43]. Plants, due to their immobility, have developed specific mechanisms to protect themselves from various environmental factors such as darkness, light, humidity, rainwater, fire, temperature, freezing, air movement, and air quality. These adaptations evolve over time and generations in response to the constantly changing environment [44].

2.4. Light in Biomimicry Architecture

Natural light in buildings is very important because not only does it have positive physical and mental effects on individuals, but it can also reduce energy consumption and help in creating sustainable buildings with minimal carbon footprint [45]. The use of natural light in architectural spaces has been very useful in various aspects such as energy efficiency, cost control, user health and well-being, and prevention of electricity [46,47]. Therefore, natural light should be the main source of illumination during the daytime. This light should create intimacy and provide a comfortable visual and thermal environment for residents [48].

Passive solar energy requires the placement and design of buildings in such a way that they interact positively with the environment and climate, and the biomimetic approach reinforces this goal [49]. Therefore, nature can be a helpful guide in utilizing solar light [50]. Biomimicry principles are modeled after nature's ways. They harness solar energy efficiently, utilize resources and energy minimally, match forms to functions, and effectively recycle and repurpose materials [51].

Plants are drawn toward sunlight as an inspiring source for architects. This shows that plants are attracted to sources of light [52] or have specific mechanisms to allow light to enter their internal organs [17]. Biomimicry offers potential solutions and inspirations for architectural lighting design projects.

Studying plant adaptation to solar conditions inspires effective design strategies to improve building performance. Integrating solar technologies with biomimetic adaptive solutions can lead to sustainable design [53,54]. Most plants need sunlight to survive and maintain their energy. They use photosynthesis to convert solar radiation into storable energy. Plants adjust their orientation and inclination according to different weather conditions [55]. Some of them are also able to actively track the sun because of their light-sensitive organs [56].

Drawing inspiration from the process of harnessing light from plants in architecture includes various aspects such as light control, light transmission, light reflection, light passage, light concentration, etc. With the knowledge of biomimicry and the help of plants, one can respond to one or more of these aspects [57]. For example, the Zehra flower basket illuminates its internal space by passing light through its translucent body, or the silver Ragwort can reflect light to a great extent and create scattered and filtered light, with many plants having the ability to utilize natural light [58,59].

Minimizing shadows through the building is another concept widely used in lighting design. Plants not only have specific ways to absorb and store sunlight, but they also have different methods to manipulate and control light. This principle is mainly seen among plants with phyllotactic geometry, which often harnesses light in building lighting design. These projects used the Fibonacci law [53]. Also, techniques in façade design for shadow casting inspired by plants can be used to modify the incoming light to the building [60]. Therefore, by incorporating plants and their potential to utilize natural light, buildings can benefit from this renewable energy in various ways.

Many studies have been conducted in recent years on biomimicry and its impact on sustainability and energy conservation, but emphasis has not been placed on the role of plants in this approach. In a study, biomimetic research within the framework of sustainable development goals has been applied. The alignment of biomimicry with key sustainable development objectives has been approved because of its interdisciplinary nature and potential to provide solutions related to healthcare [61]. In another study, the role of bionic design in promoting sustainable development has been evaluated by analyzing its applications in morphological, functional, and material aspects. The results address considerably the current challenges of bionic design in terms of environmental sustainability [22]. In recent years, there have been few studies that focus on potential solutions in nature in order to establish daylighting in buildings. In line with the importance of the potential of plants to utilize natural light to reduce energy consumption, a study has focused on designing a responsive biomimetic kinetic system inspired by the functional and adaptive principles of Gazania flowers [62]. Additionally, another paper examines the use of biomimicry to improve daylight performance in office buildings in Cairo, Egypt. Through simulating and drawing inspiration from natural processes, designs have been proposed to enhance energy efficiency with an emphasis on natural light [63]. Apart from the application of biomimicry science in the built environment, studies have been conducted on the extensive applications of plants in the fields of biomimicry and biological inspiration and their role in developing sustainable solutions in medicine, material science, and environmental technology [64]. According to the previous literature, compatible biomimetic solutions and their efficiency in the built environment in terms of daylighting concern and energy consumption are questionable. Moreover, innovative solutions based on the current literature in order to provide natural lighting and energy saving have not been presented. These concepts will be presented after analysis of the current literature in order to be evaluated in future studies.

3. Research Methodology

The methodology in this article adopts the approach of "Scientific Methods and Rationale for Systematic Literature Review (SPAR-4-SLR)" developed by Paul and colleagues [65]. SLR is a rigorous and systematic type of exhaustive review of the already existing literature on a given topic. The primary purpose of SLR is to synthesize results from various studies, which then gives researchers an opportunity to draw wider and fuller conclusions about their topic of interest. This methodology involves several fundamental steps, which include locating relevant research, selecting studies that are appropriate, and systematically evaluating the quality as well as the significance of the findings reported in those studies. The application of the SLR method cuts across several fields and disciplines; this attests to its applicability in resolving a wide variety of research problems while adequately synthesizing results, hence contributing to an enhanced understanding of intricate subjects in different areas [66]. Therefore, this research analyzes significant research works from 2005 to 2024 in reputable databases, including Scopus, Science Direct, and Google Scholar. The SPAR-4-SLR protocol is chosen due to its precise methodological framework that includes clearly defined inclusion and exclusion criteria for selecting studies, helping to maintain high standards of evidence and ensuring that only the most relevant and reliable studies are included in the review.

As shown in Table 1, the review process is divided into three distinct stages. The first stage involves identifying and reviewing the relevant literature and obtaining the related literature. This involves collecting a wide range of resources, totaling 317 sources based on the keywords of this research, to ensure that this review has been conducted at a reasonable level. Then, the collected literature is organized and refined. This includes classifying the studies, summarizing the findings, and determining which studies are most relevant to the research objectives. The final stage focuses on evaluating the quality and relevance of the literature and reporting the findings. Ultimately, 110 articles were used as references in this research. It includes analysis of results, case studies, conclusions, and presentation of information in a clear and structured manner. The aim of this study is to provide a strong foundation for understanding existing features and functions in biomimetic architecture, focusing on the potential of plants to inspire the application of nature's knowledge in this field.

Table 1. Literature review design.

Study Stage	Identification
First	Review domain: biomimicry architecture, natural light, plants. Review questions: How much research has been conducted in this field? Which are the leading countries in this field of research? Source type: journal articles Acquisition Search mechanism and material acquisition: Scopus, Science Direct, and Google Scholar database Search Period: 2005–2024 Keywords: biomimicry architecture, biomimicry design, biomimicry and sustainability, use of natural light in biomimicry architecture, light simulation, biomimicry and plants, case study of biomimicry design.
Second	Organization Organization Codes: Publication Trend Analysis, Keyword Co-Occurrence Analysis, Co-Authorship Analysis. Organization Frameworks: challenges and potentials of the desired approach Purification Article Type Excluded: proceeding and book, editorial Article Type Included: 120 articles for analysis (journal articles, review articles)
Third	Evaluation Analysis Methods: bibliometrics, analysis, and review of case examples Agenda Proposal Method: thematic analysis. Reporting Reporting Conventions: figures, tables, networks, graphs, and words. Limitations: Data accessed from dimension review are limited to bibliometric data
Overview of SPAR-4-SLR Protocol Framework (Source: Adapted from JOB Rotimi [67])	

Terminology Definitions

Researchers have been studying biomimicry for several decades. A contemporary understanding of biomimicry has been clarified through many topics. Therefore, illuminating the scope of exploration and the importance of common terms related to it in the classification of academic fields is crucial [61,68]. For this purpose, the most important words and phrases were presented in Table 2. These words were used in preparing this article, demonstrating the importance of studying biomimicry. This perspective emphasizes the interdisciplinary impact and expansion of biomimicry, creating an opportunity for specialists in various fields to collaborate and participate in discussions.

Furthermore, studies indicate that since the inception of this field in 1997 until 2024, there has been an increasing interest in biomimetic and biomimetic structures, and the esteem for this science is growing day by day. Figure 3 shows increased publication in biomimicry and biomimetic architecture, which is an indication that this field is becoming more popular not only in architecture but in other sciences as well.

Table 2. Key terminology of research keywords and definitions.

Keyword	Definition
Bionic	Bionics is a term invented to describe the prospective field, involving copying, imitating, and learning from nature [9].
Bio-morphology	It means creating structures that resemble natural organisms, meaning that the buildings have skin similar to natural shapes and are only inspired by nature in terms of appearance [31].
Phyto-mimetics	When inspiration is drawn from the structure and functionality of plants, it may be referred to as phyto-mimetics [43].
Biomimicry	Biomimicry is an approach to innovation that seeks sustainable solutions to human challenges by emulating nature's time-tested patterns and strategies [53,69].
Biomimetic	This word is made up of parts including "bio" (life) and "mimesis" (imitation). Biomicism can be defined as forms inspired by nature, adaptation, or application in architecture, which can be examined in two ways: replication of natural objects and formation process in constructions [70,71].
Biomechanic	Biomechanics is a scientific field that applies biological and engineering principles to living organisms. It studies organism movement, control of those movements, force impact, and pressure on tissues [72].
Bio-inspiration	This concept focuses on using nature for inspiration in design without copying it directly. Bio-inspiration uses nature for ideas and patterns to find creative solutions [73]. Mimicking shapes employs technology to create sustainable materials similar to natural objects [10].

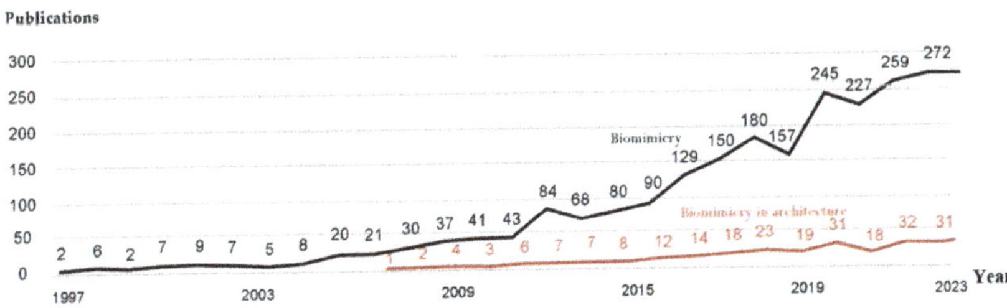

Figure 3. The total number of publications in Scopus from 1997 to 2023.

4. Analysis of Case Studies (Results)

This section analyzes and examines the five levels of biomimicry architecture (refer to Figure 2), focusing on natural solutions inspired by plants, in order to understand the effectiveness of this approach in contributing to the use of natural light for energy-efficient buildings.

4.1. Form

Forms and architectural shapes can be very effective in building lighting and benefiting from natural light. The shapes found in nature provide abundant solutions that allow designers to improve the use of daylight, both in the overall form and in the details of the building form, and control its entry into the interior space, drawing inspiration from them [30]. On the other hand, plants, due to their reliance on sunlight for growth and survival, have forms and shapes that can make the most of natural light. This dependence on natural light not only influences their physical appearance but also shapes their growth patterns and overall physiology [74].

Figure 4a,b shows the Esplanade Theatre, which is located in Marina Bay near the historic Singapore River. Architect Michael Wilford initially intended for the Marina Bay Splendid Theatre, a music hall and cultural venue, to have a transparent ceiling to provide a more accurate view from inside the building [75]. However, due to the constant heat and humidity of Singapore, a curved glass ceiling might cause the building to overheat. This led this project to draw inspiration from the durian tree in the area, a symbolic fruit (Figure 4c) of Singapore. The durian fruit is composed of three layers; the spongy middle layer has thermal properties that help preserve the durian fruit, while the outer layer has spiky features that protect the fruit from excessive heat from sunlight. The architect applies strategies of the double-layer shell by replacing the external triangular aluminum lamella fins, which are geometrically different and are calculated as a second skin above the glass dome based on the sun's path throughout the year [73].

(a) (b) (c)

Figure 4. The Esplanade Theatre, Singapore. (**a**) Distant view of the building (**b**) Close-up view of the building (**c**) Durian as a symbolic fruit in Singapore.

The design included a double-layered canopy with the second layer on top of the glass ceiling. This biomimetic design reduces the total energy consumption in the building by up to 30% and cuts down the use of artificial lighting by up to 55% [60,76]. The design utilizes a two-layer approach using triangular aluminum windows of different shapes that are determined annually by the sunlight paths on the second layer over the glass ceiling. To create maximum shade in all directions, aluminum windows are used throughout the building in such a way that their points are placed at different heights and angles depending on the locations of the daily sunlight paths annually [73,76].

The structure features a thermoplastic waterproofing membrane for single-layer waterproofing on the roof. This homogeneous elastic membrane contains a high number of solid materials with a high molecular weight that remains permanently stable over time. This consistent quality guarantees excellent durability, an optimal balance between tensile strength and elongation, as well as resistance to high temperatures and mechanical impact [77].

In another example shown in Figure 5a,b, Al Bahr Towers' integrated system is built on the concept of bio-inspired architecture, performance-oriented technology, and local architecture [78]. This design incorporates dynamic shading surfaces to meet sustainable development standards, utilizing natural daylight effectively and managing solar heat gain [79]. One of the goals of this design was to discover ideas inspired by biology. For this purpose, various natural systems were inspired by nature, from cacti and their flowers. Cacti have a canopy-like design that protects them from delicate skin, while their flowers open and close in response to weather changes (Figure 5c) [80]. This building, with its smart facade design, has improved user comfort and increased the physical and mental well being of residents. It offers better spaces that naturally brighten up through improved natural light acceptance, advanced external views, reduced use of obstructive curtains, and enhanced comfort by reducing heavy air conditioning loads.

(a) (b) (c)

Figure 5. Al Bahr Towers, Abu Dhabi, United Arab Emirates. (**a**) Distant view of the building (**b**) Close-up view of the building (**c**) Cacti and its flowers as a source of inspiration.

Figure 6a,b presents the Swiss Re headquarters in London. This building takes inspiration from a glass sponge and features a series of triangles on its exterior that resembles the shape of a sponge. This marine organism is made up of several layers of glass, and despite the small size of the threads, it creates a very sturdy skeleton (Figure 6c). This skeleton consists of networks of fibers that make up square cells, which are held together by other fibers that are arranged diagonally, thus describing spirals. The outer structure of the Swiss Re headquarters mimics the skeleton of Euplectella [60]. Its aerodynamic shape and glazing minimize wind loads while maximizing natural light and ventilation [15]. This form disperses reflected light and deflects gusting winds, thereby enhancing its environmental effects. The building structure was constructed in a grid form using the diagrid-architecture system inspired by the lattice-like exoskeleton of the Venus' flower basket for a strong assembly of lightweight steel structure [73,81]. The empty spaces are stacked to form a spiral atrium space in the building. The spiral light well inside directs fresh air up through the building and greatly reduces energy consumption with a double-glazed curtain wall [82,83]. Ultimately, these design features have led to an 80% reduction in the building's energy consumption [15].

(a) (b) (c)

Figure 6. Swiss Re headquarters in London. (**a**) Distant view of the building (**b**) Close-up view of the building (**c**) Marine organism with several layers.

4.2. Materials

Based on research conducted on plants and their mechanism, plants have a high potential to provide suitable solutions for improving building materials that meet human needs. For example, studies show that the porous internal structure of plant leaves and their biological composition of water and chlorophyll can create a spectral reflection effect on the leaves and produce materials resistant to sunlight [84]. Additionally, mimicking plants with surfaces containing defects, pores, and veins can help create semi-transparent assisting materials [80]. Plants with very fine veins on their leaves not only prevent water from sliding on their surfaces but also can be effective in reflecting sunlight and be compatible with warm and dry climates [85]. On the other hand, the tensile and flexible structure present in some plants can assist in creating suitable tensile materials in construction [53].

As shown in Figure 7a,b the implementation of a responsive facade system in a building envelope has been considered in the Media-TIC building. Figure 7c shows how the system can control the solar heat gain and improve the indoor environmental quality. ETFE is a hybrid material (Ethylene Tetra Fluoro Ethylene) with special characteristics. ETFE enhances insulation and light quality, while sensors adjust the layer's transparency according to sunlight, managing heat gain and improving indoor environmental conditions. The cushion covering system is constructed using ETFE polymer and is enclosed with lamella fins. The pneumatic mechanisms of the lamella fins are automatically activated by light sensors inspired by the plant's response to sunlight, which provides an active response to solar energy occurrence and improves thermal insulation [86]. This significantly lowers the solar factor of the building four times and reduces energy consumption by about 20% [80,85]. The mechanism used in this building was developed after comprehensive research, showing a very low economic cost compared to the project, making up 5% of the total [87].

(a) (b) (c)

Figure 7. Building façade of Media-TIC building, Barcelona, Spain. (**a**) Distant view of the building (**b**) Close-up view of the building (**c**) The porous internal structure of plant leaves.

In another design shown in Figure 8a, agave leaf mountain mint (Figure 8c) was considered as a primary concept for a 20-story office building in Lahore, Pakistan. Designing an adaptive bioclimatic façade as a practical solution to increase energy efficiency in high-gloss buildings in hot and humid areas has been considered [88]. The main facade module consists of four shadow devices that can tilt in both horizontal and vertical axes. In addition, this facade is able to generate its own electricity in addition to solar shading [89]. An adaptive biomimetic facade with adjustable shading devices that operate in horizontal and vertical positions effectively manages solar gain. As shown in Figure 8b, the shading facade, inspired by this plant, provides solar control without obstructing the user's view. This building has succeeded in reducing energy consumption by 32% and decreasing illuminance levels (500–750 lux) to 50% of the total area [90].

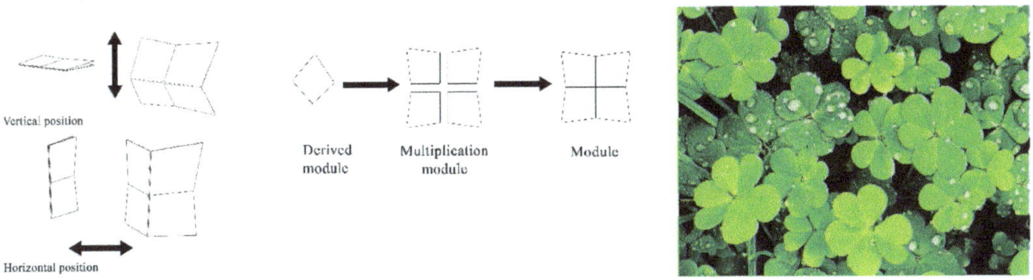

(a) (b) (c)

Figure 8. Application of vertical foldable shading devices on an office building, Lahore, Pakistan. (**a**) Shadow devices in horizontal and vertical axes (**b**) Module procedure (**c**) Agave leaf mountain mint.

4.3. Structure

The structure present in plants always provides suitable solutions to create a more stable and suitable structure. Based on studies in this area, we can leverage the ratios and structures found in plants to ensure that sufficient light reaches internal points while controlling the incoming light [44,91]. Furthermore, the structure of branches branching off tree trunks can not only help in creating columns with higher load-bearing capacity but also lead to better light absorption in indoor spaces [13].

As shown in Figure 9a,b, the Cactus Tower was designed for the Ministry of Urban and Agricultural Affairs in Doha, Qatar, which was inspired by the cactus plant (Figure 9c). This building has a shading system that is visible on its facade, with building panels resembling cactus spines drawn on the building's facade to create shade and regulate light entry. They can automatically open and close, oscillate, and adjust the amount of sunlight and heat transferred to the space to regulate the internal temperature as desired [81,92]. This solution allows this building to lower the size and amount of artificial cooling necessary for the building to 5–10%, so about 40–45% of the annual cost can be deducted, and also, the energy consumed by the HVAC can be decreased to 50% [92].

(a) (b) (c)

Figure 9. Shading system in Cactus Tower, Doha, Qatar. (**a**) Distant view of the building (**b**) Close-up view of the building (**c**) Cactus plant.

In another project, a lily flower (Figure 10c) was used to inspire the Wuhan New Energy Center in China (Figure 10a,b). This building was designed based on a structure that meets its needs through renewable energies, both in terms of form and structure. A 140 m tower is surrounded by shorter towers resembling blossoms. The inner peak extends upward in a bowl-like structure, topped by a large sun-oriented exhibition space. Much like a real plant, it effectively absorbs sunlight [93]. By collecting and recycling rainwater and using solar panels, this building presents a suitable performance for a sustainable future [94]. The building has an aluminum pipe system that not only provides natural ventilation but also brings light to indoor spaces without natural light in the building. Additionally, appropriate shading for the building facade has been used for light modulation and shadowing. The building is also designed to maximize the absorption of sunlight and obtain the necessary energy for water heating and electricity. Based on reports and studies conducted, this building has been able to reduce about 91% of its energy consumption compared to conventional buildings [95].

(a) (b) (c)

Figure 10. Natural light penetration in Wuhan New Energy Center, China. (**a**) Distant view of the building (**b**) Arial view of the building at the night (**c**) Lily flower.

4.4. Process

All plants have their own specific and complex processes that inspire architects. Biological processes such as growth, evolution, and reproduction in plants are inspiring architecture. For example, by taking inspiration from the growth process of plants, self-replicating structures can be created in buildings [96]. Plants use their exceptional capacity to harness energy from sunlight through the process of photosynthesis in nanomimic [97]. Cyanobacteria, green algae, seaweed, and some other plants have photosynthetic organisms capable of converting solar energy into electricity.

As shown in Figure 11a,b, Council House 2 (CH2) is a 10-story sustainable building situated in Melbourne, Australia. The building's design was groundbreaking and inspired by emulating tree bark (Figure 11c). Throughout the structure, biomimicry was incorporated, with the west facade resembling the tree's epidermis to regulate the external climate [1]. The north and south facades took inspiration from the tree's bronchi, acting as a protective layer that filters light and air into the ventilated wet areas. This façade is covered with timber louvers to optimize the penetration of natural lights and views [29]. CH2 boasts a rooftop garden, encouraging biodiversity and reducing the urban heat island.

(a) (b) (c)

Figure 11. Council House 2 (CH2), Melbourne, Australia. (**a**) Distant view of the building (**b**) Close-up view of the building (**c**) Bark tree.

4.5. Function

Plant performance mimicry is an important aspect in this field because the performance of structures is very important, and the more suitable the performance, the more efficient the structure will be. Therefore, utilizing plant performance, such as mimicking the sunflower rotation toward the sun, can be effective in storing natural light and improving the efficiency of solar panels [55,98]. Additionally, by creating the ability to open and close window coverings in a building inspired by the mangrove flower, one can significantly impact the building's performance in controlling daylight [99]. Moreover, controlling light from high surfaces like the plant *Fenestraria rhopalophylla*, which absorbs light through a window in its top [85], can also be utilized for shading purposes using the phyllotactic geometry present in plants to minimize self-shading of the building [53].

The Rotatable Solar House building in Germany shown in Figure 12a,b is inspired by the mechanism of the sunflower and known as the artificial sunflower (Figure 12c); it places the sun at its center and rotates the building to achieve sufficient daylighting. It features several nano-scale light absorbers and solar panels that absorb and store natural light throughout the day [81,100].

Figure 12. The Rotatable Solar, Germany. (**a**) Distant view of the building (**b**) Close-up view of the building (**c**) The artificial sunflower.

This building follows the sun's rotation to provide more clean energy and adhere to environmental sustainability principles [101]. Plants like to adjust their orientation and preferences according to different weather conditions. Reorientation is driven by the preferences of leaves and differential growth in expanding leaves, especially occurring behind the leaf, leading to leaf curvature [57]. Essentially, the building, like a plant that adjusts its status relative to changing environmental conditions, reacts to what is happening around it [102]. Additionally, the building's tracking is controlled by a computer system and can be adjusted to desired conditions and angles. The overall form of the building can be likened to a solitary tree that prevents heat loss with appropriate triple walls [100,103]. The adaptive shading of this façade provides a total energy saving of up to 82% and reduces the use of artificial lighting by 65% [104].

As shown in Figure 13a,b, in another project, Pearl River is a 71-story tower designed by Skidmore, Owings, and Merrill that was completed in 2011. The building is renowned for being the first so-called 'zero energy' building, as it was designed to produce as much energy as it consumes [82]. The architects took inspiration from sea sponges (Figure 13c), creating a sponge-like structure and using this idea to reduce energy consumption in their designs [105]. They modeled their designs after the sponge's ability to absorb gallons of water and organisms daily to reduce energy consumption. This porous tower features four openings, each housing a wind turbine, including one in the Pearl River Tower that generates electricity from strong winds. Additionally, the building harnesses solar energy through its photovoltaic system, integrated with the external solar shading and glass facade. The external solar shading system of the building and its glass facade, along with other energy-saving measures like radiant cooling, reduce the building's energy consumption by 58–60% [106].

(a) (b) (c)

Figure 13. Pearl River Tower, China. (**a**) Distant view of the building (**b**) Close-up view of the building (**c**) Sea sponges.

4.6. Case Studies Evaluation

Based on the study results and analysis of case studies inspired by biomimicry to optimize energy consumption through natural light, plants demonstrate significant potential for modeling solar energy utilization patterns. This underscores the importance of further exploration at each of the five levels of biomimetic architecture. These case studies are analyzed and compared across the five levels to identify which offers the greatest potential for aiding architects and designers in creating environmentally efficient buildings and reducing carbon footprints.

As the effectiveness of biomimicry in optimizing energy consumption depends on the specific conditions of each project, a universal guideline for drawing inspiration from plants in sustainable building design cannot be established. However, the findings aim to provide a general conclusion on the efficacy of these levels in utilizing natural light based on the case studies and research conducted. Table 3 summarizes the above case studies and their inspiration sources, and Figure 14 illustrates the energy conservation in each of the investigated buildings, which ranges from 20% to 90%.

Table 3. Application of plants for optimizing natural light.

Case	Biomimicry Level	Inspiration	Biomimicry Concept	Problem Solved	Energy Efficiency
Esplanade Theatres	Form, Material		Second layer of protection for underlying layers	- Controlling input heat - Shading	30%
Al Bahr Towers	Form, Function		A movement that controls the amount of natural light coming in and makes good use of it Daylight control	- Controlling the entry light and heat - Moving shading	50%

Table 3. *Cont.*

Case	Biomimicry Level	Inspiration	Biomimicry Concept	Problem Solved	Energy Efficiency
Swiss Reinsurance Headquarters	Form, Structure	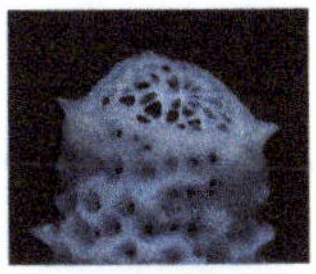	Use a series of outer triangles to create the shape of a glass sponge.	- Light control - Ventilation	80%
Media-TIC	Structure, Material		Transparent building skin insulation and light-sensitive, responsive to intense light	- Enhance light quality - Light control - Heat control	20%
Lahore office building	Material, Form		Designing a facade shadow pattern inspired by the physical structure of a plant leaf	- Visual comfort - Reduction in solar heat - Energy consumption reduction	32%
Qatar Cactus Tower	Form, Function		Reduction in heat and intense light penetration	- Adjusting the input light - Adjusting internal temperature	50%
Wuhan New Energy	Form, Structure		Transfer and storage of natural light	- Light transmission - Solar energy storage	91%
Council House 2	Process, Function, Material		It filters the light and air before entering the building.	- Controlling space light - Solar energy storage	82%
Rotatable Solar House	Structure, Function		Maximizing the use of sunlight	- Light absorption - Solar energy storage	82%
Pearl River Tower	Form, Function		Inspired by the structure of sea sponges to generate electricity and control daylight	- Electricity generation - Reduce artificial light consumption - External shading	60%

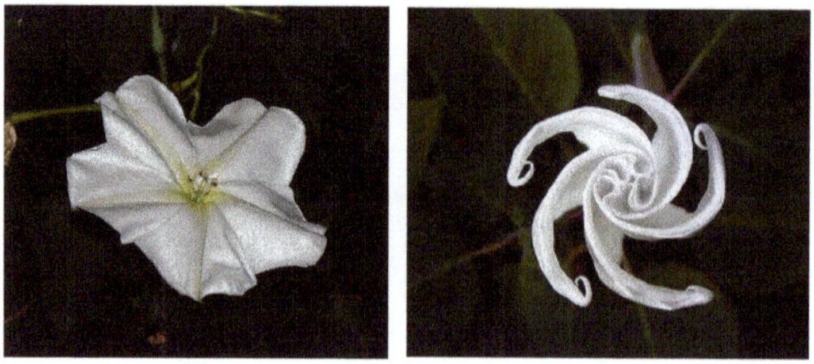

(**a**) (**b**)

Figure 14. The movement of the moonflower plant. (**a**) extended position (**b**) folded position.

The process and performance levels in biomimetic architecture often play a more pivotal role in reducing energy consumption through the utilization of light and solar energy. These levels were applied to various aspects of energy and sunlight management, such as light transmission, shading, and accounting for the sun's path, resulting in higher energy preservation compared to other levels.

Following these, the structural level demonstrated considerable potential for energy preservation by mimicking plant structures. This approach effectively aids buildings in utilizing natural light, contributing to a reduced carbon footprint. At the final levels, form and materials are discussed. These have the least impact on reducing energy consumption through light control and utilization during the day, as observed in the case studies. Additionally, specific forms and materials can pose greater challenges due to higher construction costs and the need for specialized labor, making them less practical compared to the other levels.

5. Discussion

By drawing inspiration from nature, biomimetic strategies offer innovative solutions for energy efficiency, CO_2 reduction, and climate resilience, addressing critical environmental challenges. The integration of adaptive materials, self-regulating building systems, and responsive façades can lead to more resource-efficient and low-impact construction methods. Furthermore, as climate change continues to shape building performance requirements, biomimicry provides a framework for creating resilient, self-sustaining structures that optimize natural resources like light, wind, and thermal energy. This study serves as a foundation for further research on scalable biomimetic technologies, interdisciplinary applications, and their role in shaping the future of regenerative architecture.

Innovative Design Based on Findings

Based on the findings from the various studies conducted on the specific cases highlighted in this article, it becomes evident that nature, particularly the characteristics and functions of plants, serves as an excellent model and source of inspiration for designing buildings that are energy-efficient and maximize the use of natural light. In light of this understanding, this section will delve into a range of innovative solutions that draw from the principles observed in plants, demonstrating how these natural designs can be effectively implemented to achieve optimal utilization of natural light within the built environment. By examining these plant-inspired strategies, this study aims to illustrate the potential benefits that they can bring to modern architecture in terms of sustainability and energy efficiency. Artificial lighting, as the third consumer of energy in buildings, can be controlled through optimal daylighting design through the cost-effective biomimicry architecture concepts.

The effective utilization of sunlight in architectural design hinges on the careful management of the incoming light that permeates interior spaces. It is of utmost importance that buildings are conceived in a manner that not only mitigates the risk of excessive sunlight flooding the interior but also guarantees that the inhabitants benefit from sufficient natural light. This design approach must be balanced with the need to preserve unobstructed views of the outside environment for the users. Thus, this study proposal centers around the innovative concept of a movable facade that acts as a secondary skin for the building, effectively regulating the amount of light that enters the interior spaces. This unique facade design draws inspiration from the moonflower plant, known scientifically as Ipomoea alba, which exhibits remarkable characteristics that align with design goals.

As shown in Figure 14, *Ipomoea alba* L., known as moon vine, is a night-blooming morning glory native to tropical Americas. This plant is native to the American tropics but has naturalized in many places where it was introduced. The name moonflower derives from their blooming in the evening and their being round in shape like a full moon. The flowers open at night and are up to 6 inches (15 cm) across with a slender, equally long floral tube. The flowers have five sepals, five broadly rounded corolla lobes, and a pale greenish–yellow five-pointed "star" radiating from the center of the floral tube. Even though they are night-blooming, the large, shining white flowers are easy to spot if there is any moonlight. Leaves are green, heart-shaped to slightly lobed, and stems are twining [107].

The moonflower plant is particularly notable for its ability to change shape in response to the opening and closing of its petals. This attractive feature served as a key source of inspiration for the design proposed by this study, allowing for the creation of a movable facade that is not only flexible but can be folded into two distinct positions, as shown in Figure 15a. In the first case (Figure 15b), the facade is folded once, which significantly increases the brightness inside the space and creates a bright and attractive space. In the second case (Figure 15c), the facade folds twice, increasing the amount of light entering the building, which is especially beneficial during periods when natural light is at a minimum. Moreover, this adaptability of design may provide a healthier indoor environment, whereas wind is able to freely enter the building. The folding mechanism of this innovative building facade can be designed to operate either manually or mechanically and is equipped with light-sensitive sensors. In automatic mode, the facade would open and close based on the amount of light detected by the sensors, while in manual mode, users can adjust it according to their needs by disabling the mechanical function. Additionally, the dimensions of the facades can be customized according to the specific requirements of each unit in the building. This customization allows users to adjust the facade based on their individual needs and enables them to open and close it at will to optimize their experience of natural light and the external view.

Existing and constructed buildings can significantly benefit from improvements through proper design, focusing on several key areas. By emphasizing effective design strategies and focusing on a number of essential key areas, these structures can be enhanced in ways that significantly enhance their overall performance and aesthetic appeal [108]. One of the effective solutions in this regard is the improvement of the materials used in each organization. By utilizing contemporary materials that provide enhanced performance capabilities, it is possible to significantly improve the overall durability and resilience of buildings. These advanced materials can withstand various environmental stresses and contribute to a structure's longevity and stability over time, ultimately leading to more robust construction outcomes [109].

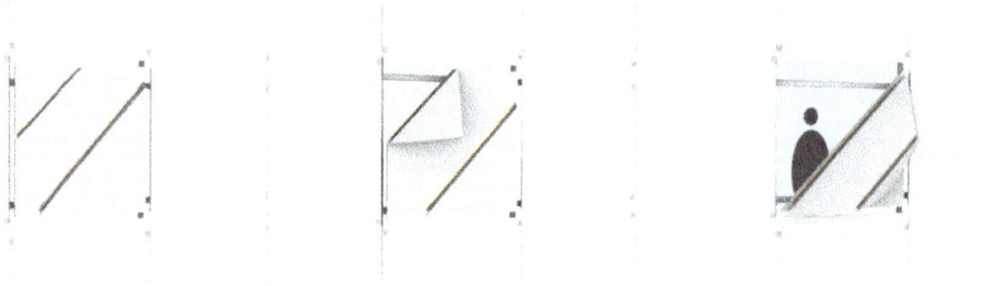

(a) (b) (c)

Figure 15. Proposed design for a double-skin facade. (**a**) Applicability of folding in two distinct positions (**b**) Folded in a time (**c**) Folded in two times.

Building shell designs inspired by the functionality of plant stomata present innovative solutions to some pressing challenges in architecture, particularly regarding energy efficiency and environmental management. Architectural shells with stomatal functions could benefit from biomimicry to better switch energy efficiency and environmental adaptability. Using those designs, the stomata opening and closing mechanism can be applied to modify shape depending on the light exposure and air quality in real time, enhancing aesthetics and functionality synergy [80,110].

As shown in Figure 16, a solution for controlling incoming light through the façade, inspired by plant apertures and pneumatic systems, involves creating a secondary skin for the building with panels containing pneumatic cushions. This system expands with increased radiation and temperature, extending the façade to block sunlight from entering.

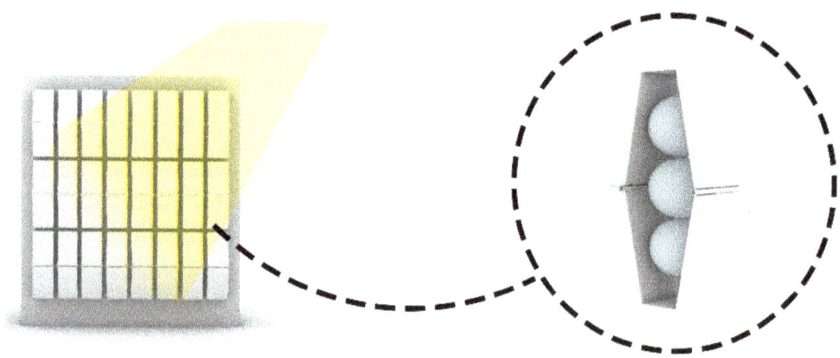

Figure 16. The shell in extended state in strong light conditions.

Conversely, when radiation decreases and temperatures drop, the structure contracts, allowing natural light to enter through the windows, as shown in Figure 17. This dynamic process effectively manages incoming light, offering precise control reminiscent of natural shading mechanisms. Ultimately, this system enhances indoor quality and optimizes energy consumption in the building.

Based on studies conducted on the selected case studies in this article, it was found that by drawing inspiration from nature, and again looking at the natural patterns found in the world, we can propose suggestions to improve the performance of these buildings in order to reduce energy consumption and carbon footprint. One of the buildings where strategies can be implemented to improve its performance in terms of energy consumption is the Al Bahr Tower. This building has a dynamic façade that helps control the intense sunlight in the area, and it can enhance its all-glass exterior with a second movable shell that opens and closes based on the sun's path. In the proposed solution, it is strongly recommended that the innovative second skin be implemented, such as this was effectively utilized in the facade of the Media-TIC building. This particular facade design incorporates semi-transparent materials that are specifically engineered to allow natural light to penetrate more deeply and evenly into the interior space. By doing so, it provides a filtered and refined quality of light that is ultimately more suitable for the interior environment. Additionally, this design approach plays a crucial role in minimizing heat gain, thus contributing to a more temperate and comfortable atmosphere within the building.

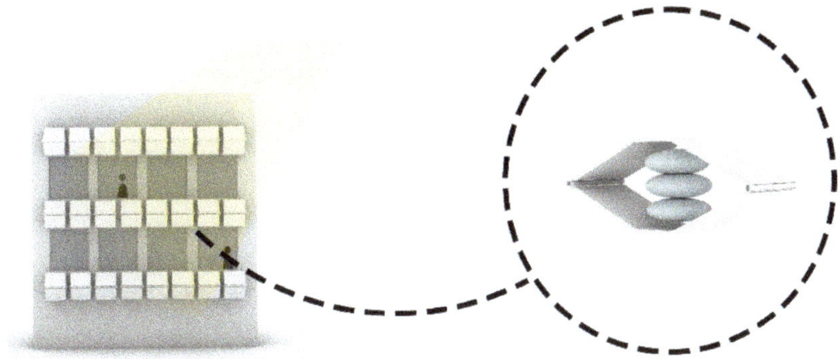

Figure 17. The shell is closed and in low light conditions.

As shown in Figure 18, in addition to the innovative design, rails can be installed all around the perimeter of the building, enabling the proposed facade to freely move along these tracks, resembling a sunflower that elegantly tracks the path of the sun throughout the day. This dynamic movement will allow residents to benefit from increased exposure to natural light during both low-light conditions and more suitable times of day, ensuring they have ample opportunities to enjoy the brightness that sunlight brings while also providing them with unobstructed views of the outdoor environment. Furthermore, akin to the energy-efficient model employed at the Wuhan Energy Center, it is also feasible to incorporate solar panels on the roof of this building. These solar panels would generate the necessary electricity to power this movable facade system as well as other essential operational needs of the building. By harnessing solar energy in this manner, we can significantly reduce overall energy consumption, contributing to a more sustainable and environmentally friendly approach to urban living. Finally, Table 4 presents suggested materials and structures, biomimicry level, inspiration and purpose of design for all proposed designs. These findings help designers to evaluate the efficiency of these applicable strategies in a real buildings.

Table 4. Potential structures of proposed designs based on study findings.

Proposed Designs	Suggested Material	Suggested Structure	Biomimicry Level	Inspiration	Purpose
Proposed design 1	PTFE	Double-skin façade, Metal support frames	Structure	Moonflower	Light control and energy consumption reduction
Proposed design 2	Composite Panel, ETFE	Double-skin façade, Pneumatic	Function, Material	Plant pore	Light control and energy consumption reduction

Table 4. *Cont.*

Proposed Designs	Suggested Material	Suggested Structure	Biomimicry Level	Inspiration	Purpose
Proposed design 3	ETFE	Pneumatic	Function, Structure	Sunflower and cactus	Utilizing solar energy and reducing energy consumption

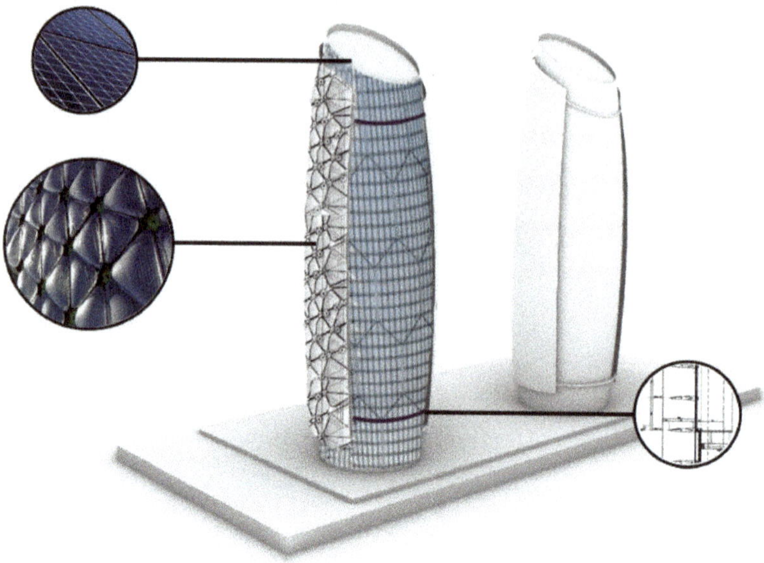

Figure 18. Proposed design combining reviewed case studies.

6. Conclusions

This article offers a comprehensive review and evaluation of biomimicry inspired by plants, focusing on building form, façade interventions, function, structure, materials, and energy. Plants have a high potential for mimicking patterns of natural light, making their use in biomimetic architecture to optimize natural lighting a sustainable and innovative solution for creating healthier and more efficient built environments. It provides valuable insights into various practical applications of this approach, emphasizing distinctive design cases and outlining future trends in this area. Additionally, this article examines key points and case studies of biomimetic applications in architecture, demonstrating capabilities for energy savings and consumption reduction in buildings. Case studies have shown how mimicking patterns, forms, and systems in nature can lead to more sustainable, efficient, and aesthetically pleasing spaces. A comprehensive review of the relevant literature from 2005 to 2024 revealed that despite numerous studies and designs in the field of biomimetic architecture, there is significant untapped potential for advancing this approach, necessitating further research in this direction. The efficiency of utilizing renewable energy sources indicates that the development of biomimicry technologies for building performance should be prioritized since this approach is critical for designing environmentally friendly buildings. The balanced and mutually beneficial relationship between green buildings and the environment, while reducing the environmental impacts of structures, is one of the important goals and advantages of biomimetic architecture. To create green structures inspired by plant biomimicry, adherence to the principles of nature and their integration with building technology, considering a comprehensive understanding of the relevant biological processes of plants is essential. By simulating the growth structure and environmental performance of plants, a healthy ecosystem can be formed to foster the development of energy-efficient bionic buildings and green structures. In addition, this study proposed theoretical solutions for future developments based on the review findings. However, further investigation into these proposals is necessary to refine, validate, and implement them effectively, ensuring their practical feasibility and maximizing their impact on advancing biomimetic architecture.

Author Contributions: Conceptualization, A.A.; methodology, A.A.; and M.B.; writing—original draft preparation, M.B.; writing—review and editing, M.E. and A.A.; visualization, M.B. All authors have read and agreed to the published version of this manuscript.

Funding: This research received no external funding.

Institutional Review Board Statement: Not applicable.

Data Availability Statement: The data presented in this study are available upon request from the corresponding authors. The data are not publicly available due to confidentiality.

Conflicts of Interest: The authors declare no conflicts of interest. The funders had no role in the design of this study, in the collection, analyses, or interpretation of data, in the writing of this manuscript, or in the decision to publish the results.

References

1. Wang, Y.; Liu, Y. Natural Building-Reflections on Bionic Architectural Design. In Proceedings of the 2021 3rd International Conference on Literature, Art and Human Development (ICLAHD 2021), Xishuangbanna, China, 29–31 October 2021; pp. 25–29.
2. Webb, M. Biomimetic building facades demonstrate potential to reduce energy consumption for different building typologies in different climate zones. *Clean Technol. Environ. Policy* **2022**, *24*, 493–518. [CrossRef] [PubMed]
3. Barbero-Barrera, M.d.M.; Tendero-Caballero, R.; García de Viedma-Santoro, M. Impact of Solar Shading on Façades' Surface Temperatures under Summer and Winter Conditions by IR Thermography. *Architecture* **2024**, *4*, 221–246. [CrossRef]
4. Mahdavinejad, M.; Nazar, N.S. Daylightophil high-performance architecture: Multi-objective optimization of energy efficiency and daylight availability in BSk climate. *Energy Procedia* **2017**, *115*, 92–101. [CrossRef]
5. Oh, S.; Choi, G.-S.; Kim, H. Climate-Adaptive Building Envelope Controls: Assessing the Impact on Building Performance. *Sustainability* **2023**, *16*, 288. [CrossRef]
6. Kong, T.; Luo, G.; Zhao, Y.; Liu, Z. Bioinspired superwettability micro/nanoarchitectures: Fabrications and applications. *Adv. Funct. Mater.* **2019**, *29*, 1808012. [CrossRef]
7. Varshabi, N.; Arslan Selçuk, S.; Mutlu Avinç, G. Biomimicry for energy-efficient building design: A bibliometric analysis. *Biomimetics* **2022**, *7*, 21. [CrossRef]
8. Faivre, N.; Fritz, M.; Freitas, T.; De Boissezon, B.; Vandewoestijne, S. Nature-Based Solutions in the EU: Innovating with nature to address social, economic and environmental challenges. *Environ. Res.* **2017**, *159*, 509–518. [CrossRef]
9. Verbrugghe, N.; Rubinacci, E.; Khan, A.Z. Biomimicry in Architecture: A Review of Definitions, Case Studies, and Design Methods. *Biomimetics* **2023**, *8*, 107. [CrossRef]
10. Speck, O.; Speck, D.; Horn, R.; Gantner, J.; Sedlbauer, K.P. Biomimetic bio-inspired biomorph sustainable? An attempt to classify and clarify biology-derived technical developments. *Bioinspiration Biomim.* **2017**, *12*, 011004. [CrossRef]
11. Baumeister, D.; Zari, M.P.; Hayes, S. Biomimicry. An opportunity for buildings to relate to place. In *Ecologies Design*; Routledge: London, UK, 2020; pp. 85–95.
12. Speck, O.; Speck, T. Biomimetics in botanical gardens—Educational trails and guided tours. *Biomimetics* **2023**, *8*, 303. [CrossRef]
13. Contreras, G.S.; Lezcano, R.A.G.; Fernández, E.J.L.; Concepción, M.; Gutiérrez, P. Architecture Learns from Nature. The Influence of Biomimicry and Biophilic Design in Building. *Mod. Appl. Sci.* **2023**, *17*, 1–58. [CrossRef]
14. Benyus, J.M. *Biomimicry: Innovation Inspired by Nature*; Morrow: New York, NY, USA, 1997.
15. Dash, S.P. Application of biomimicry in building design. *Int. J. Civ. Eng. Technol.* **2018**, *9*, 644–660.
16. Goldstein, J.; Johnson, E. Biomimicry: New natures, new enclosures. *Theory Cult. Soc.* **2015**, *32*, 61–81. [CrossRef]
17. Awadalla, R. Biomimicry as an Innovation Behavior in Architecture and Interior Design. *J. Des. Sci. Appl. Arts* **2023**, *4*, 81–92. [CrossRef]

18. Chaurasia, M.; Srivastava, S. Biomimicry and its applications—A Review. *Int. J. Eng. Appl. Sci. Technol.* **2020**, *4*, 545–549. [CrossRef]
19. Białkiewicz, J.J. Biomimetics and biomimicry. Their role as a tool and ideology in contemporary architecture. *Architectus* **2024**, *79*, 93–100. [CrossRef]
20. Pugalenthi, A.; Khlie, K.; Hassan Ahmed, F. Harnessing the power of biomimicry for sustainable innovation in construction industry. *J. Infrastruct. Policy Dev.* **2024**, *8*, 6663. [CrossRef]
21. Yıldız, A.; Avinç, G.M. Nature Inspired Practices in Super Tall Building Designs: Similarities of Form. *Black Sea J. Eng. Sci.* **2024**, *7*, 37–38. [CrossRef]
22. Jiang, M.; Deng, W.; Lin, H. Sustainability through Biomimicry: A Comprehensive Review of Bionic Design Applications. *Biomimetics* **2024**, *9*, 507. [CrossRef]
23. Uchiyama, Y.; Blanco, E.; Kohsaka, R. Application of biomimetics to architectural and urban design: A review across scales. *Sustainability* **2020**, *12*, 9813. [CrossRef]
24. Pohl, G.; Nachtigall, W. *Biomimetics for Architecture & Design: Nature-Analogies-Technology*; Springer: Berlin/Heidelberg, Germany, 2015.
25. Mohamed, N.A.; Bakr, A.F.; Hasan, A.E. Energy efficient buildings in smart cities: Biomimicry approach. In Proceedings of the 24th International Conference on Urban Planning, Regional Development and Information Society, Karlsruhe, Germany, 2–4 April 2019; pp. 2–4.
26. Shahda, M.; Elmokadem, A.A.E.; Abd Elhafeez, M.M. Biomimicry levels as an approach to the architectural sustainability. *Port Said Eng. Res. J.* **2014**, *18*, 118–121. [CrossRef]
27. Othmani, N.I.; Mohamed, S.A.; Abdul Hamid, N.H.; Ramlee, N.; Yeo, L.B.; Mohd Yunos, M.Y. Reviewing biomimicry design case studies as a solution to sustainable design. *Environ. Sci. Pollut. Res.* **2022**, *29*, 69327–69340. [CrossRef] [PubMed]
28. Othmani, N.I.; Yunos, M.Y.M.; Ismail, N.A.; Rahman, K. Review on biomimicry levels. *Am. J. Humanit. Soc. Sci. Res. (AJHSSR)* **2018**, *2*, 55–58.
29. Lipholt, N. Biomimicry-Where Nature Is Changing Innovation. Bachelor's Thesis, University of Twente, Enschede, The Netherlands, 2019.
30. Mohamed Abd El-Rahman, S.; Ibrahim Esmail, S.; Bakr Khalil, H.; El-Razaz, Z. Biomimicry inspired Adaptive Building Envelope in hot climate. *Eng. Res. J.* **2020**, *166*, 30–47. [CrossRef]
31. Amer, N. Biomimetic Approach in Architectural Education: Case study of 'Biomimicry in Architecture' Course. *Ain Shams Eng. J.* **2019**, *10*, 499–506. [CrossRef]
32. Elmeligy, D.A. Biomimicry for ecologically sustainable design in architecture: A proposed methodological study. In Proceedings of the 6th International Conference on Harmonisation between Architecture and Nature, Alicante, Spain, 13–15 July 2016; pp. 45–57.
33. Zari, M.P.; Storey, J. An ecosystem based biomimetic theory for a regenerative built environment. In Proceedings of the Sustainable Building Conference, Lisbon, Portugal, 12–14 September 2007.
34. Harsha, M.S.; Lakshmi, V.S. An analytical approach to sustainable building adaption using biomimicry. *Mater. Today Proc.* **2020**, *33*, 514–518. [CrossRef]
35. Levit, G.S.; Hossfeld, U. Ernst Haeckel in the history of biology. *Curr. Biol.* **2019**, *29*, R1276–R1284. [CrossRef]
36. El Ahmar, S.; Fioravanti, A. *Botanics and Parametric Design Fusions for Performative Building Skins*; Department of Architecture and Built Environment, Faculty of Engineering and Environment, Northumbria University: Newcastle upon Tyne, UK, 2014; p. 595.
37. Kader, J.-C.; Delseny, M. *Advances in Botanical Research*; Academic Press: Cambridge, MA, USA, 2011.
38. Bantserova, O.L.; Kasimova, A.R. Bionic approach to the organization of architectural objects in the sustainable development paradigm. *Civ. Eng. Archit.* **2023**, *11*, 939–947. [CrossRef]
39. Haque, O. Biomimicry Inspired Design for Daylighting Through Roof of Multipurpose Hall. Master's Thesis, Bangladesh University of Engineering & Technology, Dhaka, Bangladesh, 2019.
40. Pearcy, R.W.; Muraoka, H.; Valladares, F. Crown architecture in sun and shade environments: Assessing function and trade-offs with a three-dimensional simulation model. *New Phytol.* **2005**, *166*, 791–800. [CrossRef]
41. Di Salvo, S. Advances in research for biomimetic materials. *Adv. Mater. Res.* **2018**, *1149*, 28–40. [CrossRef]
42. Jiang, N.; Wang, H.; Liu, H.; Hu, X.; Li, W. Application of bionic design in product form design. In Proceedings of the 2010 IEEE 11th International Conference on Computer-Aided Industrial Design & Conceptual Design 1, Yiwu, China, 17–19 November 2010; pp. 431–434
43. Durai Prabhakaran, R.; Spear, M.J.; Curling, S.; Wootton-Beard, P.; Jones, P.; Donnison, I.; Ormondroyd, G.A. Plants and architecture: The role of biology and biomimetics in materials development for buildings. *Intell. Build. Int.* **2019**, *11*, 178–211. [CrossRef]
44. Yetkin, E.G. Effects of Biomimicry on Architecture. *Eur. J. Nat. Sci. Med.* **2020**, *4*, 100–113. [CrossRef]

45. Velásquez, C.; Espín, F.; Castro, M.Á.; Rodríguez, F. Energy Efficiency in Public Lighting Systems Friendly to the Environment and Protected Areas. *Sustainability* **2024**, *16*, 5113. [CrossRef]
46. Maghlakelidze, M.; Ben Alon, L. Biophilic Daylight Design in Healthcare Environments: A Critical Review. In Proceedings of the SimAUD 2021, Virtually Hosted, 15–17 April 2021.
47. Ruggiero, S.; Iannantuono, M.; Fotopoulou, A.; Papadaki, D.; Assimakopoulos, M.N.; De Masi, R.F.; Vanoli, G.P.; Ferrante, A. Multi-objective optimization for cooling and interior natural lighting in buildings for sustainable renovation. *Sustainability* **2022**, *14*, 8001. [CrossRef]
48. Atamewan, E.E. Appraisal of Day-Lighting in Sustainable Housing Development in Developing Countries. *J. Stud. Sci. Eng.* **2022**, *2*, 59–75. [CrossRef]
49. Dikou, C. Sustainability, through passive solar building envelope design, based on the principles of bioinspiration. *E3S Web Conf.* **2023**, *436*, 01013. [CrossRef]
50. Presti, D.; Delbrück, M. Photoreceptors for biosynthesis, energy storage and vision. *Plant Cell Environ.* **1978**, *1*, 81–100. [CrossRef]
51. Mohammed, A.B. A Systematic Design Technique of Biomimicry to Correlate and Integrate Architecture and Biology to Attain Green Buildings. *JES. J. Eng. Sci.* **2023**, *51*, 239–259. [CrossRef]
52. Helliwell, R. Daylight in relation to plant growth and illumination of buildings. *Arboric. J. Int. J. Urban For.* **2013**, *35*, 202–219. [CrossRef]
53. Jamei, E.; Vrcelj, Z. Biomimicry and the built environment, learning from nature's solutions. *Appl. Sci.* **2021**, *11*, 7514. [CrossRef]
54. Jalali, S.; Nicoletti, E.; Badarnah, L. From Flora to Solar Adaptive Facades: Integrating Plant-Inspired Design with Photovoltaic Technologies. *Sustainability* **2024**, *16*, 1145. [CrossRef]
55. Badarnah, L. Light management lessons from nature for building applications. *Procedia Eng.* **2016**, *145*, 595–602. [CrossRef]
56. Gay, C.; Yerkes, J.; Wilson, J. Performance advantages of two-axis tracking for large flat-plate photovoltaic energy systems. In Proceedings of the 16th Photovoltaic Specialists Conference, San Diego, CA, USA, 27–30 September 1982; pp. 1368–1371.
57. Badarnah, L. Form follows environment: Biomimetic approaches to building envelope design for environmental adaptation. *Buildings* **2017**, *7*, 40. [CrossRef]
58. Gu, Z.-Z.; Wei, H.-M.; Zhang, R.-Q.; Han, G.-Z.; Pan, C.; Zhang, H.; Tian, X.-J.; Chen, Z.-M. Artificial silver ragwort surface. *Appl. Phys. Lett.* **2005**, *86*, 201915. [CrossRef]
59. Aizenberg, J.; Weaver, J.C.; Thanawala, M.S.; Sundar, V.C.; Morse, D.E.; Fratzl, P. Skeleton of *Euplectella* sp.: Structural hierarchy from the nanoscale to the macroscale. *Science* **2005**, *309*, 275–278. [CrossRef]
60. Chayaamor-Heil, N.; Guéna, F.; Hannachi-Belkadi, N. Biomimétisme en architecture. État, méthodes et outils. *Les Cah. De La Rech. Archit. Urbaine Et Paysagère* **2018**. [CrossRef]
61. Raman, R.; Sreenivasan, A.; Suresh, M.; Nedungadi, P. Mapping biomimicry research to sustainable development goals. *Sci. Rep.* **2024**, *14*, 18613. [CrossRef]
62. Sommese, F.; Hosseini, S.M.; Badarnah, L.; Capozzi, F.; Giordano, S.; Ambrogi, V.; Ausiello, G. Light-responsive kinetic façade system inspired by the Gazania flower: A biomimetic approach in parametric design for daylighting. *Build. Environ.* **2024**, *247*, 111052. [CrossRef]
63. Amr, H.A.E.-A.; Essam, N. Designing adaptive facades using biomimicry for improving daylighting in office building: A case study of Cairo, Egypt. *J. Eng. Res. Eng. Res.* **2024**, *8*, 31.
64. Barbinta-Patrascu, M.-E.; Bita, B.; Negut, I. From nature to technology: Exploring the potential of plant-based materials and modified plants in biomimetics, bionics, and green innovations. *Biomimetics* **2024**, *9*, 390. [CrossRef]
65. Fahimnia, B.; Sarkis, J.; Davarzani, H. Green supply chain management: A review and bibliometric analysis. *Int. J. Prod. Econ.* **2015**, *162*, 101–114. [CrossRef]
66. Carrera-Rivera, A.; Ochoa, W.; Larrinaga, F.; Lasa, G. How-to conduct a systematic literature review: A quick guide for computer science research. *MethodsX* **2022**, *9*, 101895. [CrossRef] [PubMed]
67. Rotimi, J.O.B.; Moshood, T.D.; Rotimi, F.E. The Potential Challenges and Limitations of Implementing Modern Office Design Features in Residential Spaces: A SPAR-4-SLR Approach. *Buildings* **2024**, *14*, 3037. [CrossRef]
68. Fried, E.; Martin, A.; Esler, A.; Tran, A.; Corwin, L. Design-based learning for a sustainable future: Student outcomes resulting from a biomimicry curriculum in an evolution course. *Evol. Educ. Outreach* **2020**, *13*, 22. [CrossRef]
69. Lee, J.; Kang, J.; Han, S.; Lee, T. Biomimicry Method for Implementing Natural Lighting Performance in Underground Buildings. *Int. Proc. Chem. Biol. Environ. Eng.* **2016**, *93*, 32–38.
70. Maksudovich, P.M.; Yusufovich, R.Y. Applying biomimetic approach in architecture. *Достижения науки и образования* **2019**, *49*, 15–16.
71. Kapoor, R.T.; Rafatullah, M.; Qamar, M.; Qutob, M.; Alosaimi, A.M.; Alorfi, H.S.; Hussein, M.A. Review on recent developments in bioinspired-materials for sustainable energy and environmental applications. *Sustainability* **2022**, *14*, 16931. [CrossRef]
72. Knudson, D.V.; Knudson, D. *Fundamentals of Biomechanics*; Springer: Berlin/Heidelberg, Germany, 2007; Volume 183.

73. Chayaamor-Heil, N. From Bioinspiration to Biomimicry in Architecture: Opportunities and Challenges. *Encyclopedia* **2023**, *3*, 202–223. [CrossRef]
74. Altomonte, S. Daylight for energy savings and psycho-physiological well-being in sustainable built environments. *J. Sustain. Dev.* **2008**, *1*, 3–16. [CrossRef]
75. San Ha, N.; Lu, G.; Shu, D.; Yu, T. Mechanical properties and energy absorption characteristics of tropical fruit durian (Durio zibethinus). *J. Mech. Behav. Biomed. Mater.* **2020**, *104*, 103603.
76. Radwan, G.A.; Osama, N. Biomimicry, an approach, for energy effecient building skin design. *Procedia Environ. Sci.* **2016**, *34*, 178–189. [CrossRef]
77. ElDin, N.N. Biomimetic Approach for Building Envelope Adaptation in Hot and Dry Regions. *Green Build. Constr. Econ.* **2023**, 367–383. [CrossRef]
78. El-Abedin, E.E.Z.; Ali, S.; Saleh, A.; Foad, M. Biomimicry As a Catalyst to Improve the Behaviour of Building Performance. *Int. J. Innov. Res. Sci. Eng. Technol. (IJIRSET)* **2022**, *11*, 1111–1122.
79. Xuereb, K. The Al Bahr towers-enhancing sustainability through innovation. In Proceedings of the Sustainable Energy 2014: The ISE Annual Conference, Qawra, Malta, 20 March 2014.
80. Altahhan, R. Saving Energy by Using Biomimicry Facade to Adapt to Climate Change. Master's Thesis, Lisansüstü Eğitim Enstitüsü, Altınbaş Üniversitesi, Bağcılar, Turkey, 2022.
81. Makram, A. Nature-based framework for sustainable architectural design-biomimetic design and biophilic design. *Archit. Res.* **2019**, *9*, 74–81.
82. Al-Kodmany, K. Eco-iconic skyscrapers: Review of new design approaches. *Int. J. Sustain. Des.* **2010**, *1*, 314–334.
83. Al-Kodmany, K. Green towers and iconic design: Cases from three continents. *ArchNet-IJAR Int. J. Archit. Res.* **2014**, *8*, 11. [CrossRef]
84. Yang, Y.; Liu, Z.; Hu, B.; Man, Y.; Wu, W. Bionic composite material simulating the optical spectra of plant leaves. *J. Bionic Eng.* **2010**, *7*, S43–S49. [CrossRef]
85. López, M.; Rubio, R.; Martín, S.; Croxford, B. How plants inspire façades. From plants to architecture: Biomimetic principles for the development of adaptive architectural envelopes. *Renew. Sustain. Energy Rev.* **2017**, *67*, 692–703. [CrossRef]
86. Nasr, Y.T.A. Using Smart Materials to Mimic Nature in Architecture. Doctoral Dissertation, Graduate School, Faculty of Engineering, Alexandria University, Alexandria, Egypt, 2017.
87. de Vroom, P. *Media-TIC Building*; Time-Based Architecture International, 2009; Volume 7. Available online: https://openurl.ebsco.com/EPDB:gcd:14:11328446/detailv2?sid=ebsco:plink:scholar&id=ebsco:gcd:56537213&crl=c&link_origin=scholar.google.com.my (accessed on 11 February 2025).
88. Sohail, M.; Qureshi, M. Energy-efficient buildings in pakistan. *Sci. Vis.* **2011**, *16*, 27–38.
89. Nagy, Z.; Svetozarevic, B.; Jayathissa, P.; Begle, M.; Hofer, J.; Lydon, G.; Willmann, A.; Schlueter, A. The adaptive solar facade: From concept to prototypes. *Front. Archit. Res.* **2016**, *5*, 143–156. [CrossRef]
90. Sheikh, W.T.; Asghar, Q. Adaptive biomimetic facades: Enhancing energy efficiency of highly glazed buildings. *Front. Archit. Res.* **2019**, *8*, 319–331. [CrossRef]
91. Sadri, M.; Kavandi, M.; Jozepiri, A.; Teimouri, S.; Abbasi, F. Bionic architecture, forms and constructions. *Res. J. Recent Sci.* **2014**, *2277*, 2502.
92. Yassin, H.E.A.E.; Fathallah, A.M.; Elshapasy, R.A.I. Biomimicry in Managing Daylighting and Energy Consumption in Office Buildings. *Int. J. Eng. Res. Technol. (IJERT)* **2017**, *6*, 521–526.
93. Haider, Z.; Salman, M.; Girma, A. Biomimetic Architecture: An Innovative Approach to Attain Sustain-ability in Built Environment. *Ethiop. Int. J. Eng. Technol.* **2023**, *1*, 39–49. [CrossRef]
94. Bajaj, G. Comprehensive Analysis of Application of Biomimicry in Architecture to Enhance Functionality. Doctoral Dissertation, World University of Design, Sonepat, India, 2020.
95. Ali Hussein Mohamed, N.; Alaa Mandour, M.; Marey, S. Applying innovative criteria to achieve Biomimicry in Architecture. *Eng. Res. J.* **2022**, *176*, 90–109. [CrossRef]
96. Gruber, P. *Biomimetics in Architecture: Architecture of Life and Buildings*; Springer: Berlin/Heidelberg, Germany, 2011.
97. Prasad, R. *Advances in the Understanding of Nanomaterials Research and Applications*; Springer: Berlin/Heidelberg, Germany, 2019.
98. Ersanli, E.T.; Ersanli, C.C. Biomimicry: Journey to the Future with the Power of Nature. *Int. Sci. Vocat. Stud. J.* **2023**, *7*, 149–160.
99. Hassan, F.H.F.; Ali, K.A.Y.; Ahmed, S.A.M. Biomimicry as an Approach to Improve Daylighting Performance in Office Buildings in Assiut City, Egypt. *J. Daylighting* **2023**, *10*, 1–16. [CrossRef]
100. Fan, X.; Zhang, Z.; Yang, F.; Cao, L.; Xu, J.; Yuan, M. Analysis on design and performance of a solar rotary house. *IOP Conf. Ser. Earth Environ. Sci.* **2017**, *61*, 012085. [CrossRef]
101. Marins, K.R.d.C.C. Análise comparativa multicriterial de estratégias em sustentabilidade urbana aplicada aos bairros de Cidade Pedra Branca (Palhoça, SC) e Vauban (Freiburg, Alemanha). *Ambiente Construído* **2017**, *17*, 393–408. [CrossRef]

102. Li, J. Research on the Application of Bionics in Modern Architectural Design. In Proceedings of the 2018 International Workshop on Advances in Social Sciences (IWASS 2018), Qingdao, China, 29–30 December 2018; pp. 676–679.
103. Zuo, J.; Zhao, Z.-Y. Green building research–current status and future agenda: A review. *Renew. Sustain. Energy Rev.* **2014**, *30*, 271–281. [CrossRef]
104. Srisuwan, T. Applications of biomimetic adaptive facades for enhancing building energy efficiency. *Int. J. Build. Urban Inter. Landsc. Technol.* **2022**, *20*, 7–18. [CrossRef]
105. Frechette, R.; Gilchrist, R. Towards zero energy: A case study of the Pearl River Tower, Guangzhou, China. In Proceedings of the CTBUH: Proceedings of the Council on Tall Buildings and Urban Habitat's 8th World Congress, Dubai, United Arab Emirates, 3–5 March 2008; pp. 7–16.
106. Mirniazmandan, S.; Rahimianzarif, E. Biomimicry, an approach toward sustainability of high-rise buildings. *Iran. J. Sci. Technol. Trans. A Sci.* **2018**, *42*, 1837–1846. [CrossRef]
107. Srivastava, D.; Rauniyar, N. *Medicinal Plants of Genus Ipomoea*; LAP Lambert Academic Publishing: Beau Bassin, Mauritius, 2020.
108. Kośny, J.; Kośny, J. Short history of PCM applications in building envelopes. In *PCM-Enhanced Building Components: An Application of Phase Change Materials in Building Envelopes and Internal Structures*; Springer: Cham, Switzerland, 2015; pp. 21–59.
109. Belleri, A.; Passoni, C.; Marini, A.; Riva, P. Hinged-wall solutions for the structural strengthening of existing RC buildings. In Proceedings of the Fib Symposium, Cape Town, South Africa, 21–23 November 2016.
110. Le Ferrand, H.; Riley, K.S.; Arrieta, A.F. Plant-inspired multi-stimuli and multi-temporal morphing composites. *Bioinspiration Biomim.* **2022**, *17*, 046002. [CrossRef] [PubMed]

MDPI AG
Grosspeteranlage 5
4052 Basel
Switzerland
Tel.: +41 61 683 77 34

Biomimetics Editorial Office
E-mail: biomimetics@mdpi.com
www.mdpi.com/journal/biomimetics

www.ingramcontent.com/pod-product-compliance
Lightning Source LLC
Chambersburg PA
CBHW061507210926
51850CB00008B/1844

* 9 7 8 3 7 2 5 8 6 9 0 6 0 *